# CAMBRIDGE ENERGY STUDIES

Edited by Richard Eden
*Professor of Energy Studies, University of Cambridge*

## INTRODUCTION TO ENERGY

# INTRODUCTION TO ENERGY
## Resources, technology, and society

**EDWARD S. CASSEDY**
*Polytechnic University*

**PETER Z. GROSSMAN**
*Washington University*

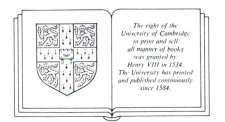

The right of the
University of Cambridge
to print and sell
all manner of books
was granted by
Henry VIII in 1534.
The University has printed
and published continuously
since 1584.

## CAMBRIDGE UNIVERSITY PRESS

Cambridge

New York   Port Chester   Melbourne   Sydney

Published by the Press Syndicate of the University of Cambridge
The Pitt Building, Trumpington Street, Cambridge CB2 1RP
40 West 20th Street, New York, NY 10011, USA
10 Stamford Road, Oakleigh, Melbourne 3166, Australia

First published 1990

Printed in the United States of America

*Library of Congress Cataloging-in-Publication Data*

Cassedy, Edward S.

Introduction to energy: resources, technology, and society
Edward S. Cassedy and Peter Z. Grossman.
p.   cm. – (Cambridge energy studies)
ISBN 0-521-35091-3 (Cambridge University Press). – ISBN
0-521-35941-4 (pbk.)
1. Power resources.   2. Power (Mechanics)   I. Grossman, Peter Z.,
1948–       II. Title.   III. Series.
TJ163.2.C4   1990                                89-70826
333.79 – dc20                                          CIP

*British Library Cataloguing in Publication Data*

Cassedy, Edward S.

Introduction to energy: resources, technology, and
society. (Cambridge energy studies)
1. Technology. Social aspects.
I. Title   II. Grossman, Peter Z.
306.46

ISBN 0-521-35091-3 hardback
ISBN 0-521-35941-4 paperback

For Sylvia

# Contents

# Preface

This book was developed for a course on the philosophical and social implications of modern energy technology that we taught at Polytechnic University in Brooklyn, New York. The course was created as part of a pioneering program in Science Technology and Society (STS) that Polytechnic established in the early 1980s with the support of the Andrew W. Mellon Foundation. Polytechnic's program, in turn, is part of a growing STS movement encompassing more than 100 colleges and universities. The goal of the movement is to help humanities and social science students achieve a measure of what is sometimes called *technological literacy* – an ability to think critically about modern technology.

This book, which is intended primarily as a text for STS courses, has the same general goal: to develop critical thinking skills about technology. Accordingly, we present not only the problems posed by energy technology, but also the technical and philosophical principles that might be brought to bear in evaluating any solutions. For the most part, we avoid advocacy and prescriptions; instead we bring out the underlying values surrounding technological choices so that students are challenged to find their own answers. Because energy issues are complex and have effects that may stretch many generations into the future, we believe that pat answers would only limit the usefulness of this book.

As a result, technology is neither the hero nor the villain. Critical analysis is incompatible with simple technical-fix solutions because all technological solutions have social consequences. At the same time, a critical approach must regard with equal skepticism a view of technology as a socially malevolent force, where technical advance is seen as self-determining and synonymous with dehumanization. Technology does, of course, have a major effect on society, and people may be justifiably concerned that in the process of technical development they may lose autonomy and freedom. Our book delves into these issues and asks how we can use technology without being used by it, how our institutions can be adapted to control technology rather than be destroyed by it, how the benefits of technology can be understood and evaluated against its costs. These questions are at the heart of the debate over energy technology and will confront today's students in the years to come.

To be able to appreciate fully the issues raised in this book, the reader should know basic physical science at a level that is taught in science courses offered to nonscience majors. Nevertheless, the book contains an appendix on scientific principles (Appendix A) that starts its review with beginning definitions. We believe that with this appendix a layperson with some appreciation of science will be able to grasp the essential concepts without taking a prerequisite course.

ix

The book also employs concepts from ethical and social philosophy. Most students who took our course at Polytechnic had taken a prior course in philosophy, but those who had not were able to keep up with the class discussion with the help of some basic philosophical texts. Suggested readings as well as a brief review of basic principles of ethics and justice are given in Appendix B.

In addition, energy issues require some understanding of concepts from the social sciences – especially economics, but also sociology and even psychology. We do not, however, assume prior knowledge of any of these areas. A class whose students have had some economics will probably be able to cover the material more rapidly, but technical concepts in economics are explained in the text. Lack of formal training in economics should not pose a barrier to an appreciation of the issues of energy technology.

The primary aim of the book is to raise the level of understanding of technology in students of the humanities and social science. At the same time, we have found that the course at Polytechnic is valuable for students in engineering and physical science. Naturally, they find the science material relatively simple, although frequently they seem to gain a greater insight into thermodynamics through the direct application of the scientific principles to specific energy issues. More important, these students gain an important philosophical and social perspective on the technology they will one day help create.

Although this book is primarily intended as an STS text, we have found it a useful supplement for other courses. We have used portions of it successfully in a graduate course in energy policy that was taught to engineering, economics, and management students. We think it would be useful for courses related to technology management or safety and health regulation.

As an introduction to energy, the book necessarily covers a broad range of topics. It is divided into three sections. In the first section, the book considers basic issues connected with resources, energy demand, and the principles of conventional thermal energy conversion. (We have a chapter on thermal conversion because 95 percent of our energy is derived from the thermal processes.) In the next two sections, the book delves into specific technologies – present and future – and the important issues surrounding them. In all cases, we seek to deepen the student's understanding of how these technologies work as well as the impacts they may have on society.

The book stresses the value of a quantitative as well as a qualitative appreciation of energy issues. While we try to avoid complex mathematical formulation, we introduce appropriate units (such as quads to measure total national energy consumption or dollars per MBTU as a measure of cost) to familiarize students with quantitative measures. This emphasis not only enables the reader to grasp more fully the technological issues, but it is also crucial to the economic and philosophical debate, particularly with regard to the measurement of social risks and benefits (Chapter 10).

The timeliness of facts is often a concern when dealing with issues such as the world oil market, nuclear power, or the development of new technologies. Our approach has been to keep the discussion as timeless as possible. The 1970s and early 1980s witnessed drastic and sudden changes in perception about energy issues. (As you will see, much of the data we use comes from that period. For example, in the United States, government cutbacks in the 1980s not only reduced research in energy technology, they lessened information gathering as well.) In the late 1980s, as this book was written, events did not move so rapidly, but the rapidity of change in the past suggests how precarious it can be to rely on the latest headlines for a perspective on energy questions.

Several sources have been used to update information on resources, economics, and technology developments. In such instances we have only briefly indicated those sources and given dates – the responsibility for the selection of data from them is ours.

We would like to express our appreciation to a number of people for their help in creating this book. Above all, we wish to thank Donald Hockney and Carl Mitcham who, as administrators of the Contemporary Liberal Arts Program at Polytechnic University, provided us with encouragement, support, and time to see this work to completion.

The authors are also indebted to a number of their Polytechnic colleagues who graciously read and commented on portions of the book while it was in progress: Donald Hockney, Walter Kiszenick, Carl Mitcham, Werner Mueller, Romualdas Sviedrys, and Nancy Tooney. In addition, we wish to thank others who read part or all of the book: John Byrne and Paul Durbin of the University of Delaware; Peter M. Meier of SUNY Stony Brook; Steven Hoffman of the College of St. Thomas; Theodor N. Kapiga of the UN Development Programme; Joshua M. Kay, formerly of Pacific Gas and Electric; and two anonymous reviewers. Their comments were always welcomed and extremely valuable.

We would like to acknowledge the help of people and organizations in providing information and answering questions relating to issues discussed in this book. We would especially like to thank Chaim Braun of the Electric Power Research Institute, Ted Flanagan of the Rocky Mountain Institute, Mudassar Imran of the World Bank, and Charles Komanoff of Komanoff Associates in New York. Recognition is also due Kunmo Chung of the Korean Electric Power Company, a former colleague at Polytechnic, who stimulated our interest in energy policy dilemmas during the 1970s, and thanks are due to Laurence Pringle for guidance in locating illustrations.

We also wish to acknowledge the help of the Department of Electrical Engineering and Computer Science and the Department of Humanities and Communications at Polytechnic for logistical and secretarial support. Word processing of the several drafts was expedited by Joseph Washington, with the assistance of Le Dung and Tien Nguyen. Computer graphics were ably carried out by Leah Zabar and Lucille Kalmbacher.

Finally, we would like to express our appreciation to our editor, Peter-John Leone, at Cambridge University Press, for his guidance and understanding in bringing this book to publication.

# 1

# Introduction

## Energy crisis

October 1973.

The energy crisis began in dramatic fashion. In response to a war between Arabs and Israelis, Arab nations suddenly cut off all shipments of oil to Israel's ally, the United States. For the next few weeks, Americans experienced a discomfort that they had not known before; across the nation, millions of people were compelled to sit each day in their cars waiting in line for gasoline. In the United States since World War II, people had taken an abundant supply of cheap energy resources – oil especially – for granted and had created an energy-consuming society. Now, there were shortages everywhere.

But the embargo had far more serious implications than the discomfort and inconvenience the gas lines suggested. Americans began to realize just how much energy they needed. The press was suddenly filled with startling statistics about the millions of barrels of oil the United States used each day – and how much more oil the nation used than did the rest of the world. People also realized, with shock, that too much of that oil was imported, that America's energy resources were controlled by others, and that supplies could be shut off instantaneously at any time. Perceptions about energy and the world had started to change.

And it was only the beginning.

The embargo ended, but the crisis did not.

Throughout the 1970s, there seemed to be one energy crisis after another, not only in the United States but all over the world. There were more shortages, most notably a natural gas shortage in the United States in 1976. We began to hear ominous predictions that some resources would be virtually exhausted in the very near future.

There were continual price hikes, too. Oil, which was $3 a barrel at the beginning of the decade, reached more than $30 per barrel at the end of the 1970s, and there was no end to rising prices in sight. The Organization of Petroleum Exporting Countries (OPEC), which controlled the majority of oil reserves in the world and appeared to control the price of oil as well, seemed determined to raise prices indefinitely. Few energy experts saw how OPEC could be prevented from doing so. Analysts predicted that oil would cost $80 per barrel or more in the 1980s. Electricity rates soared as well, partly from rising oil, gas, and coal prices, partly from the cost of building electric generating plants that would not need oil, gas, or coal – but that ran instead on nuclear fuel. Once regarded as a cost saver, nuclear power increasingly appeared to be an economic disaster.

To make matters worse, energy use, the means by which industrial society had been created and had persevered, was now widely recognized as a problem in itself. Not only did our demand for energy bring us closer to the day resources would be exhausted, but our use – through combustion of fuels – created pollution that affected our land, our air, our water, and our health. By the end of the decade, President Jimmy Carter declared that the energy crisis was the "moral equivalent of war" and challenged the nation to combat it and find answers.

The energy crisis period had already challenged the entire world in important ways. For Americans especially, the crisis had posed a wide range of dilemmas. Prior to the crisis period, not only did individuals take cheap supplies for granted, they also took the smooth and consistent running of technological society as a given. People were optimistic about the ability of technology to make everyone's life better and richer. But the reality was not so simple. Suddenly, technology was uncertain, potentially dangerous, and costly. It was not clear whether technology was the problem or the solution.

Energy crises also created a period of economic dislocation. The need of industrial nations for energy resources had resulted in the massive shift of wealth to oil exporters and in economic instability for the rest of the world as a result of rising energy costs. Two significant economic recessions took place within the crisis period, and there were fears that economic hardship would only grow in the years ahead.

The energy crises, however, were more than technological and economic phenomena. They were political and social phenomena as well. The crises changed political perceptions. The U.S. government faced political blackmail with these terms: no support, no oil. In the crisis period of the 1970s, the United States resisted that pressure, endured some hardships as a consequence, but it had the advantage in that not all OPEC members supported the embargo and in fact used it – thereby undercutting the Arab states – to gain a greater share of the U.S. market. But this did not lessen the sense of vulnerability Americans felt. Next time, what if OPEC nations acted in concert? The United States might have to choose between running its industry and maintaining control of its

foreign policy. It was not a choice people wanted to face.

Finally, the energy crisis period challenged people of the industrial world to think about how we lived and would live in the future. Could we continue to use as much energy as we were using? What would it mean to use less? Were we ready to change our life-styles? Were we ready to drive smaller cars, turn down our thermostats, make do with fewer gadgets, and live a technologically simpler life?

Such choices, as many philosophers noted at the time, were not simply matters of practicality. Our patterns of energy use carried moral implications, as did any solutions a society adopted to forestall future energy crises. For example, were the consumption-oriented life-styles of the industrial West a violation of world distributive justice? That is, did we have the moral right to use more fossil fuel per capita than the rest of the world? Was the pollution we generated morally defensible, a threat not only to all people in the world today but a threat perhaps to the generations to come? But if government demanded a change of life-style or an end to polluting industries, would that not violate individual freedom? These questions were debated throughout the 1970s in many public forums. Although there appeared to be no easy answers at the time, it seemed of utmost urgency that society reach a consensus on such questions as soon as possible.

Then the situation turned upside down.

## Postcrisis to . . . ?

Though as we will see in Chapter 4 the industrial world began for a time to use less energy, consensus on the larger moral issues was not achieved. Instead, the debate tapered off. Almost as abruptly as it had begun, the energy crisis period ended. It seemed to vanish, in fact. Oil prices that had been expected to hit $80 per barrel fell below $15, and there was far more supply than demand – a glut of oil and gas. OPEC lost its grip on the market and even on its own members. OPEC production quotas designed to prop up prices were routinely violated by member nations afraid of losing their markets altogether to non-OPEC producer states like Mexico and Norway. For all practical purposes, the OPEC cartel lost its power.

There was some feeling, in the United States particularly, that the crisis had been an illusion. The Reagan administration appeared to regard it, at most, as an insignificant interlude. Government programs to develop alternative energy resources were cut or eliminated. There was no new policy approach to the larger social and moral questions and indeed it did not appear necessary. By the late 1980s, consumption of energy resources began rising again, but there were still low prices and a glut of supply.

The crisis seemed to be over, and the issue settled.

But was it? The immediate crisis certainly did end, but there was disagreement as to how settled the matter had become. By all appearances, it seemed that the old days of abundant cheap energy had returned. Questions that seemed so important in the 1970s seemed out of step in a world in which there was a glut of oil and gas.

Yet many people were unconvinced by appearances. Articles began appearing in both scientific journals and in the popular media arguing that a new crisis was looming (see, for example, Crawford 1987 and Hirsch 1987). According to this view, fossil fuels were becoming exhausted (albeit more slowly than the press had suggested in the 1970s), and growing consumption was destined to lead to a new crisis in the 1990s or thereafter. These observers argued that people should think of the postcrisis period as an opportunity to ensure that we would not face a crisis again. They urged us to learn from the last crisis and act to forestall the next one. If society became complacent, they declared, it guaranteed that another crisis would be upon us soon. And the next one, some experts believed, would be worse than the one we had already faced.

## A look backward

Who is right? Are energy crises only behind us, or are they in front of us as well?

Although we cannot answer these questions definitively, we do know that even though the crisis period of the 1970s was unique in our experience, it was not unique to history. Energy crises have happened before, sometimes lasting the better part of a century, sometimes having a dramatic and lasting result on the world.

The ancient Egyptians, for example, apparently experienced a long-term energy problem. In an attempt to maintain supplies of wood fuel, they extended their empire further and further south. Their search for energy resources led them away from the Mediterranean, the center of their civilization, and moved deeper and deeper into the jungles of Africa. They extended themselves too far, some historians argue. This move led to the decline of Egyptian preeminence in the ancient world (see LeBel 1982).

A more recent energy crisis, by constrast, may well have contributed to the *rise* of a nation and indeed of modern technological society itself. In the sixteenth century, England was becoming rapidly deforested. Of course, wood had uses beyond energy – in building especially – but it was also the primary fuel of the time, and the need for fuel contributed significantly to the crisis. Wood fuel was used in several key industries such as glassmaking, metal smelting, and extracting salt from seawater. This energy crisis lasted about a century, and in the process of ending it, the English paved the way for the Industrial Revolution (see Nef 1977).

The English had an alternative energy resource to wood: coal. Coal was abundant, in many places coal seams reached the surface, and coal had been used as a fuel since Roman times. But it also had drawbacks that limited its applications. Above all, it burned dirty smoke. This led to such bad pollution that coal was banned in London during the fourteenth century. Even more important, it made a poor industrial fuel. Components of the smoke led to industrial products that were at best impure, at worst unusable.

The need to find an alternative to wood, however, led to two notable technological advances. First, the reverbatory furnace was developed. Essentially a hollowed sphere, its

**FIG 1.1.** The Newcomen engine. *Source*: Courtesy of Babcock & Wilcox Co.

shape caused the heat to reflect back onto itself — to reverberate — raising the temperature and resulting in a cleaner burn. Now coal could replace wood for most industrial purposes, and by the end of the 1600s, it had.

The second innovation grew out of the need to boost supplies of coal. The surface supplies were soon exhausted, and new supplies required underground extraction. The deeper the mines went, though, the harder the task became. One especially difficult problem was water seepage into mine shafts. Better ways of pumping were sought. In 1705, an English inventor, Thomas Newcomen, developed a pumping device powered by steam, which itself was produced by burning coal (see Fig. 1.1). The Newcomen engine was widely utilized for the rest of the eighteenth century. More important, it was the first practical employment of a steam engine. It was, of course, the steam engine (albeit in a different configuration) that later powered the Industrial Revolution.

What lessons do the crises of the past have for us? The English experience paradoxically can provide us with two very different lessons. On the one hand, the long duration of the English crisis could make us fearful about expecting quick solutions to difficult energy problems. But the ultimate outcome might encourage us to be confident that technology will find answers. The story of the English energy crisis may suggest a future that provides an opportunity for the employment

and expression of human ingenuity. At the same time, the Egyptian experience might suggest that without successful innovation, scarcity of energy resources can doom a civilization.

Of course, there is no perfect historical analogy. There are important differences today. We are far more dependent on abundant energy and we must have it now. We cannot wait a century for innovation and maintain current life-styles. Technology may hold promises, but it also provides no guarantees of when or whether it will succeed. There may be good cause for optimism but, as one oil executive put it, technology cannot be expected to come to the rescue "like the cavalry riding over the hill." There also is danger in technology as well as promise. Even in the face of an apparent energy resource glut, scientists expressed fears that our vast appetite for energy resources is creating intolerable levels of pollution and is risking global climatic change.

## Thinking about energy issues

There is no single answer to the questions about energy in the years ahead; we cannot even be sure of the questions themselves. But there are most certainly going to be challenges and, probably, crises as well. Some questions will require active and definitive answers from everyone in our society and perhaps in our world. These issues will affect our wealth, our life-styles, our health, and our children.

Ultimately, these issues engage us all. For the individual who is not an energy expert, however, the question arises: How can one provide answers to technological problems without being an engineer or a scientist or an energy analyst? As we will discuss in the chapters that follow, energy technologies, though intricate at times, are not beyond the average individual's comprehension. More important, we will see that any analysis of energy issues must go beyond the strictly technical. The broader context of energy issues and energy policy is social and philosophical. Technology itself does not stand in isolation from the world. Judging its place in, and value to, society is something that everyone can, and arguably must, do, since such judgments will directly affect the way we all live.

The specific answers to the questions on energy we face now and will face tomorrow may not be knowable. But such answers must come from a full analysis, a sense of the social and philosophic context in which energy technology and resources are used, and a keen appreciation of what energy issues mean to the way we live and to the world we live in. This book does not provide a prescription for the future. Instead, our quest is for a *critical appreciation* of energy issues, including resources, technology, safety, economy, behavior, and so on. Only in this way are we likely, as a society, to be able to answer – equitably and clearly – the questions that will arise in the years ahead.

## Bibliography

Crawford, M. 1987. "Back to the Energy Crisis." *Science*, February 6, pp. 626–7.

Gever, J., R. Kaufmann, D. Skole, and C. Vorosmarty. 1986. *Beyond Oil*. Ballinger, Cambridge, MA.

Hirsch, R. L. 1987. "Impending United States Energy Crisis." *Science*, March 20, pp. 1467–73.

LeBel, P. G. 1982. *Energy Economics and Technology*. Johns Hopkins Press, Baltimore, MD.

Nef, J. U. 1977. "An Early Energy Crisis and Its Consequences." *Scientific American*, October, pp. 141–51.

# Energy resources and technology

# 2

# Energy resources

## Introduction

The use of natural resources has caused controversy since the Industrial Revolution in the nineteenth century. Questions about who controlled these resources, who benefited from their exploitation, and what were the environmental consequences of this exploitation have been raised repeatedly. Energy resources, and fossil fuels in particular, have been the focus of intense and protracted controversy throughout the industrialization of the United States and Europe.

In the United States, these disputes have raged for nearly a century and have common threads that can be followed to the present day (see Wildavsky & Tenenbaum 1981). Some of these threads are part of the broader historic fabric of this country, such as the antimonopolism of the last decades of the nineteenth century that grew into the Progressive movement of the early twentieth century. A symbol of these political struggles was John D. Rockefeller and his Standard Oil Company. Following the discovery of oil in Pennsylvania in 1859, Standard's rapid acquisition of oil production and refining operations provoked the charge that the company sought monopoly power. Whatever the truth of the charge, Rockefeller's company was following a larger trend of the era whereby conglomerates of industry, for example in steel, were being formed.

The popular reaction against business conglomeration led to, among other consequences, the passage of the Sherman Antitrust Act (1890) and the founding of the Progressive movement. Progressivism extended into the twentieth century with the trust-busting of President Theodore Roosevelt (1901–1908) and the presidential candidacy in 1924 of Robert La Follete, who campaigned on antitrust platforms. The underlying sentiment in these movements was mistrust of big business, especially those companies that had accumulated sufficient power through conglomeration to exercise monopolistic control of the market for that industry's product. Standard Oil, for example, attempted to control the market not only for the extraction of oil, but also sought to dominate the petroleum refining and distribution businesses. Even in the late nineteenth century, these markets were very significant and included such oil-based products as kerosene (for lighting).

Standard Oil was broken up into separate operations, but that did not settle all of the issues that the oil trust had raised. The most crucial was the question of control of the extraction and use of resources. A company like Standard had the power to determine how much oil and its associated products reached the market. In theory, they could always keep the amounts low with respect to demand in order to keep prices and profits high and to maintain supplies for the longer term. After Standard was broken up, no one company had the power to control the supply of oil any longer. But it remained in the interest of the new companies to maintain policies that would preserve the policy goals of the oil trust.

Of course, in time, if the companies were seen to be acting in collusion and, for example, deliberately withholding supplies from the market, they risked new government actions. Alternatively, if they did not collude, they might face competition from firms willing to supply oil when they were not. But the latter case would assume that an additional supply could be found in sufficient quantities and at a cheap enough cost to be worth producing. In other words, controlling the *information* about resources could be important – crucial both in terms of the market and of government behavior. It is not surprising therefore that much of the controversy about resources has to do with information – what is it, how to determine it, who should control it.

It also is not surprising that throughout the twentieth century, antimonopolists have challenged estimates of oil (and later natural gas) reserves. Charges have been made as recently as the 1970s that the oil industry[1] had underestimated reserves and even had contrived shortages to force prices upward. There were data to support the industry's case. But as we will see, the issue has hardly been settled.

## Controversies over resource estimates

How much oil (or gas or coal) was available? How much is available today? We must emphasize at the outset that determining what is available and where is no easy matter. Often estimates of resources have come from the industries themselves and have been routinely challenged. Two groups whose origins go back to the last century have been especially notable in resisting industry estimates of resources.

The first – the consumer movement – is rooted in the Progressivist tradition, although the emphasis of the two groups is different.

Consumerists are more concerned with the quality and price of products than with the structure of companies. Nevertheless, there is an implicit sense of mistrust of large corporations, including the major oil companies, underlying this activism. As at the turn of the century, the fear of massive transfers of wealth to big business, and the political control that is perceived to go with it, is a force underlying this social/political movement.

Another contemporary movement concerned with resources – the ecology movement – has its antecedents in the conservation movement of the last century. Conservationism (or preservationism, as it was often called) was distinct from Progressivism and in fact did not embrace a distinct political ideology. Like today's ecological advocates, preservationism united individuals across the political spectrum who shared a belief in the conservation of the world's resources. Interestingly, although they often are advocated by the same public figures, *consumerism* and *conservationism* have exerted opposing pressures on the exploitation of resources. In general, conservationism advocates the preservation of natural resources and thus slow rates of extraction of fossil fuels. Consumerism, on the other hand, advocates maintaining low prices, which encourages increased extraction of fossil fuels in an unrestrained market. With respect to consumerism, however, free market advocacy is often inadvertent since these activists often ignore market questions or explicitly oppose greater market freedom. Still, it would appear that freer markets would be the logical outcome of their program. Some consumer groups have noted this outcome at least tacitly and have favored outright government control of the market: first, to restrain production and, at the same time, to control prices. Conservationists would tend to endorse greater market control – if not by government then by the major oil companies who would be inclined to reduce production, albeit for their own ends. It has been said that "the monopolist is the conservationist's best friend," and, we might add, the consumerist's worst enemy.

Though they have what in some sense are

contrary interests, consumerists and conservationists have been allied in their mistrust of the public information supplied by the resource industries. Specifically, both have felt that industry-supplied data on fossil fuel supplies are self-serving. According to this view, companies will say that oil and gas supplies are lower than they are in order to keep prices high, when such statements are in the companies' interest. Or they will declare that there is more oil or gas when they wish to encourage demand or influence favorable legislation.

Perhaps the most acute public expression of mistrust occurred during the gasoline shortage of the summer of 1979, which prompted a special government report[2] on the practice of using industry data. In this report, consumer groups challenged the confidentiality of oil industry resource information. Individual oil companies claimed that they needed to protect proprietary resource information from their competitors. The issue raised a fundamental question for democratic government: When (and why) must a private concern make public disclosures that may be against its own interests? It is a question that has important legal and ethical implications.

Whatever the validity of industry data, it should not be forgotten that resource industries are not the only groups with agendas that can affect their choice of data. The same groups that challenge industry have their own biases and predispositions. As we will see, others besides the oil companies have put forward estimates of resources that were later proven to be far from the mark.

## The estimation of fossil resources

The main problem in resolving disagreements about fuel resources is that estimating them is very difficult. In fact, there is no known method for making accurate and certain predictions of future amounts of fossil resources that will be available as energy fuels. This has been true for oil, natural gas, and coal and also is expected to be the case for other fossil resources such as shale and tar sands.

Geologists have tried various techniques for estimating these resources based on physical evidence and statistical data. It is prohibitively expensive to prospect for such natural resources exhaustively over vast land areas by definitive determinations such as drilling. Therefore, various broad survey methods have been tried to keep this cost of information reasonable. One method, the *volumetric method*, relied on surface reconnaissance (often from the air) and mapping to find geological formations that were likely to be rich in resources. If the reconnaissance of the earth's surface were reliable for such identification, the theory goes, then it would be relatively simple to estimate the volume of the resources below these areas. However, surface reconnaissance data have turned out to be highly uncertain in the identification of resource-bearing areas. Earlier in the history of oil exploration, for example, surface identification was quite successful in locating new resources in Louisiana, but a few years later it proved quite poor in Texas.

The most widely used method of large-scale resource estimation is the *logistics curve*. This method simply takes the historic pattern, over time, of extraction and extends it to fit a classic shape appropriate for any finite resource. This result is an S-shaped curve for cumulative production, or, equivalently, the bell-shaped curve for production rates.[3] Examples of such curves may be seen in Figure 2.1a and b, which would be appropriately used for large regions to determine the cycle of exploration of an exhaustible natural resource over time.

The curves are interpreted in three major segments: the *early-time* segment, the *midrange* segment, and the *late* segment (Fig. 2.1). During the early segment (*a*), when the resource is just beginning to be exploited, the rise in the amount extracted each year is a fixed fraction of the cumulative amount extracted to date – that is, the growth is compounded. During this period, the rate curve and the cumulative curve are exponential in time. (Were it not for the finite nature of the resource, exponential growth could continue, if there were a demand for it.)

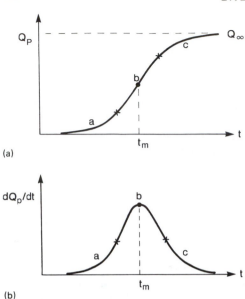

(a)

(b)

**FIG. 2.1.** Logistics curves. (a) Cumulative production versus time. (b) Production rate versus time. $a$ = early-time segment, $b$ = midrange segment, $c$ = late-time segment.

Segment ($b$) of the S-shaped curve is the earliest indication of the finiteness of the resource. Here the curve has reached its maximum slope versus time midrange, and the rate curve passes through a peak (at $t = t_m$) and thereafter drops. This peak in the *rate* of production is a significant signal that the resource will become increasingly scarce.

The late-time segment ($c$) is the period when the cumulative production (Fig. 2.1a) is reaching the limit of the resource: $Q_\infty$ (*ultimate cumulative production* or *ultimate recoverable resource*). Here, as the cumulative curve approaches its asymptotic limit[4] ($Q_\infty$), the rate curve (Fig. 2.1b) falls to zero (zero production). Note also that the total area under the production rate curve must equal the ultimate cumulative production ($Q_\infty$).

This description of the logistics curve shows its appropriateness for describing the time history of the use of a finite resource. Actually, it is but one of many uses of this classic curve. In biology, for example, the growth of a population (e.g., protozoa) in a limited environ-

ment is known to follow the S-shaped curve. In areas of human activity, such as the historic adoption of new technologies (e.g., steamships), the S-shaped curve may be observed in the pattern of development over time. (The correlation of the curve for fossil resources to economic activity will be given later in this chapter.)

Now, in order to understand how the logistics curve is currently used to estimate oil, natural gas, and coal resources, we must add two more curves: one for *cumulative discovered reserves* and another for *remaining proved reserves*. First, we should explain what these terms mean and define their relationship to the production curves above. In terms of formulas:

$Q_p(t)$ = cumulative production
$Q_d(t)$ = cumulative discovered reserves
$Q_r(t) = Q_d(t) - Q_p(t)$ = remaining proven reserves (or simply *reserves*)

where cumulative discoveries are simply assumed to precede production with the same relative shape versus time, as follows:

$$Q_d(t) = Q_p(t + \Delta t)$$

with

$\Delta t$ = the time lag between discovery and production

Also

$$Q_\infty = Q_d(\infty)$$
$$= Q_p(\infty)$$

= ultimate recoverable resource or ultimate cumulative production

Figure 2.2 shows the use of all three curves to describe the time history of a particular resource. Note that the vertical axes of the cumulative (and reserve) curves will be labeled in the appropriate units for the particular resource: billions of barrels (*BBℓ*) of oil, billions of metric tons (*Bmt*) of coal, or trillions of cubic feet (*TCF*) of natural gas. The rates of extraction, on the other hand, should be

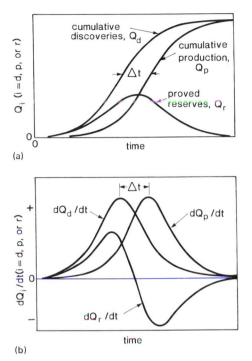

**FIG. 2.2.** Time histories of resource use. (a) Cumulative quantities versus time. (b) Quantity rates versus time. Adapted from Penner & Icerman (1974).

expressed as measures per unit of time: billions of barrels of oil per year ($BB\ell/yr$), billions of metric tons of coal per year ($Bmt/yr$), and trillions of cubic feet of natural gas per year ($TCF/yr$). Special note should also be made of the fact that $reserves - Q_r(t)$ – vary with time. The popular impression is often that reserves are a fixed amount of resources held in reserve for later use, and even the use in the oil and gas industry of the reserve/production ratio:

$$\frac{R}{P} = \frac{Q_r}{dQ_p/dt} \quad [year]$$

gives the impression of a static prediction, whereas its strict interpretation is: "If this year's reserves ($R$) were to be depleted at the rate of this year's production ($P$), those reserves would be exhausted in $R/P$ years." Reserves can, of course, vary due to changes of either discovery

or production. In fact, the shape of the reserve curve can corroborate salient features of the cycle of exploitation.

The estimation of fossil fuel resources is controversial, because no one knows what the ultimate recoverable resource ($Q_\infty$) is for any of them. This is so, at least in part, because when we deem a resource recoverable, we mean that it is *economically recoverable with current extraction technology*. These resources are *conventional*. Resources do exist that are called *unconventional* when they cannot be extracted either for economic or technical (or both) reasons. Technical breakthroughs and price changes could cause an unconventional resource to become conventional and as a result raise the level of the ultimate recoverable resource. But at present there is uncertainty; some unconventional resources may never be technically or economically recoverable.

We can see, then, that judgments must come into play at two levels: first as to the size of the total resource in the ground (*in place*, that is, whether it is economically recoverable or not) and second as to what fraction of it is economically recoverable. When these judgments are combined with the social and economic pressures associated with fossil fuels, controversies ensue. New estimates are accepted or disputed according to whether they support or contradict the current point of view of a given interest group. An optimistic or pessimistic outlook on the part of the estimator can even color the predictions. The various predictions arising from these controversies will usually be represented by production rate curves where the size of the areas under the curves corresponds to various estimates of $Q_\infty$.

Figure 2.3a illustrates how such variations in estimated $Q_\infty$ show up on resource production curves. These curves were given by the eminent geologist M. King Hubbert and illustrate the range of estimates on ultimate, recoverable, world-oil resources ($Q_\infty$). Hubbert is credited as the person who brought home the realization of the imminent limitations of our oil and natural gas resources. His estimate for world oil (made in 1969) is: $1,350 < Q_\infty < 2,100 \, BB\ell$. A more recent study (Wilson 1977) used a value

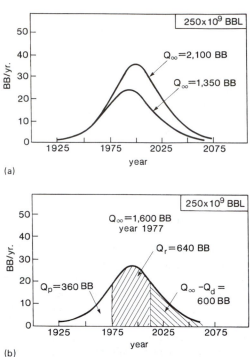

(a)

(b)

**FIG. 2.3.**  World oil – projected exploitation. (a) $Q_\infty = 2,100$ BB$\ell$ and 1,350 BB$\ell$ estimates. (b) $Q_\infty = 1,600$ BB$\ell$ estimate. Adapted from Hubbert (1973) and Wilson (1977).

Finally, it is possible to divide the total area under the production curve ($Q_\infty$) into various cumulative quantities. Figure 2.3b, for example, gives the world-oil production rate for the year 1977 and therefore a vertical line is drawn under the curves at that year. The area under the (rate) curve to the left of the year's line must be the cumulative production ($Q_p$) to that year. Of the remaining area under the curve, the cross-hatched area marked $Q_r$, (Fig. 2.3b) represents the cumulative reserves ($Q_r$) that were estimated for the year 1977. Furthermore, the sum of the two areas $Q_p$ and $Q_r$ should represent the cumulative discoveries ($Q_d$) up to that year, since:

$$Q_d = Q_p + Q_r$$

The last segment represents the remaining resources that are as yet undiscovered. Starting logically from the cumulative-discovery area ($Q_d$), the remaining area under the production rate curve to the right of the $Q_p$ and $Q_r$ areas must be $Q_\infty - Q_d$, the *remaining undiscovered resources*. This uncertain quantity is the object of controversial judgments concerning the size of the remaining resource and the fraction that will be economically recoverable.

of $Q_\infty = 1,600$ BB$\ell$, which is about midway between Hubbert's extremes.

The effect of these different estimates of total resources on production rates, as predicted by the logistics curves, is interesting. The first feature is that the asymptotic time to exhaustion is *not* proportional to the ultimate size of the resource. Hubbert's two world-oil curves, for example, represent a variation of over 50 percent in the estimated resource, but the (asymptotic) times to exhaustion are nowhere near that ratio. The logistics curve predicts that for an increase in the estimated $Q_\infty$, there is an increase in the level of production as well as extension in time. The reasons for an expected rise in production, as well as an extension in time, have to do with the effects of perceived scarcity or abundance on the demand and price of the resource (see Chapter 4).

## Estimates of U.S. and world resources

### U.S. oil resources

The 1982 estimate of U.S. domestic oil resources is represented in Figure 2.4. This logistic curve is an update of Hubbert's estimate of 1980 and is consistent with other recent estimates. Note that the 1980 $R/P$ ratio for U.S. oil was about 10 years and that the remaining undiscovered resources ($Q_\infty - Q_d$) were estimated at about 35 BB$\ell$.

At present, there are widely accepted oil re-source estimates for the United States, although they have been preceded by decades of disputes. The first estimates of U.S. fossil resources were attempted by the U.S. Geodetic Survey (USGS) from 1908 to 1918 using what would now be

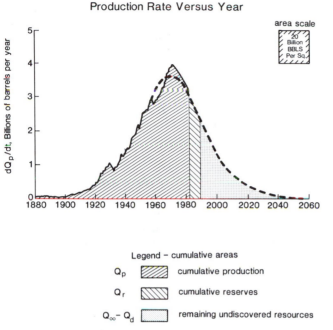

Production Rate Versus Year

FIG. 2.4. U.S. oil – production rate versus year. $Q_\infty = 170\,\mathrm{BB}\ell$, $R = Q_r = 26.5\,\mathrm{BB}\ell$, $P = dQ_p/dt = 3.0\,\mathrm{BB}\ell/\mathrm{yr}$; $Q_\infty - Q_d = 35\,\mathrm{BB}\ell$
*Sources*: World Petroleum Congress (1983), *Oil & Gas Journal* (1987), and U.S. Geodetic Survey (1989). Adapted from IEEE (1977).

considered primitive methods. The intent of these estimates was to aid the then-emerging oil industry, but the industry reacted hostilely to the USGS's conclusions. The estimates were considered too low and raised what was then termed the exhaustion bogey. In retrospect it appears that the estimates were approached from what we will call a *limitationist*[5] viewpoint and the criticisms arose from *expansionist* views. The limitationist geologists at the USGS believed that oil was basically running out and they feared wasting it. They wanted the government to regulate the oil companies' production rates. The expansionist companies, on the other hand, did not want any government restrictions on their extraction at that time.

History soon proved the geologists wrong. Their estimates were made just before large new deposits of oil were discovered in the U.S. Southwest and therefore took place within a period when experience in the cycle of discovery was limited. Perspectives and the nature of debate changed radically with the big findings in Texas and Oklahoma. Glut replaced scarcity as the focus of concern, especially for the major oil companies,[6] who worried about collapsing

prices and profits in the long term. Then in the late 1920s, because of large estimates of the resource, the same companies reversed themselves and advocated production restrictions either voluntarily or by regulation to change and stabilize the market in their favor. Actually, the oil companies faced a dilemma; they wanted production restricted, but were loath to have it regulated by the government.

Meanwhile, the USGS was still issuing pessimistic estimates of remaining oil resources, now motivated apparently by a different conservationist concern – long-term exhaustion. During the 1920s, the oil companies tried to preserve their interests by forming their own committees for collecting data and estimating reserves.

Following World War II, reserves rose undisputedly until about 1967, but exploratory drilling efforts fell off after 1957. Some critics (see Commoner 1979) claimed that this drop-off of drilling effort had much to do with the growing supply of imported oil. This alternative source of cheap oil cut the incentive to drill in the United States and, since drilling is part of the process of discovery, additions to reserves slowed down. Thus, according to the critics, estimates of reserves (and even of recoverable resources) were too low.

There were also questions during this period about the finding rate of oil, which is the rate at which oil is discovered per unit of exploratory drilling. Evidence was growing that during the postwar era oil was becoming more difficult to find in the heavily worked U.S. oil basins. However, even these finding rate statistics, which are used by geologists to extrapolate discovery rates, were not immune to interpretation.

In 1963, A. D. Zapp, a USGS geologist, published an estimate of $Q_\infty = 590$ BB$\ell$ for U.S. oil resources. M. K. Hubbert meanwhile had already made a lower estimate and in 1969 published a figure of 165 BB$\ell$. Whatever the underlying basis of Zapp's optimism may have been, his predictions were supported by the industry (see Blair 1978). The underlying motivation for the industry's enthusiasm for higher resource predictions was said to be a desire for oil import quotas (see Wildavsky & Tennenbaum 1981). These quotas, the critics said, helped to hold off the downward pressure on domestic oil prices caused by increasing supplies of cheap imported oil. Predictions of abundant oil resources at home supported the notion that the United States could afford to restrict oil imports. Shortly thereafter, however, the import quotas were revoked as a result of

consumerist pressure, and imports increased to supply growing demands that domestic production could not supply.

In any event, Zapp's optimism was not borne out by subsequent estimates, which tended to confirm Hubbert. At the heart of the matter has been the assumption of the rate at which new oil would be discovered per unit of effort of exploration (the *finding rate*). Zapp had assumed a constant finding rate, suggesting that oil discoveries would continue at the same rate of previous years. Hubbert, however, assumed that the finding rate would decline, which for the most part it did.

In recent years, government and industry have accepted Hubbert's viewpoint that the rate of production of domestic oil resources had peaked. Unless dramatic changes occur in either the technology or the estimates of the resources in place, production rates will continue to decline.

This conclusion is corroborated if we examine the time history of (cumulative) reserves for U.S. oil, as shown in Figure 2.5. Here we see a curve starting from 1945 and rising to a maximum in the early 1960s. The abrupt increase in 1970 is due to the inclusion of the Alaskan discoveries into the data base, since the curve would have continued in a more smooth decline had it included only the lower 48 states. In general, reserves peaked ahead of the production rate, thus fitting the pattern of a logistics curve.

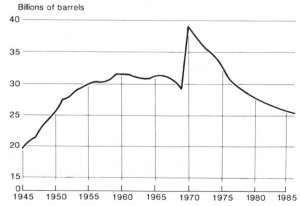

**FIG. 2.5.** U.S. oil – proved reserves versus year. *Sources*: American Petroleum Institute (1977) and *Oil & Gas Journal* (1987–1988).

## U.S. natural gas resources

A similar picture in time is presented by domestic natural gas (Fig. 2.6). Here also we see that the production rate has passed its peak, a fact again corroborated by an earlier peak in reserves (Fig. 2.7). The pattern of yearly discovery subtracted from yearly production also supports this view (the bar graph of Fig. 2.7). Observe also in these results the one-year jump in 1970 due to concurrent discoveries of natural gas associated with oil in Alaska.

The current $R/P$ ratio for U.S. (including Alaskan) natural gas reserves is about 10 years and the remaining undiscovered resources $(Q_\infty - Q_d)$ were estimated to be about 1,118 TCF in 1987. These overall estimates appear to have been relatively non-controversial, except during the period leading up to the passage of the Natural Gas Act of 1978, when widely varying estimates of gas resources were offered to the public.[7]

One source of confusion was a strong public relations effort by the industry aimed at passage of the act, which notably deregulated gas prices. The industry claimed, for example, that "deregulation is sure to bring impressive additional supplies" (*Energy Users News*, Fairchild Business Newspapers, New York, October

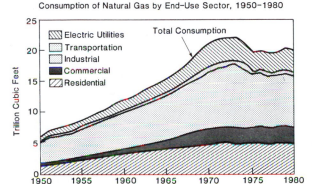

FIG. 2.6. Consumption of natural gas by end-use sector versus year. $Q_\infty \simeq 1,700$ TCF, $Q_\infty \simeq 1,700$ TCF, $R = Q_r = 160$ TCF; $P = dQ_p/dt = 20$ TCF/yr; $Q_\infty - Q_d = 1,188$ TCF. *Sources*: Department of Energy (1982, 1988). *Oil & Gas Journal* (1985–1989) and CONOCO (1989).

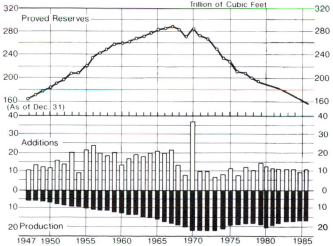

FIG. 2.7. U.S. natural gas reserves. (above) Reserves versus year. (below) Discovery additions and production for each year. *Sources*: American Gas Association (1980), CONOCO (1989), *Oil & Gas Journal* (1982–1988), and U.S. Department of Energy (1988).

1978). Some of these public statements gave a vague impression of large resources without actually issuing data that could be challenged. At the same time, other statements offered estimates that were grossly larger than most others of the period. What these latter sources failed to note, either by omission or by design, was that their estimates were not made on the same basis as conventional estimates, that is, they did not follow the accepted definition of recoverable resources as that portion of the resource (*in place*) that is economically recoverable with current technology. The industry sources referred to gas resources in Devonian shale and geopressurized formations that are *not* extractable economically by current technology. Indeed, such unconventional gas resources (see Chapter 5 and 11), although known to exist, have been incompletely explored and estimates of their size are largely unconfirmed hypotheses.

Throughout the period of the deregulation legislation, there was also a confusing series of fluctuations in gas supplies. In 1976 there was a shortage in several sections of the country; in the Midwest, for example, there were curtailments of supplies to industrial users, causing worker layoffs. In 1979, an excess of supply (termed a bubble) occurred. Although the same partisans of deregulation claimed that the bubble proved the existence of vast new reserves in natural gas, it has since become evident that the excess was merely a short-term fluctuation due to effects on the market of price deregulation.

## U.S. coal resources

A startlingly different picture is presented for U.S. coal resources (Fig. 2.8). While the general shape of this logistics curve is the same as those for oil and natural gas, the scales are very much larger, both in magnitude and time. The peak production rate (almost 4 bmt/yr), for example, is almost an order of magnitude larger than the current rate (about 0.8 Bmt/yr in 1985). The peak is not predicted to occur until the twenty-third century.

This logistics curve is currently in that early time segment where growth is exponential in time – a situation that will continue well into the next century. In short, this resource is so large, compared to present demands, that none of the concerns or controversies over scarcity exist, as they do for oil and natural gas. Note, incidentally, that this logistics curve is drawn assuming that 50 percent recovery is economic; in other words, the total resource in place is twice as large as the amount assumed to be recoverable. The proven reserves area for the coal production curve (Fig. 2.8) is only about 13 percent of the area, indicating that about 87 percent of this huge resource that can be exploited economically using present technology remains undiscovered. The current $R/P$ ratio is almost 200 years.

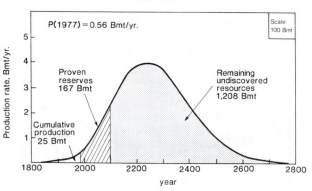

U.S. Coal

P(1977) = 0.56 Bmt/yr.

Scale: 100 Bmt

Proven reserves 167 Bmt

Remaining undiscovered resources 1,208 Bmt

Cumulative production 25 Bmt

Production rate, Bmt/yr.

year

**FIG. 2.8.** U.S. coal – projected cycle of exploitation. *Source:* Adapted from Hubbert (1973) and Wilson (1980).

Despite this huge abundance of coal, it currently supplies only about one-fifth of U.S. annual energy demand. Coal use is not expected to grow rapidly in the coming decades because of the limitations and constraints it places on the environment, public health, transportation, and technological development, as we shall see in following chapters.

# World fossil-fuel resources

Figure 2.3 shows a logistics curve of world oil production. We see that the peak of world oil production is predicted to occur somewhere in the last decade of this century, although the exact prediction will depend on what estimate of ultimate recoverable resource ($Q_\infty$) is used.

The recent history of world fossil-fuel production and projections of resources may be summarized as follows:

*World oil*

Production (1984):    $P = \dfrac{dQ_p}{dt} \simeq 21\ \mathrm{BB}\ell/\mathrm{yr}$

Reserves (1984):    $R = Q_r \simeq 649\ \mathrm{BB}\ell$

Ratio: $R/P = \dfrac{649}{21} \simeq 31$ years

Ultimate recoverable resources (1984):
$Q_\infty \simeq 1{,}866\ \mathrm{BB}\ell$
Remaining undiscovered resources (1984):
$Q_\infty - Q_d \simeq 1{,}314\ \mathrm{BB}\ell$

*World natural gas*

Production (1980):    $P = \dfrac{dQ_p}{dt} \simeq 48\ \mathrm{TCF}/\mathrm{yr}$

Reserves (1979):    $R = Q_r \simeq 2{,}470\ \mathrm{TCF}$

Ratio: $R/P = \dfrac{2{,}470}{48} \simeq 52$ years

Ultimately recoverable resources (1979):
$Q_\infty \simeq 6{,}950\ \mathrm{TCF}$
Remaining undiscovered resources (1979):
$Q_\infty - Q_d \simeq 4{,}480\ \mathrm{TCF}$

*World coal*

Production (1980):    $P = \dfrac{dQ_p}{dt} \simeq 2.45\ \mathrm{Bmt}/\mathrm{yr}$

Reserves (1980):    $R = Q_r \simeq 663\ \mathrm{Bmt}$

Ratio: $R/P = \dfrac{663}{2.45} \simeq 270$ years

Ultimately recoverable resources (assuming 50 percent recovery; 1980): $Q_\infty \simeq 5{,}380\ \mathrm{Bmt}$
Remaining undiscovered resources (1980):
$Q_\infty - Q_d \simeq 4{,}610\ \mathrm{Bmt}$

The long-term world fossil-fuel situation is somewhat similar to the situation in the United States. Coal is again an enormous resource, but will be slow to develop because of technological and environmental limitations. Oil and natural gas, worldwide as in the United States, have more immediate resource limitations.

The most interesting comparisons of national resources are for oil, since the impact of the world market in this commodity is so pervasive. We see that not only is world oil production expected to peak about two decades after the U.S. peak, but that the present world $R/P$ ratio is about three times higher than the U.S. $R/P$ ratio. Not surprisingly, much of this differential is found in the OPEC countries. The most outstanding examples of large oil holdings are found in the Middle East, with Saudi Arabia as the largest holder and producer. This region has close to half of the world's recoverable oil resources, identified reserves of about 500 BB$\ell$, and a current $R/P$ ratio over 50 years, in sharp contrast with the United States. Thus, it is obvious why these countries have been in a position to export oil to countries such as the United States over the past decade, and they are likely to maintain that position.

Nevertheless, conventional oil and natural gas resources worldwide appear to be in the process of exhaustion. The time scales for the predicted production peaks in these two fossil resources is a matter of decades, after which world production rates are expected to slowly decline as they have already in the United States. If these predictions hold true, a worldwide scarcity of these most heavily used fossil fuels will become all too likely. Such a reality and the resulting escalations in the price of energy could have a severe economic impact during the decades when the scarcities occur. However, the finiteness of natural resources also raises major societal and philosophic considerations today.

## Conflicting views concerning resource depletion

If the predictions of impending depletion of energy resources are true, then what should we do now to forestall the consequences? The answer to that question is by no means clear-cut. There is widespread disagreement on what to do and even on whether there really is a problem in the first place. In part, as we have noted before, these attitudes and proposed solutions may reflect the influence of special interests. For instance, a consumerist might argue that resources should be exploited fully and rapidly to keep prices down; a stockholder in an energy company might want to see a policy that leads to tight supplies and high prices to keep company profits up. But the question of resource depletion has led to the advocacy of very different approaches for philosophic reasons as well as those of self-interest. Most of these philosophical positions are variants on what we call the *limitationist* and the *expansionist* perspectives.

The limitationist view stresses that natural resources are finite and that we must deal now with the reality that they will run out. Furthermore, rapid exploitation would be bad even if there were no danger of depletion. It causes pollution and despoils the land; it risks

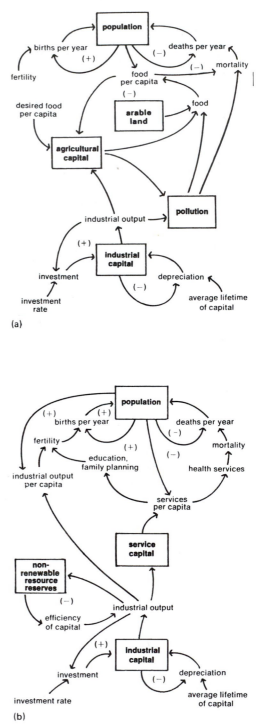

(a)

(b)

**FIG. 2.9.** Dynamic system diagrams. *Source*: Donella H. Meadows, Dennis L. Meadows, Jorgen Randers, and William W. Behrens, III, 1972, *The Limits to Growth: A Report for the Club of Rome's Project on the Predicament of Mankind*. A Potomac Associates book published by Universe Books, N.Y. Graphics by Potomac Associates.

all life by destroying the ecosystem itself. Limitationists are basically pessimistic about the ability of technology to solve these problems quickly and, in fact, are pessimistic generally about what lies ahead for the world because of resource-related problems. Some see truly ominous possibilities for the future if we do not take drastic steps now to limit resource exploitation. They project the possibility of global catastrophes – catastrophes that could at the very least destroy industrial societies and lead to hardship for millions of people.

In the early 1970s, a series of Massachusetts Institute of Technology (MIT) studies on world dynamics and the limits of growth (see Meadows, Meadows, Randers, & Behrens 1974) supported the limitationist point of view. These studies were computer-generated projections of future developments of the worldwide ecological/economic system, and they produced disquieting results. In any one of several scenarios, the MIT group's results (depicted as time-dynamic curves) showed catastrophes, with ultimate saturation levels reached in either food production, population, pollution, or capital investment (see Fig. 2.9). Furthermore, the studies suggested that technology offered no likely solution: "We have shown that in the world model the application of technology to apparent problems of resource depletion or pollution or food shortage has no impact on the essential problem, which is exponential growth in a finite and complex system" (Meadows et al. 1974). These conclusions represent the basic thrust of limitationist thinking.[8] Not surprisingly, the limitationist view often calls for policies that limit (forcibly if necessary) resource exploitation, population, and even world economic growth.

If these projections are correct, it is hard from a moral perspective to argue against a policy response along limitationist lines. How can we commit acts that will lead to the collapse of the world's economy or the destruction of the ecosphere? Destruction of civilization – even if it is a generation or two away – would have to be considered a great evil, and action to forestall and prevent it an equally important good. Yet it also must be noted that to take such actions

would have immediate consequences that are not necessarily beneficial. For example, limiting economic growth would deprive people of material goods; government restrictions on consumption of resources would limit individual freedom. However, these deprivations and restrictions would be based not on an evil that exists, but on one that has been hypothesized for the indeterminate future.

Nevertheless, the Limits to Growth studies had supporters who vigorously defended the conclusions and urged action to prevent catastrophes from occurring. But while the studies added food for thought, they certainly did not settle the issue of whether limitation would be the best or the right policy for the future. Critics (see Cole, Freeman, Jahoda, & Pavitt 1973) challenged the fundamental assumptions of the studies concerning total recoverable resources of all kinds; others attacked the methodology, noting that changing some assumptions made the forecasts far less perilous. It was also pointed out that catastrophe theory was nothing new and previous theories were not borne out by experience.

Perhaps the most famous historical example was the catastrophe theory of Thomas Malthus, a British philosopher and economist (1766–1834). In the early nineteenth century, he predicted that population would outstrip agricultural production and mass starvation would follow. His argument was in part mathematical; population would expand at a geometric rate, he maintained, while agricultural production would expand arithmetically. Malthus had good reasons for his arguments, but they were wrong because he had overestimated population growth and had underestimated the productive potential of agriculture. In other words he had underestimated technology.

Optimism about the potential of technology underlies the expansionist approach (see, among others, Kahn, Brown, & Martel 1976 and Simon 1981). This view holds that resources should be exploited as rapidly as needed for economic growth or, even more strongly, that resources should be exploited at whatever rates are required for economic growth. Expansionists are not necessarily blind to the possibility of resource depletion, but they are not worried about the outcome. As one expansionist put it, "In effect, technology keeps creating new resources" (quoted in Simon 1981).

Limitationist and expansionist positions embody the ambivalence that industrial society has shown for the technology that created it. The expansionist belief in a *technical fix*[9] for problems that beset us is deeply rooted in the experience of the twentieth century. We have seen technological progress at work and believe in its potential. But at the same time, many see technology as a villain. Although a machine may seem morally neutral, its existence is inseperable from its uses. The limitationist view reminds us that employment of technology creates moral and social problems: pollution, population pressures, resource depletion, and so on. The limitationist contends that we need to gain control of technology before we are destroyed by its effects.

To the expansionists; technology will provide the answers to today's problems, as it has to those of the past. An expansionist may grant that oil and gas are finite, but he or she is confident that technology will provide solutions or alternatives before exhaustion occurs. Economist Julian Simon (1981) argues that as we begin to run out, we will soon find other ways of meeting our needs. Pollution, too, will be fixed by technological solutions. As for population growth, the expansionist does not even see it as a problem. Rather it expands the world's pool of labor and means more people to apply their ingenuity toward the problems of the future (Simon 1981).

This line of argument sees strong economic growth as a spur to solutions, since a robust economic system will sustain itself. There will be, the growth advocates say, a great demand for new energy technology, and consequently innovative entrepreneurs will have incentive (high profits) to create and develop the means to solve resource depletion. In the meantime, we can enjoy economic growth and all nations can raise their standard of living.

The expansionist view is optimistic on both the future of technology and the capacity of the free market economic system to adjust itself to cope with problems unrelated to near-term profits and losses (Simon 1981).[10] Technology in the context of the free market becomes not just a more efficient approach, but a way to better the lot of all humanity. Like the limitationist argument, the expansionist approach is clearly a moral position, but the focus is changed. It seeks to maximize both economic growth and liberty simultaneously, now and in the future, although it is based on an assumption – not a certainty – that the future will find a way to take care of itself.

The issue of uncertainty is important. The expansionist and the limitationist offer estimates and projections of the future. Neither can offer proof. Technological progress, as we will see in later chapters, cannot be assumed at all, much less within the appropriate time frame. As for the Limits to Growth studies, they cannot be proven right or wrong. Historic failures of catastrophe theories do not mean that new ones will be wrong. It can be argued that these projections are more likely to prove valid than those of the past because they are based on a larger body of information than Malthus or other early theorists possessed. Actually, the limitationist scenario is right in one respect; fossil fuel resources – oil, natural gas, and coal as well as other unrenewable resources – are finite and must run out at some point. But where is that point and what will happen in the meantime? We might ask: Is it morally acceptable to infringe on human freedom now when the outcome is highly uncertain?

It is interesting to note that the Limits to Growth studies initially were denounced as reactionary. Whereas the limitationist view would seem to be associated with a liberal political outlook – government intervention and regulation of corporate activity, the environment, and so on – the studies were interpreted to advocate, by implication, the status quo. In other words, the conclusions suggested that the rate of industrialization be reduced, serving to delay advances in underdeveloped countries, while leaving the industrialized world at close to a current (and at a high) level of prosperity. Thus, it was charged that advocates of slowed expansion were not liberal or farsighted, but rather against progress.

The debate subsequently entered U.S. politics, where advocacy followed a more expected ideological path. Political liberals took up the limitationist line; the conservatives followed the expansionist point of view. This is apparent in comparing the views of Jimmy Carter's administration with those of his successor, Ronald Reagan. The Carter *Global 2000 Report* expressed concerns over the capacity of resources and agriculture to supply expanding populations, and it advocated long-term global analysis leading to vigorous public policy to deal with future problems. On the other hand, the Reagan administration emphasized expansion of production. Consistent with this view were the programs pursued by the Reagan Department of Interior and Environmental Protection Agency, which sought to deemphasize environmentalism and conservationism on the grounds that they restrain production and retard economic growth.

However, these positions reflect something else: the immediate state of the world. In the late 1970s, oil price hikes, natural gas shortages, and other signs of resource problems made the Limits to Growth projection appear all too plausible, all too soon. But the oil and gas glut – and falling prices for energy – in the 1980s made the same scenario appear, if not completely absurd, at least far in the future. The plausibility of one scenario at any given time is likely to affect policy. This may be a poor way to approach the issue because perceptions of the problems of resource depletion may change from year to year, making long-term policy choices in the United States or the rest of the world unlikely. Policymakers will typically find it difficult to take actions that appear to be contradicted by current conditions.

Of course, there is no way to determine which side is right in this debate. Both sides make ethical, social, economic, and political points. Both also claim allegiance to similar values – personal freedom, justice, protecting the environment. They would differ, however, in

the system of those values, that is, the hierarchy in which they would rate values would differ. Thus, for the expansionist, personal freedom might have a higher value than protection of the environment, while for the limitationist the situation would be reversed. Both viewpoints also differ in their assessment of the risks involved, or at least the acceptability of those risks – although that in itself suggests a system of values. In other words, to say the risk of doing $x$ is acceptable is to place the benefit from doing $x$ above the possibility of $x$ turning out badly. But then, any conclusions we are likely to arrive at will be made on the basis of weighing all the factors according to our own value systems.

If we adopt a limitationist view, we should note that the actual formulation and implementation of policy becomes enormously difficult. The issue of resource depletion is global; every nation could suffer if we do nothing now, and every nation would probably play a part in creating or forestalling the catastrophes. Unlike a policy on nuclear power, we can go only a small way in solving it within this or any one country. No one has yet begun to tackle the enormity of trying to set global energy policies, where conflicting interests, sovereign prerogatives, and old enmities inevitably arise. In the meantime, industrial society is not trying to limit growth or resource exploitation. It has in effect adopted the expansionist line. Thus, no policy is in itself a policy decision.

## Notes

1. The industry, as represented by the American Petroleum Institute (API), has maintained committees for estimating oil reserves and production capacity since 1920. A similar activity for natural gas has more recently been instituted by the American Gas Association (AGA). The reserves data and production statistics provided by the API/AGA are the principal source of such information to the U.S. government today on domestic oil and natural gas.
2. "An Evaluation of the Adequacy and Reliability of Petroleum Information," U.S. Department of Energy, Office of Consumer Affairs, January 1980.

3. Since cumulative production varies with time, it is described as a mathematical function of time: $Q_p(t)$ and the production rate is described as the derivative in time: $P = dQ_p/dt$.

4. The *asymptotic limit* is a strict mathematical property of the curve as it approaches the ultimate value $(Q_\infty)$ as time increases to infinity. For practical purposes, a curve reaches the ultimate (asymptotic) limit when the amount remaining is negligible (e.g., less than 0.1 percent).

5. Historically, this is synonymous with the preservationist or conservationist outlook (see Wildavsky & Tennenbaum 1981).

6. Wildavsky and Tennenbaum (1981) have an interesting analysis of the markedly different perspective on prices and profits of the smaller independent oil companies throughout the 1920s and 1930s.

7. Representatives of the AGA gave reserve data and remaining-resource estimates close to those cited here (advertisements appeared in *Energy Users News*, Fairchild Publications, New York, July 17 and October 23, 1978). However, some business journals (for example, *Business Communications*, BCR Enterprises, Inc., Hinsdale, IL, Summer 1977) offered reserve estimates that were several times higher than the figures cited here and spoke of "additional reserves" in Devonian shale and geopressurized methane that were an order of magnitude larger than the $Q_\infty - Q_d$ estimate given above. Also, one group of gas producer companies (Pitts Energy Group, Dallas, Texas) in an advertisement appearing in *Energy Users News* (1978) cited "deep, high-pressure, saltwater" sources on the Texas Gulf Coast that were an order of magnitude larger than most $Q_\infty - Q_d$ estimates.

8. Notable limitationist arguments, especially those by Erhlich and Hardin, preceded the Limits studies. See discussion and bibliographic references in Chapter 5. Since the Limits study, World Bank economist Herman Daly has been a leading proponent of limitationism. In his view, an economy should develop in a steady state to preserve the world's ecosystem. The steady state would limit or disallow continued growth, but he distinguishes growth from development, which to him means improvement.

9. An energy policy study report by the Ford Foundation in 1974 used the phrase "technical fix" to apply to a particular energy scenario. Here we mean that society will develop appropriate technological solutions for many of its problems.

10. Technological optimism as a philosophic position has had a long tradition. In the nineteenth century, for example, Scottish philosopher Andrew Ure (see Bibliography) wrote of the "blessings which physio-mechanical science has bestowed on society," and for its ability for further "ameliorating the lot of mankind." Ure also believed that technology required a free market economy to flourish.

# Bibliography

Barber, W. J. 1967. *History of Economic Thought.* Penguin, New York.

Blair, J. 1978. *The Control of Oil.* Random House, New York.

Burnett, W. M., and S. D. Ban. 1989. "Changing Prospects for Natural Gas in the United States." *Science*, Vol. 244, pp. 305–10, April 21.

Cole, H. S. D., C. Freeman, M. Iohoda, and K. L. Davitt. 1973. *Models of Doom – A Critique of Limits to Growth.* Universe, New York.

Commoner, B. 1979. *The Politics of Energy.* Knopf, New York.

Dorf, R. C. 1978. *Energy, Resources, and Policy.* Addison-Wesley, Reading, MA.

Forrester, J. W. 1971. *World Dynamics.* Wright-Allen, Cambridge, MA.

Heiman, E. 1974. (reprint). *History of Economic Doctrines*, Oxford University Press, New York.

Hubbert, M. K. 1973. "Survey of World Energy Resources." *Canadian Mining and Metallurg. Bull.*, Vol. 66, pp. 37–53, July. (Also reprinted in M. G. Morgan, ed. 1975. *Energy and Man.* IEEE Press, New York.)

Kahn, H., W. Brown, and L. Martel. 1976. *The Next 200 Years: A Scenario for America and the World.* Morrow, New York.

Meadows, D. H., D. L. Meadows, J., Randers, and W. W. Behrens. 1974. *The Limits to Growth,* Signet, New York. (Second edition in 1975. Signet Books, New York.)

Penner, S. S., and L. Icerman. 1974. *Energy. Volume I. Demands, Resources, and Policy,* Addison-Wesley, Reading, MA.

Schurr, S., J. Darmstadter, H. Perry, W. Ramsey, and M. Russell. 1979. *Energy in America's Future: The choices before us.* Johns Hopkins, Baltimore, MD. (See Chapter 7: "Mineral Fuel Resources.")

Simon, J. L. 1981. *The Ultimate Resource.* Princeton University Press, Princeton, NJ.

Stobaugh, R., and Yergin, D. 1979. *Energy Future: Report of the Harvard Business School.* Random House, New York. (Third edition in paperback. 1983. Vintage Books, New York.)

U.S. Department of Energy, Energy Information Agency, *Monthly Energy Review.*

Ure, A. 1967 *The Philosophy of Manufactures* Augustus M. Kelley, New York.

Wildavsky, A., and E. Tennenbaum. 1981. *The Politics of Mistrust.* Sage, Beverly Hills, CA.

Wilson, C. L. project director. 1977. *Energy: Global Prospects 1985–2000, Report of the Workshop on Alternative Energy Strategies (WAES).* McGraw-Hill, New York.

Wilson, C. L. project director. 1980. *Coal – Bridge to the Future. Report of the World Coal Study (WOCOL).* Ballinger, Cambridge, MA.

# 3

# Conventional conversion of energy

## Energy conversion

To power industrial society, energy resources have to be converted into useful work. In fact, the technical wonders of the modern world, by and large, depend on a handful of what we will call *conventional* energy conversion technologies, the principles of which have been known for a century or more. Most of these technologies are *heat engines*. Hydropower (water power) also has long been a source of energy; it currently provides about one-fifth of the electric generating capacity in the United States. But that represents only about 4 percent of energy consumption of all types, and it is the only nonheat conversion technology widely used. Some alternative energy technologies, such as solar conversion and wind conversion, are also not of the heat-engine type. These alternatives, however, have achieved only a small penetration of the energy market to date, despite the attention they received during the 1970s (See Chapters 11 and 12).

In fact, energy technologies have been based primarily on conversion of thermal energy – heat – from the start of the Industrial Revolution to the present day. This concentration on thermal conversion seems likely to continue into the future. Not only have there been huge investments in current technology, but some prospective alternatives, notably synthetic fuels or *synfuels*, are also based on thermal

conversion. Even nuclear technologies, both nuclear fission and nuclear fusion, are forms of the heat engine. Only a major technological – and commercial – breakthrough in solar-direct conversion (see Chapter 11 and Appendix C) would change this historic reliance on thermal conversion.

## Conventional heat conversion

The category of conventional heat engines includes two particularly important varieties: the steam power plant, by far the major means of generating electricity, and the internal-combustion engine, the principal power source for transportation in industrial society. Indeed, two slightly different forms of internal-combustion engines – the gasoline and diesel engines – provide about nine-tenths of all passenger travel and about one-third of all freight transport in the United States. The diesel engine has also found a place in small-scale electric generation. Another form of internal-combustion engine, the gas turbine, figures prominently in electric generation too.

But electricity is still primarily generated by external-combustion steam engines. And we will begin our discussion of conventional energy technologies with a consideration of steam engines and of the steam electric power plant.

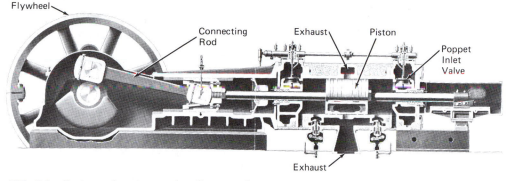

**FIG. 3.1.** Reciprocating steam engine. Courtesy of Skinner Engine Company, Erie, PA.

## Steam engines

Steam engines operate on the Rankine cycle[1] and follow thermodynamic paths that follow changes in pressure and volume of the steam as the machine goes through its cycle of operation. In part, the paths of the Rankine cycle are determined by the particular characteristics of the working substance as it changes from water to steam or from steam to water. (See Appendix A for a review of thermodynamic principles.)

A classic example of the Rankine cycle is the reciprocating steam engine, which powered factory machinery in the nineteenth century and is similar to those used on railroad locomotives earlier in this century. In Figure 3.1, we can see a piston inside a cylinder that is connected to a flywheel by a drive shaft. There are also inlet and exhaust valves on the cylinder. The basic

action of the engine is probably familiar. First, the steam enters from the right into the cylinder. As the steam expands, the piston is driven horizontally to the left. But after the steam thrust, the motion imparted to the flywheel forces the piston back to the right. In the process, the spent steam is expelled. The motion of the piston and the flywheel also controls the valves that regulate steam input and exhaust. As long as the steam is fed to the piston chamber, the reciprocating motion will continue. The reciprocating steam engine is an example of an *open* thermodynamic cycle in that the working substance – steam – is exhausted from the system after its use.

The changes that take place in the pressure and volume of the steam as the cycle of reciprocating motion takes place can be described on a pressure-volume (p-V) diagram. A path on a p-V diagram describes the changes

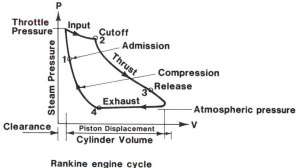

**Rankine engine cycle**

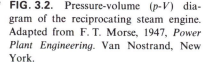

**FIG. 3.2.** Pressure-volume (*p-V*) diagram of the reciprocating steam engine. Adapted from F. T. Morse, 1947, *Power Plant Engineering.* Van Nostrand, New York.

in pressure that accompany changes in volume of the working substance (in this instance, steam) of a heat engine (see Appendix A).

In Figure 3.2, we see the four Rankine cycle p-V paths:

Path 1–2: steam (heat) input
Path 2–3: steam expansion (thrust)
Path 3–4: steam rejection (exhaust)
Path 4–1: compression

We can also consider this diagram in terms of the positioning of the valves that control steam input and exhaust.

Point 1. *Admission* of steam starts as the input valve is opened; the exhaust valve remains closed.
Point 2. *Cut-off* of the input valve occurs and thus both ports are closed for path 2–3.
Point 3. *Release* as the exhaust valve opens while the input valve remains closed.
Point 4. *Compression* starts as the exhaust valve closes; again, both valves are closed.

Modern steam plants no longer use reciprocating engines to deliver mechanical work. Rather they use steam turbines that usually operate in a *closed* thermodynamic cycle. In a closed cycle, the working substance is constantly recycled without any exchange of its mass. In terms of its thermodynamic description, the p-V paths are simply retraced as the cycle is repeated over and over again. A schematic diagram of the closed cycle for a modern steam-turbine power plant is shown in Figure 3.3. The water-steam working substance is shown there to be circulating around a physically closed path: from the *boiler* to the *turbine* to the *condenser* to the *pump* and back to the boiler again. In contrast, in the reciprocating engine, the steam is discharged at the end of each cycle.

The thermodynamically closed cycle of the steam plant is depicted on the p-V diagram in Figure 3.4. Starting at point 1, water under pressure is fed into the boiler, where heating changes it to steam. The boiler output (point 2) is in fact high-pressure, high-temperature steam that is fed to the turbine. The steam expands through the turbine, causing it to spin, and yield useful work output. After the temperature and pressure drop from expansion, the steam exits the turbine (point 3) and is fed into the condenser. There the steam's latent heat of vaporization (the energy required to turn water to steam) is given up to cooling water, without any exchange of mass from the working fluid to the cooling water. At the condenser the working fluid turns to water – point 4 in the cycle – at essentially atmospheric pressure. A pump then raises the water pressure to that in the boiler, and the cycle repeats, starting again from point 1. This cycle of operation is a closed Rankine cycle.

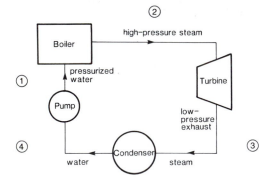

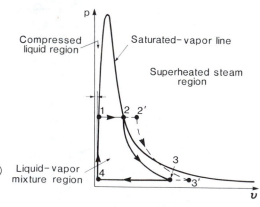

**FIG. 3.3.** Schematic diagram of a steam–turbine power plant.

**FIG. 3.4.** P-V diagram of a closed-cycle steam plant. Adapted from Faires (1948).

The p-V paths on Figure 3.4 for the closed Rankine cycle are similar, but not identical, to those for the open reciprocating steam engine. The heat input path (1–2) of both is a constant-pressure process that results from the liquid-to-vapor transition. Over most of this p-V path, the boiler is adding the necessary heat to change the *phase* of the working substance from liquid to vapor. The p-V expansion path (2–3) for the turbine is similar to that of the reciprocating engine. The exhaust path (3–4) is a constant-pressure process for the closed cycle, as it was for the open-cycle steam engine. Finally, the compression (path 4–1) in the closed Rankine cycle takes place at essentially constant volume through the action of the separate pump.

In the closed Rankine cycle, the working substance is typically brought to a superheated condition by continuing to add heat beyond the point where the water turns to steam. When this is done (in superheater coils in the boiler), the heat–input path is extended. Point 2′ in Figure 3.4 is such a superheated state. The path extension 2–2′ is merely a horizontal line, since the superheated extension is still at constant pressure. However, the steam temperature rises in this superheated extension, which results in an increase in the thermal efficiency. That is, the ratio of energy available for useful work to input energy increases (see Appendix A).

The thermal efficiency of the modern steam turbine plant is often enhanced by other means. For example, some of the steam may be diverted back to the boiler before it passes through the turbine. The steam is then reheated and sent back to the turbine, where it reenters near the point where it has been diverted. Reheating can improve efficiency by as much as 3 percent. (See Appendix A for further discussion of thermal efficiency.)

## Steam power plants

In a steam power plant, we can see the physical realization of the Rankine steam cycle. The entire process of energy conversion is depicted in Figure 3.5, starting with the fuel input (coal) at the top and ending with the electric output (the high-voltage transformer) on the right.

If we examine the components of the thermal core (Fig. 3.5), beginning with the boiler, we see:

1. *Fuel loading* – a conveyor carries the coal across the top and drops it into a *storage bunker*.
2. *Fuel input* – a *traveling grate* moves the coal into the furnace, where it combines with a *forced draft* of heated air to achieve combustion in the *fire box*.
3. *Water input/steam output* – boiler tubing in the furnace fire box contains the working substance, which enters as water and leaves as steam on its way to the steam turbine.

The combustion of fuel – together with the transfer of its heat – in a modern steam plant takes place at high temperatures. As we note in Appendix A, the temperature at which the input heat is supplied has a critical effect on the thermodynamic efficiency (that is, the amount of energy output from a given energy input) of the cycle. The higher the input temperature, the higher will be the efficiency that can be attained. The highest temperature in the fossil-fired power plant is at the flame itself, typically around 3,000°F. The hot combustion gases in the furnace chamber in the immediate vicinity of the flame, however, range from 2,100 to 2,300°F. These combustion gases actually transfer the heat into the working fluid in the boiler tubes. But as this transfer takes place, there is a considerable temperature drop; the working fluid will only be in the 500°F range for pressurized water and about 950°F for superheated steam.

Tracing the remainder of the closed cycle of the plant, we find:

4. *Steam drum* and *superheater* feeding the turbine.
5. The *turbine* itself.
6. The *condenser* directly below the turbine.
7. *Boiler feed pump* feeding the pressurized water back to the boiler.

The turbine, of course, is where heat energy is converted into mechanical work. A simple

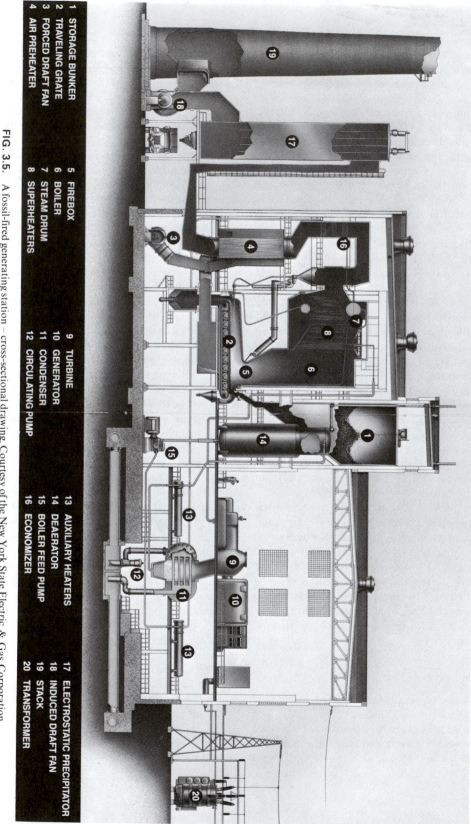

**FIG. 3.5.** A fossil-fired generating station – cross-sectional drawing. Courtesy of the New York State Electric & Gas Corporation.

1 STORAGE BUNKER
2 TRAVELING GRATE
3 FORCED DRAFT FAN
4 AIR PREHEATER
5 FIREBOX
6 BOILER
7 STEAM DRUM
8 SUPERHEATERS
9 TURBINE
10 GENERATOR
11 CONDENSER
12 CIRCULATING PUMP
13 AUXILIARY HEATERS
14 DEAERATOR
15 BOILER FEED PUMP
16 ECONOMIZER
17 ELECTROSTATIC PRECIPITATOR
18 INDUCED DRAFT FAN
19 STACK
20 TRANSFORMER

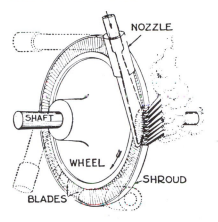

**FIG. 3.6.** Simple impulse steam turbine. Adapted from Severns & Degler © John Wiley & Sons (1948).

example of a turbine is shown in Figure 3.6. Called an impulse turbine, it consists of a bladed wheel set on a shaft. Essentially, steam is forced out through the nozzle at high pressure and is directed to the blades of the turbine wheel. The impulse of the jetting steam turns the wheel, which turns the shaft for useful work. In a second type of turbine, the reaction turbine, steam enters through passages in stationary blades and turns other sets of blades as it builds up velocity and pressure. A turbine can be a composite of both types, making use of impulse and reaction processes, or it can be a simple impulse or a simple reaction turbine, depending on its design.

The next major component in the closed Rankine cycle is the *condenser* (Fig. 3.7). This condenser is an early version of a modern surface condenser, in which the steam and condensate do not come into direct contact with the cooling water, but rather only undergo heat transfer. In the condenser, steam enters through the top and passes over metal tubes carrying the cooling water. Upon contact with the cool tubes, the steam turns to water, which drops to the bottom of the condenser and flows through the outlet. The condensing of the steam is accompanied by a pressure drop, which creates a vacuum and thereby reduces the *back pressure* into the turbine exhaust (see Appendix A for a discussion of the Rankine cycle). The heat of condensation from the steam is transferred almost entirely to the cooling water circulating in the tubes and is carried away.

The final major component in the closed Rankine cycle is the *boiler feed pump*, which returns condensed water to the boiler at the proper pressure. In its simplest concept the feed water pump merely raises the pressure of the liquid condensate. However, the pressures required by modern boilers can be quite high – in a range of 1,000 pounds per square inch (psi).

The condenser not only transfers heat from steam into cooling water, it also transports waste heat to the outside environment. If we return to Figure 3.5, we see at the bottom the loop of cooling water, with the circulating

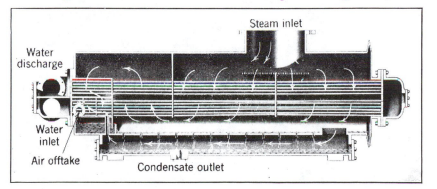

**FIG. 3.7.** Longitudinal section of a small surface condenser. *Source*: Severns & Degler © John Wiley & Sons (1948).

pump, that passes the water through the condenser. When the cooling water is discharged, its temperature can be sufficiently high to raise the temperature of the body of water it drains into. In the early days of condenser-equipped power plants (late nineteenth and early twentieth century), there was little concern about the discharge of waste heat. However, with the development of large-scale plants, biologists discovered that temperature changes could have noticeable effects on aquatic life. Spurred by the environmental movement of the last few decades, governments have established regulations on thermal pollution, limiting the permissible temperature increase of a body of water, be it a river, a bay, or a lake. These regulations sometimes require electric utilities to construct cooling ponds or towers to reduce the temperature of the cooling water being discharged. At some sites, particularly in arid regions, cooling towers are needed, regardless of regulations, in order to recycle the coolant back to the condenser.

The operating principle of cooling ponds or towers is the same, namely partial evaporation, a commonly known cooling mechanism. In cooling ponds or towers, a portion of the water evaporates, taking heat from the remaining water and thereby cooling it. Cooling in a pond is achieved from evaporation by exposing a large surface area of a shallow body of water to the air, thus maximizing the ratio of cooling area to volume of water. Even more effective cooling is achieved in a tower (Fig. 3.8), in which water descends in sprays or sheets over baffles through an upward draft of air. The air draft can be achieved through a natural convective flow or can be forced with fans.

One last component of the thermal plant is worth noting – the air preheater. The purpose of a preheater is to improve the plant's efficiency by recovering some of the heat exhausted to the stack after combustion. This partial recovery of heat is accomplished through a heat exchange from the hot stack gases to the incoming flow of air. (The two streams of gas are physically separate.) Figure 3.9 illustrates the actual equipment used in a large-scale power plant.

Air preheating is external to the closed Rankine cycle and therefore should not be thought of as an improvement of the Rankine cycle's efficiency, as is the case with the reheating of steam. However, air preheating eliminates part of the need for heat from combustion. That means less fuel can supply the same heat input to the working substance.

FIG. 3.8. Power plant cooling tower. *Source*: The Marley-Mouchel Co. *Note*: A photo of this particular power configuration has been included because the cooling tower depicted here has become the symbol of the contemporary power plant and particularly the nuclear power plant. But it represents only one part of thermal plant operation and is not distinctive of nuclear plants.

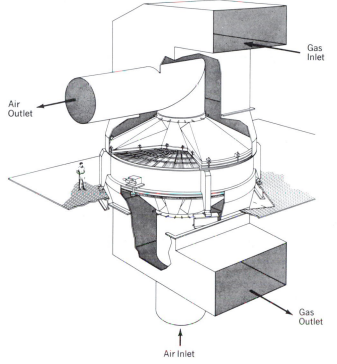

Gas
Inlet

Air
Outlet

Gas
Outlet

Air Inlet

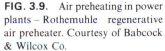

**FIG. 3.9.** Air preheating in power plants – Rothemuhle regenerative air preheater. Courtesy of Babcock & Wilcox Co.

## Hydroelectric generation

Although we will focus most of our attention on heat engines, we should not entirely overlook water power. The conversion of falling or flowing water to useful work is an ancient technology. Before the development of the steam engine, water wheels were used as engines during the early industrialization of Europe. In the United States, the availability of water determined the locations of the earliest factories, such as the textile mills in New England.

Even though hydrocapacity grew in absolute terms during the nineteenth century, the huge demand for mechanical power brought about by the Industrial Revolution was far beyond the capacities of small hydropower sites in the Northeast. Consequently, coal used with steam engines supplied industrial power at first. Later in the century, coal powered electric plants and the electricity they produced supplied an increasing proportion of the energy burden. Electric generation grew rapidly (usage doubled every decade for more than five decades) during the twentieth century and supplied most of the power needed by industry, as well as a great deal of the energy used for residential and commercial purposes.

Much of electric power generation required thermal-type plants. Today, about four-fifths of the electric energy is generated by thermal plants (fossil-fired and nuclear). Still, the remaining 20 percent is generated by water power. Most hydroelectric development in the United States has either utilized sites of tremendous power potential, such as the Niagara River or Hoover Dam, or has been part of large regional development schemes, as in the Tennessee Valley or Pacific Northwest watersheds. In the expansion of electric power earlier in this century, not only were smaller water power sites ignored for hydroelectric

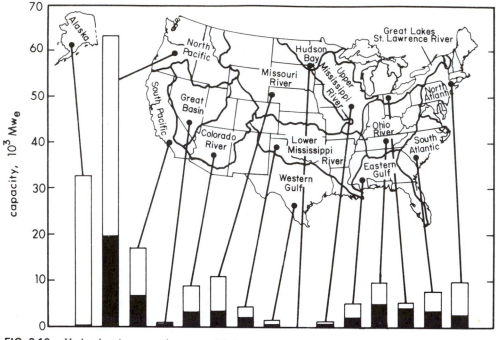

**FIG. 3.10.** Hydroelectric generating potential in the United States. *Source*: U.S. Federal Power Commission (1972).

development, but in many cases existing small-dam facilities were closed down in favor of the larger fossil-fired plants.

But actually, these small sites if added together could provide a great deal of power. This potential remains undeveloped, as Figure 3.10 shows. In some regions less than half the potential hydro generating capacity has been tapped. The scale for these bargraphs is important. Each unit represents $10^3$ megawatts (MW) of electric power, which is approximately equal to the output capacity of a large, present-day thermal power plant. In other words, full development of hydropower in the United States could replace several dozen coal or nuclear power plants. A similar situation exists in other parts of the world, particularly in underdeveloped nations (see Chapter 5).

Prior to the energy crisis era of the 1970s, the conventional wisdom of the electric power industry was that hydropower was fully developed in the United States. What was

meant, of course, was that large hydrosites had been developed. However, with the advent of expensive fuels, rapidly escalating nuclear plant construction costs, and controversies over the environmental consequences of both coal and nuclear power plants, small hydrosites came to be viewed in an entirely different light, and a number of small hydroprojects were undertaken, but the potential remains for considerably more development.

## Hydroelectric basics

Hydroelectric generation is based on the principle of *potential* energy. Natural processes of evaporation and rain place volumes of water at various altitudes. Bodies of water at elevations above some point of use can provide energy for useful work. As the water falls from the higher elevation, the potential energy of the water is transformed into kinetic energy (see Appendix A for a discussion of the forms of

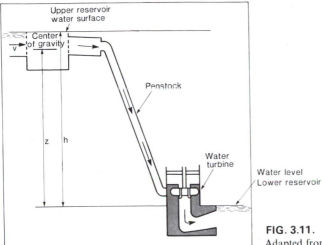

FIG. 3.11. Principles of hydroplant operation.
Adapted from Davis & Sorenson (1969).

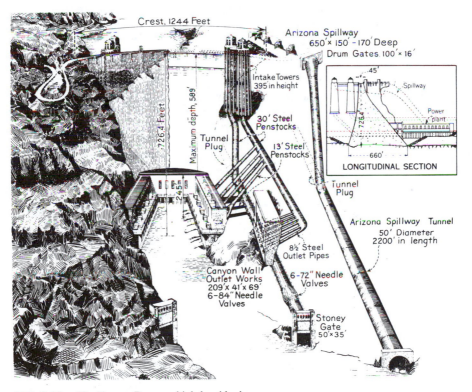

FIG. 3.12. The Hoover Dam – a high-head hydroelectric plant. *Source*: U.S. Department of Interior
(1972).

energy). This kinetic energy is then converted into mechanical form by a hydraulic turbine. The conversion to kinetic energy may occur gradually over a downhill slope or over a sharp drop but the overall principle is the same. This general principle is illustrated in Figure 3.11.

Hydrofacilities are grouped according to their *head*, which is the altitude of the fall before the water does its work. At a *high-head plant*, the drop is in the range of 500 feet or more; a *low-head plant* usually has a drop under 65 feet; those in between are termed *medium-head plants*. A clear example of a high-head plant is the Hoover Dam on the Colorado River in Nevada (Fig. 3.12).

Hydroplants, like other types of generating plants, are also grouped according to the size of their electric power output. The Hoover Dam is both a high-head and a large-scale plant, having an electric power output capacity of 1,344 MW. The enormous James Bay Complex in Quebec, if fully developed, will have an output capacity close to 25,000 MW.

Large-scale output is not achieved using high heads alone. Rather, high-output capacity is the product of the head (height of the reservoir measured in feet) and the rate of water flow (gallons per second). Accordingly, large-scale output power can be achieved with a low-head

plant having a large volume of flow. An example is the Robert Moses hydrostation on the St. Lawrence River, which has a generating capacity of 800 MW, but a head of only 30 feet.

Small-scale hydroelectric plants typically operate in the range of a few kilowatts (KW) to about 5 MW. Most small hydroplants have low heads, such as the dam shown in Figure 3.13. The flow of the river as well as the pressure of the height of the head behind the dam supply the power. The output obtained for the normal volume of flow is called the *run of the river*. The hydroplant for the dam in Figure 3.13 operates essentially on the flow volume of the run of the river, with virtually no head.

## Conversion to electricity

The conversion of hydropower and steam power to electricity is based on the same concept. Energy either in the form of falling water or steam turns a turbine that runs an electric generator. Since most electricity is generated in thermal plants, we will focus on the conversion of thermal energy to electricity. To generate electricity in a steam power plant, a turbine is connected to an electric generator by a common rotation shaft (Fig. 3.14). The turbine is the *prime mover*, meaning that it supplies the mechanical work to rotate the shaft. (In the case of hydrogeneration, the hydraulic turbine performs the same task.)

Electricity is generated according to the principles of *electromagnetic induction*, by which an electric voltage is induced in a coil of wire by a rotating magnetic field (see Appendix A). Two types of electricity can be generated – direct current (d.c.) or alternating (a.c.). But since virtually all centrally generated electric power is a.c., we will focus on the conventional a.c. generator as depicted in Figure 3.15.

Essentially, an a.c. generator has two parts: a stator and a rotor. A stator (Fig. 3.15 a) is a fixed structure that holds the coils of wire in which voltages will be induced. The stator structure is made of a laminated steel, which channels the magnetic flux within the machine.

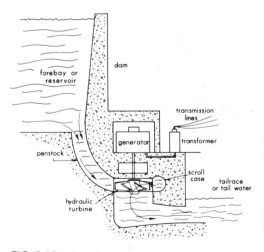

**FIG. 3.13.** Low-head hydroelectric generation. Courtesy of City of Seattle, Light Department.

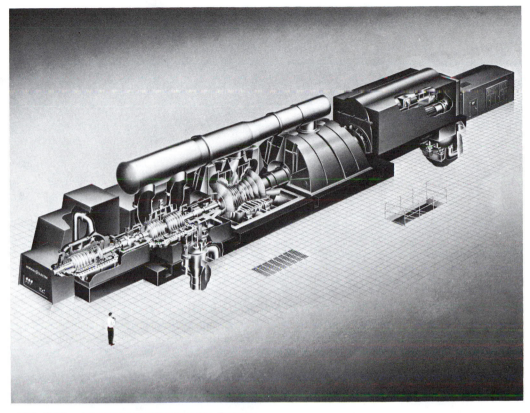

**FIG. 3.14.** A large steam turbogenerator. Courtesy of General Electric Company.

Steel concentrates and guides the magnetic flux of the rotor poles as they pass from one stator coil to the next at the periphery of rotation. The purpose of the laminations is to reduce magnetically induced *eddy currents*, which cause electrical losses (see Tarboux, 1946).

The rotor (Fig. 3.15 b), which is driven by the turbine, contains electromagnets. At the periphery of the rotor are magnetic *poles* that are magnetized by *field coils* placed just inside the radius of the pole faces. This particular example is a *four-pole machine*, in which two pairs of north-south (N-S) poles are created by winding field coils on the rotor (shown as single turns in Fig. 3.15 b). The field coils receive electric current from a separate d.c. source, which actually can be produced by a d.c.

generator on the same shaft as the main a.c. generator.

As the magnetic poles turn on the rotor past the coils embedded in the stator, the a.c. voltage in each coil alternates in polarity, positive and negative. A smooth alternation of polarity ideally traces out a sine waveshape in time (see Appendix A), with a repetition rate of sixty times per second (60 Hertz[2]). If a current flows from the stator coil, it too has a smooth waveform with a repetition rate of 60 Hertz (Hz). These two waves together determine the power output, because the output of a generator at any moment equals the product of voltage times current (and is measured in watts, see Appendix A). Of course, the power output will then also alternate as a sine wave. The *average*, over time,

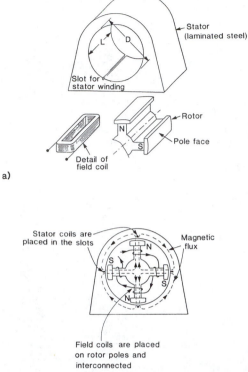

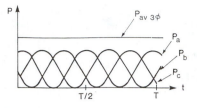

**FIG. 3.16.** Three-phase power waveforms versus time.

**FIG. 3.15.** Alternating current generator. (a) Disassembled parts. (b) An assembled four-pole machine. Adapted from Elgerd (1977).

of a generator's power output waveform represents the useful energy the generator delivers. That is, the energy that can be converted back into mechanical work by a motor or can be converted back into heat for customers connected to the power grid.

But if we look at these alternating waveforms of voltage and current, we would surmise the power also will vary in time. In order to avoid having a fluctuating supply of power, electric systems are usually operated on a *three-phase basis*, resulting in three distinct power waveforms ($P_a$, $P_b$, and $P_c$), each one staggered equally in time from the others as shown in Figure 3.16.[3] These voltages are applied to circuits having three conductors (in addition to a possible ground connection). The three-phase operation makes the sum of the power outputs of all three phases *constant*, not fluctuating (Fig. 3.16). A constant electric output is a major advantage of the multiphase a.c. power system.

After the three-phase power is generated, the voltage is raised for transmission. Electrical transformers change low voltage to high voltage and reduce it again at the point of use (see Appendix A for a discussion of the transformation of electricity). High voltages are the best way to transmit electricity over long-distance transmission because they reduce the electrical losses in the transmission

**FIG. 3.17.** High-tension transmission line. Courtesy of Carolina Power & Light Company.

line conductors. Electrical losses can be minimized by using high-tension transmission (see Fig. 3.17).

Electric power demand fluctuates over daily cycles, rising to peaks when people are active during the daytime and falling to a minimum at night. Electric generating stations are therefore designed for two major phases of operation – *base load* and *peak load*. Base-load service supplies that minimum level of power demand that continues night and day. Peak-load service helps to supply the maximum power needed during the short period of the day when it is demanded. The electric utility is required to supply power at whatever level demanded and at the lowest possible cost.

Peak-load demand is best supplied by generators with relatively low capital costs but high running costs (such as the gas turbines described later in this chapter). Since the high running costs are incurred only a small fraction of the time, they are more than offset by the low capital costs.

## The efficiency of steam electric power plants

As we have indicated, a typical steam power plant (schematically shown on Fig. 3.18) does not use all or even close to all of its energy input. In the figure, useful output – electricity –

is only 35 energy units, while input equals 100 units. Therefore the plant efficiency (the ratio of useful energy output to the required fuel input energy) is only 35 percent. A small amount of power ( ∼ 2 percent) is needed for plant operation, such as the boiler feed pump, but the rest of the energy is lost as waste heat – some at practically every step along the way. Stack losses, for example, account for 12 units, and boiler radiation and convection loss take away 1 unit. The largest loss, however, is from condenser waste heat – 47 units – more energy than is delivered as output. Yet we know from thermodynamics that most of this loss is unavoidable (see Appendix A).

This 35 percent efficiency is representative of current conventional steam power plants. Plant efficiencies may be increased up to 40 percent through energy conservation and technical efficiency improvements, such as regeneration, reheating, and preheating. (See Chapter 4 for the economic and policy implications of *technical efficiency* for energy technologies in general.) Even higher efficiency improvements appear possible with technologies under development. Some of these, such as *combined-cycle*, *cogeneration*, or *magnetohydrodynamics* (MHD), are discussed in Appendix C. These new technologies may have the potential for efficiencies of at least 50 percent. But conventional steam power plants today operate near their practical limits of conversion efficiency.

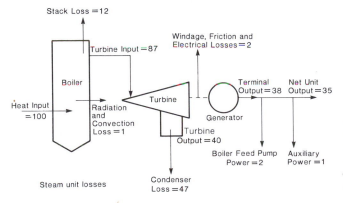

FIG. 3.18. Steam power plant: input/output energies. Adapted from W. D. Marsh, 1980, *Economics of Electric Utility Power Generation.* Clarendon Press, Oxford University Press, New York.

## Internal combustion

Most of modern transportation uses a different heat engine technology from the one employed in electric generation. Automobiles, trucks, and aircraft use forms of internal combustion engines. Automobile and truck engines are mainly of two types: the gasoline engine, operating on the Otto cycle,[4] and the diesel engine, burning a lower petroleum distillate and operating on the diesel cycle. While Otto-cycle engines are used almost exclusively in powering cars and trucks, diesel engines also run small generating facilities and most present-day railroad locomotives.

Jet aircraft use a third type of internal combustion machine, the combustion or gas turbine. Gas turbines, however, are also used as prime movers for electric generators in specific and important applications. Most notably, they provide extra power during times of the day when electric demand reaches peak load. A gas turbine has special advantages. First, it can be started quickly. (A boiler needs considerable time for steam buildup; a turbine does not.) Second, although it burns high-cost fuels like petroleum distillates or natural gas, the turbine equipment itself is a much less costly investment than a steam generator or a hydroelectric dam. Since it is operated for small durations of time throughout the year, a gas turbine uses little fuel in total and so it is more cost effective, when both capital and operating costs are considered, than most other alternatives for peak-load service.

The term *internal* combustion refers to the fact that the fuel is burned *within* the engine. In steam-cycle conversion, the fuel is burned externally, inside a furnace. The heat of combustion is then transferred into the working substance and the working substance does the work for useful output. In internal combustion engines, the combustion takes place within the working machine and the hot combustion gases then make up the working substance, delivering the useful work. Also unlike modern steam-cycle operations, internal-combustion technology is all thermodynamically *open cycle*. In gasoline and diesel engines and gas turbines, the combustion gases are exhausted from the machine once they have delivered their work.

## Otto and diesel cycles

Almost all gasoline and diesel engines operate with systems of pistons and the thermodynamic cycles follow paths similar to those of the open-cycle steam engine. The Otto cycle for conventional gasoline engines follows the thermodynamic paths indicated on Figure 3.19. The cycle requires four *strokes* (one-way motions). One of these strokes, however, is not strictly a path in the thermodynamic cycle, but rather a fuel suction stroke (0–1). Also one of the mechanical strokes combines two thermodynamic processes. The thermodynamic cycle therefore consists of the paths 1–2–3–4, as indicated on the p-V diagram.

The four strokes of the Otto cycle are:

*Stroke 0–1.* During the *suction stroke*, the fuel mixture[5] is drawn into the cylinder through the open intake valve as the piston moves downward.

*Stroke 1–2.* During the *compression stroke*, the fuel mixture is compressed as the piston moves up and both valves close. This stroke, a path in the thermodynamic cycle, is ideally an *isentropic* compression process (see Appendix A).

*Stroke 2–3–4.* The *power stroke* consists of two thermodynamic paths (2–3 and 3–4). The first, path 2–3, is the heat input process, where the compressed fuel mixture is ignited by a spark at point 2. Path 2–3 is nearly a constant volume process; the combustion process raises the pressure while the piston starts its motion from minimum volume. Expansion process 3–4 delivers work to the engine crankshaft. It starts once maximum pressure (point 3) has been reached.

*Stroke 4–0.* During the *exhaust stroke*, the heat is rejected, corresponding to the heat rejection stroke of the steam engine. Just as the steam engine ejects steam in this stroke, the gasoline engine ejects the products of combustion (gases) from the cylinder through an exhaust valve.

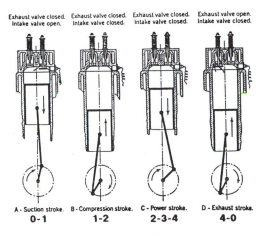

A - Suction stroke.   B - Compression stroke.   C - Power stroke.   D - Exhaust stroke.
      **0 - 1**              **1 - 2**              **2 - 3 - 4**            **4 - 0**

a). **Cycle of operation of an Otto four-stroke cycle engine.**

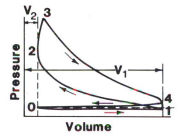

b). **Indicator diagram of an Otto four-stroke cycle engine**

**FIG. 3.19.** Otto-cycle internal-combustion engine. (a) Cycle of operation of an Otto-cycle four-stroke engine. (b) Indicator p–V diagram of an Otto-cycle four-stroke engine. Adapted from Severns and Degler (1948).

Whereas Otto-cycle engines use electrical ignition to initiate combustion in the cylinders, diesel-cycle engines do not. Ignition is achieved in the diesel engine when the compression stroke raises the fuel mixture to a sufficiently high temperature for spontaneous ignition (*auto ignition*) to take place. Also unlike the Otto cycle, the diesel heat input path is a constant-pressure process, rather than constant volume.

The diesel cycle (Fig. 3.20) is for a two-stroke machine, meaning that the entire thermodynamic cycle is completed in two strokes (up and down) of the piston. (Four-stroke diesel operation is also possible.) In the two-stroke

operation, four thermodynamic processes are incorporated:

*Stroke 0–1–2*

*Path 0–1.* With top valves open, compressed air forces products of combustion of the previous cycle out of open exhaust valves on the side, called the *scavenging* operation.

*Path 1–2.* During the *compression* process, which is ideally isentropic, ignition temperature is reached at point 2. All valves are closed.

*Stroke 2–3–4–0*

*Path 2–3. Injection* of heat takes place. Ideally, this is a constant-pressure thermodynamic

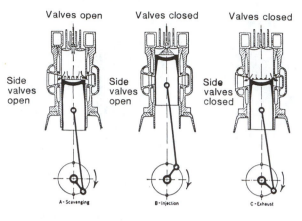

a). Cycle of operation of a Diesel two-stroke
cycle engine

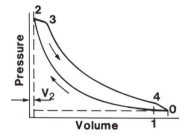

**FIG. 3.20.** Diesel-cycle, internal-combustion engine. (a) Cycle of operation of a diesel-cycle two-stroke engine. (b) Indicator p–V diagram of a diesel-cycle two-stroke engine. Adapted from Severns and Degler (1948).

b). Indicator diagram of a Diesel two-stroke
cycle engine.

process (position B in Fig. 3-20a), with all valves closed.

*Path 3–4.* The *expansion* process delivers work. All valves are closed.

*Path 4–0.* During the *exhaust* operation, the side valves open at the end of the expansion stroke and heat rejection is accomplished.

Figure 3.21 shows the major parts of an automobile engine. Whereas the engine shown is a gasoline engine, many of the parts are similar to diesel engines. Figure 3.21a is a photograph of the engine block showing the *cylinders* in which the pistons move. Figure 3.21b is a photograph of the *pistons* themselves.

Finally, in Figure 3.21c we have a drawing of the engine assembly in which the major parts of a gasoline engine are labeled.

The diesel cycle holds several advantages over the Otto cycle. First, it requires no separate electrical ignition system and no carburetor, as does a gasoline (Otto cycle) engine (see Fig. 3.21). Secondly, the diesel cycle can be operated at temperatures and pressures yielding higher ideal thermal efficiencies than the Otto cycle. Finally, because it is a lower petroleum distillate than gasoline, diesel fuel is cheaper than gasoline. Earlier disadvantages of the diesel engine, such as special starting procedures and excessive engine weights, have been largely

(a)

(b)

OIL FILLER CAP

STROMBERG CARBURETTOR

COOLING WATER THERMOSTAT

VALVE CHAMBER COVER

INLET MANIFOLD

EXHAUST MANIFOLD

CARBURETTOR THERMOSTAT

WATER IMPELLER

FLYWHEEL

FAN

VIBRATION DAMPER

OIL PUMP DRIVE

TIMING CHAIN SPROCKETS

DRILLED CRANKSHAFT    CAMSHAFT

FUEL PUMP  OIL PUMP

OIL SUMP

(c)

FIG. 3.21. Automobile engine – major parts. (a) Engine block. (b) Crankshaft and pistons. (c) Engine assembly. Adapted from Judge (1972).

overcome in recent years. As a result, the diesel engine, long a standby for trucks, has been more widely adopted for automobile use.

Thermal efficiencies of internal-combustion engines can ideally be higher than those of the external-combustion (steam-cycle) plants. In part this is due to the elimination of some of the losses, such as the stack and radiation losses. It is also possible to achieve higher working-substance temperatures in internal-combustion processes, which has implications for higher thermal efficiencies (see Appendix A).

A typical gasoline engine, however, is far from the ideal. The average gasoline-powered automobile, as indicated in Figure 3.22, has a thermal efficiency of 26.5 percent at a compression ratio of about 5:1. This measure for automotive engines is called a *brake* efficiency, since it is for an engine at a constant rotational speed (rpm) working against a friction brake. This value is in the midrange of efficiencies for various gasoline and diesel engines (Fig. 3.22). Some can achieve efficiencies as high as 44 percent, but each brake efficiency applies to one operating speed and power output only. Even though brake efficiency measures do not reflect the range of performance of driving demands, they are nonetheless good measures of the basic technical capabilities of engine designs.

Because the working substance in internal combustion engines is not retained within the system, as it is in the steam-turbine power plant, the opportunities for reusing some of the waste heat for improvement in thermal efficiency are limited for automotive engines. Attempts to improve the basic thermodynamic cycles of these engines have focused on the *compression ratio*, which is the ratio of cylinder volume before compression to that after compression. In general, as this ratio increases, so does the available energy per cycle and therefore the

thermal efficiency also increases. As we can see in Figure 3.22, at high compression ratios, Otto and diesel cycle engines can ideally have efficiencies of 60 to 70 percent.

Although we see thermal efficiencies increasing with compression ratios, they do so with diminishing returns as the ratio approaches 20:1. Most practical engines today operate in the midrange of this curve on either side of 10:1. One of the reasons for an upper limit in the ratio for gasoline engines is *pre-ignition* (familiarly known as *knock* or *ping*) – really an autoignition as in a diesel engine. What happens is that higher compression ratios induce higher temperatures, which lead to autoignition or preignition. Lead additives in gasoline were formerly used to inhibit preignition, but these are now being phased out to reduce air pollution.

There are other ways besides thermal efficiency to measure the performance of automotive engines, ways that better reflect real-life driving conditions. The most important is *fuel economy*. Fuel economy measures, such as miles per gallon (mpg), are overall indicators of how a given vehicle performs its designed task. However, this measure is merely an indicator of the fuel cost to get the vehicle to its destination and is averaged over ranges of speeds, accelerations, and road conditions. More specific measures exist for trucks in terms of *ton-miles per gallon*, which indicate the fuel costs of transporting given weights of freight over given distances. These data again reflect an average of speed, road conditions, and driving habits, and so have limitations. Nevertheless, measures such as miles per gallon or ton-miles per gallon tell a driver more about the fuel input requirement than does a measure of thermal efficiency.

Automotive fuel economy, in fact, involves more than maximizing the thermal efficiency of the engine (see Horton and Compton 1984). After engine efficiency, weight is the principal factor for lowering fuel requirements. With every reduction in the overall weight of the vehicle, (Fig. 3.23), significant improvements in fuel economy can be achieved. This occurs because a lower inertial weight requires less

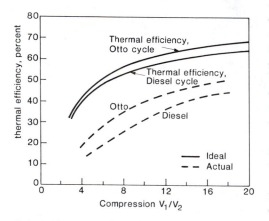

**FIG. 3.22.** Internal combustion engine: thermal efficiency versus compression ratio. Adapted from Severns & Degler (1948) and Judge (1972).

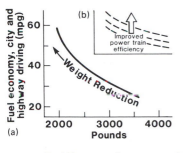

**Fuel Economy Improvements**

**FIG. 3.23.** Fuel economy improvements. Adapted from Horton & Compton (1984).

power to achieve the same performance on acceleration. Weight reductions in automobiles have been achieved (and more are possible) through the use of lighter materials such as cast aluminum, plastics, and low-alloy steels. The average U.S. automobile manufactured in 1985 weighed about 2,700 pounds, compared with 3,800 in 1975, which according to Figure 3.23, would imply a 72 percent gain in fuel economy for highway driving conditions.

An aerodynamic design can also affect fuel economy by reducing the drag of air resistance. Air resistance cannot, of course, ever be eliminated entirely. A practical limit for drag reduction currently appears to be about 25 percent below present designs. Since air drag is only part of the total rolling losses of an automobile, the resulting improvements in fuel economy are only a fraction of the percent reduction in air drag: a 10 percent reduction in drag, in fact, results in only about a 2 percent improvement in mpg.

Further improvements are expected through power train designs, which include combustion and exhaust innovations as well as transmission and gearing developments (see Horton and Compton 1984). Electronic controls for optimizing engine performance are also yielding measurable gains. As with aerodynamic improvement, each gain cannot be expected to be dramatic; rather individual innovations probably will add only a few percent improvement to gasoline mileage, although, as we will

see in Chapter 4, taken together these gains can be sizable.

Whatever the potential for improved technology, it is important to recognize that fuel economy data is based on average driving conditions and that variations around this average will be substantial. Indeed, individual driving habits and personal tastes in choice of models exert an enormous influence on automobile fuel economy. It is fair to say that such subjectively determined factors are at least as important as engineering design in questions of national energy policy. (We will consider human behavior in all sectors of energy usage in Chapter 4.)

## Gas turbines

The third and final internal combustion engine is the combustion or gas turbine, in which the thrust of expanding combustion gases drives the blades, instead of steam, and thus converts thermal energy to work (Fig. 3.24a). As with other engines, we can follow the paths of the thermodynamic cycle (called the Brayton cycle) on a p-V diagram. Starting at point *a*, the air input is compressed to point *b*. The heat input is supplied in a separate combustion chamber, fed by the compressed air in a constant pressure process (path *b–c*). This is followed by expansion (path *c–d*). The expansion, in this case, is of the hot, high-pressure combustion gases (Fig. 3.24b).

This process is significantly different from the process that drives the steam-cycle turbine. First, in the gas turbine, the hot combustion gases themselves are the working substance; they enter and expand directly through the blades of the turbine, in contrast to the steam cycle where the working substance derives its heat from the combustion gases in a furnace. Second, the Brayton cycle[6] for the gas turbine is an open cycle – unlike the steam-turbine cycle – since the hot gases simply exhaust to the atmosphere. Finally, the working substance of the gas-turbine cycle does not undergo a phase change as it traverses its thermodynamic paths.

Gas turbine cycles have a brake efficiency as high as 30 percent. Better utilization of the

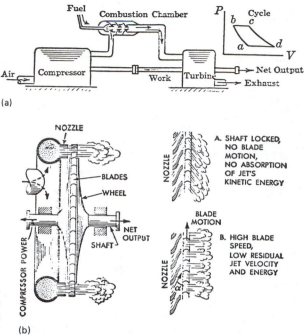

**FIG. 3.24.** Gas turbines. (a) Basic combustion in a gas turbine plant. (b) Principle of the gas turbine. Adapted from Morse (1947).

petroleum derivative or natural gas fuel can be achieved by operating the gas turbine in a *combined cycle* with a steam-turbine system. That is, the hot exhaust gases from the gas-turbine system are used to heat (or preheat) steam for the steam cycle. Thus, the gas-turbine waste heat is utilized in a combined thermodynamic cycle, resulting in a higher thermal efficiency. Combined-cycle operation yields thermal efficiencies approaching 50 percent and many plants that use this method are in operation today.

## Conclusion

At this point, we want to reemphasize the importance of these conventional conversion technologies. As we will see in the next chapter, they account for virtually all of our transportation and electric generation, and well over half of our consumption of energy resources generally. The technologies discussed in this chapter are the main reasons we need and desire

energy resources. They are the principal engines of the modern industrial world.

## Notes

1. Named after W. J. M. Rankine (1820–1872), engineer, scientist, musician, and professor at Glasgow, Scotland. He made notable contributions in steam engines, mechanics, and ship building.
2. Hertz (Hz) is a unit of *frequency*, the number of repetitions or *cycles* per second. The frequency of a.c. voltages and currents in the United States is 60 Hz. In other countries it is 50 Hz.
3. In Figure 3.16, $P_a$, $P_b$, and $P_c$ are the time variations of power output of each of the three phases, taken over the basic period of repetition ($T$) of the voltages and currents. $P_{av2\varphi}$ is the sum of the three powers $(P_a + P_b + P_c)$ for every instant, which turns out to be constant over time.
4. These thermodynamic cycles are named after Nikolaus A. Otto (German engineer, 1832–1891) and Rudolf Diesel (German engineer, 1858–1913), respectively.

5. Combustible air-gasoline vapor mixture prepared by the carburettor (see Judge 1972).
6. Named after George B. Brayton (1830–1892), American inventor and creator of the Brayton engine.

## Bibliography

Faires, V. M. 1948. *Heat Engines*, MacMillan. New York. Elementary engineering level, but with much understandable to the layman; good figures.

Horton, E. J., and W. D. Compton. 1984. "Technological Trends in Automobiles," *Science*, August 10, pp. 587–93. Review of contemporary trends in automotive technology.

Judge, A. W. 1972. *Automobile Engines*. Robert Bentley, Cambridge, MA. Detailed descriptions of engines, with numerous figures; also, design and operation.

Lamme, B. G. 1984. "High Speed Turbo-Alternators – Designs and Limitations," American Institute of Electrical Engineers Transactions (1919). Reprinted with commentary in *Proceedings of the IEEE*, Vol. 72, No. 4, April, pp. 493–526. History of the modern steam-driven turbine electrical generator; excellent for gaining understanding of operation.

Park, H. E. 1975. *Auto Engines of Tomorrow*. Indiana University Press, Bloomington, IND. A popular level exposition of automotive engines.

Rose, D. J. 1986. *Learning about Energy*. Plenum Press, New York. The insights of a longtime researcher in energy technologies.

Setright, L. J. K. 1975. *Some Universal Engines*. Mechanical Engineering Publications, The Institution of Mechanical Engineers, London. An interesting history of internal combustion technology, also helpful to understanding operation.

Singer, J. G. ed. 1981 (or latest edition). *Fossil Power Systems*. Combustion Engineering, Windsor, CT. A reference book on fuel combustion and steam generation with photos and figures of the latest contemporary designs.

Tarboux, T. G. *Electric Power Equipment*. McGraw-Hill, New York. Good descriptions of electrical machines with photos and drawings; the layman can ignore the theoretical sections.

Woodruff, E. B., and L. B. Lammers. 1977. *Steam Plant Operation*. McGraw-Hill, New York. Good descriptions with clear figures on furnaces, boilers, and combustion processes; the layman can ignore the engineering data in text.

# 4

# The demand for energy

## Introduction

Modern industrial society uses enormous quantities of energy each moment of the day. It needs energy to run its machines, to power its transportation, and to provide heat, light, and refrigeration. We depend on the instant response of the electric switch on the wall and the reliability of the internal combustion engine. People base their life-styles on the certainty of readily available sources of energy. If that certainty no longer existed, the developed world would change radically.

As individuals, we are all consumers of energy – a part of a market that is both national and international in scope. Each individual's demand for energy probably differs from the needs of another individual, and each society's aggregate demand will be somewhat different from the demand of another society. Yet we all – individually and together – demand certain quantities of energy, in the form of fuels and resources, from the energy market. How much we demand and what factors affect our demand decisions are important questions. The answers will tell us not only how much we use, but also whether we can continue to use as much as we do. On that question hangs the continuation of the life-styles we have grown accustomed to and the economic health of industrial society itself.

These questions have especially important implications for the more distant future. As we saw in Chapter 2, crucial fossil fuel resources are finite. The obvious point is that if we use less now, we will have more for the future. But here we must ask: What will happen to the economy if we use less? Can we cut consumption (or demand) now without great sacrifice, without radically altering the way of life people have grown to expect? And if we can cut, by how much and by what means can we do so? Or in essence we can reduce the issue to a single, controversial question: Do we really need what we consume?

## Energy demand: review of the market model

This chapter will focus on the demand for *energy*. But from an economist's standpoint, the demand for energy is no different in its basic structure from the demand for any other commodity or group of commodities. Indeed, consumer demand is a fundamental component of economics per se – a concept inseparable from economic descriptions of a market economy or even, to some extent, of a socialist economy.

There is not simply an idea of demand in economic theory. Demand has the status of economic "law."[1] Although this law does admit possible exceptions and requires considerable discussion and amplification, it has substantial empirical verification.

The law of demand is a relationship between the quantity of any commodity (any good) demanded by consumers and the commodity's price. Essentially it states that the quantity demanded of a commodity will be inversely related to its price, or simply, the higher the price, the less consumers will demand.

There is an important qualification to this law. That is, it applies assuming that all other things are equal. Since a market economy is extremely complex and continually changing, the qualification becomes very large, since "all" must truly include all. For example, in considering demand, we must include such unquantifiable noneconomic variables as taste. Yesterday's hot fad (such as pet rocks) may not be desirable today even at half the price. Demand will also be affected by nonprice elements such as population growth.

Two nonprice determinants of demand are especially important with respect to the energy market: national income (both per capita and aggregate) and the prices of related goods. The first determinant reflects the state of economic growth generally, which in turn can affect how much we consume of any commodity. The second determinant involves prices among interchangeable or mutually dependent commodities. (We can readily imagine how the demand for a commodity might be affected by the price of an obvious substitute – margarine and butter for example. But demand for a commodity may also be affected by prices of not so obvious substitutes, for instance, the demand for labor may be affected by the cost of capital.[2]) These are significant elements and will be discussed here in the context of energy demand. But even if the price/quantity relationship does not provide the whole picture, it is nonetheless fundamental and must be the starting point for any economic discussion of energy demand.

Demand can be thought of as a schedule. At any specific time – a demand schedule must be limited to a given time frame – the price of a commodity, gasoline for example, will call forth a demand for a certain quantity from

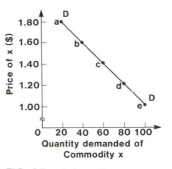

FIG. 4.1.  A demand curve.

consumers. At $1.50 per gallon, U.S. consumers might demand 2 billion gallons per week. However, at $2, they might demand only 1.5 billion gallons, while at $1, they would purchase 2.5 billion. It is important to keep in mind that in this specified time period, all of these possible price/quantity relationships are true simultaneously. A shift in price means that the quantity demanded will change. But the basic structure of demand, as depicted by the schedule, has not changed. A consumer demand schedule is usually depicted by economists as a downward sloping curve (Fig. 4.1). Though the curve shown here is linear, the shape of the curve and the steepness of its slope may vary from market to market or even within the same market from time period to time period. Still, in the energy market, the demand for individual resources, as well as the aggregate demand for energy, can always be depicted by some such curves.

Price changes lead to changes in the quantity demanded, but price does not change *demand* (that is, the schedule). A change in demand occurs for nonprice reasons and the phrase "change in demand" means strictly that the curve itself has shifted; price changes will then have a different impact on quantity. Consider, for instance, an increase in population due to immigration (Fig. 4.2). We can see that the curve shifts to the right. Now, according to this new schedule, at every price, more energy is being demanded.

The shape and slope of any demand curve are important features not only because they depict the schedule of quantities and prices, but

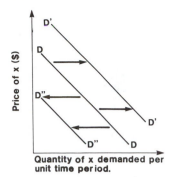

Quantity of x demanded per
unit time period.

**FIG. 4.2.** A change in demand. The shift from *D* to
*D'* results from an increase in population. The shift
to D″ represents a population decrease. Adapted from
R. L. Miller, 1982, *Intermediate Microeconomics*.
McGraw-Hill, New York.

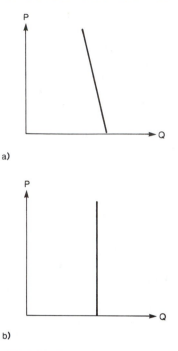

a)

b)

**FIG. 4.3.** Demand curves. (a) A relatively inelastic
demand curve (high slope). (b) A totally inelastic
demand curve (vertical slope).

also because they suggest the responsiveness of
demand to changes in price – what is called the
*price elasticity of demand*. So a steep slope would
suggest that prices have to change a great deal
before there is a significant change in the
quantity demanded (see Fig. 4.3a). Indeed it is
possible to imagine a nearly vertical curve
(Fig. 4.3b), where a certain amount of a good
will be demanded regardless of price. For short
periods of time, demand for certain energy
resources has seemed relatively *inelastic*, that is,
it exhibited a nearly vertical curve. This
occurred with respect to gasoline after prices
rose dramatically in the 1970s. In rural areas
of the United States where gasoline was needed
to run farm machinery, there was little change
in the quantity of gasoline demanded despite
the price rise. However, it is crucial to keep in
mind that a short-run curve can have a vastly
different shape and slope from a long-run curve
for the same commodity.

Although the slope depicts the elasticity of a
demand curve, the visual impression can be
misleading. A steep slope is usually a sign of
unresponsive, or inelastic, demand; a flat slope
indicates a highly elastic demand. But elasticities
can change from one part of the curve to the
next, and a curve that looks highly elastic may
be the opposite, depending on which section of
the curve one measures.

The price elasticity (*E*) of demand can be

defined in a simple ratio:

$$E = \frac{\% \text{ change in quantity} (Q)}{\% \text{ change in price} (P)} \quad \text{or}$$

$$E = \frac{[dQ/Q]}{[dP/P]}$$

where *dQ* is the change in quantity and *dP* is
the corresponding change in price.

Take the example of $E = 1$. It means that for
a percentage change in price (*dP/P*) there will
be an identical percentage change in the
quantity (*dQ/Q*) demanded. This is called *unit
elasticity*. If *E*, on the other hand, is less than
one, the change in demand will be smaller than
the percentage change in price – or relatively
inelastic. A vertical curve would have an
elasticity of zero, where no change in quantity
demanded occurs regardless of price (as in
Fig. 4.3b). Where *E* is greater than one (very

elastic), then changes in quantity demanded will be greater in percentage terms than price changes, meaning that the market is very responsive to price changes.

How much a price change affects the quantity demanded may depend in part on whether consumers can *substitute* one good for another. For example, if the price of coal rises, industrial users of coal can substitute natural gas or oil. (There are indeed some industrial furnaces that can adapt easily to different fuels.) A rise in the price of one fuel thereby increases the demand for another fuel and the price/quantity relationship between coal and natural gas is then said to be positive. In the energy market, substitution is not limited to choosing among the various resources. If energy prices rise too high, a consumer may choose to substitute labor-intensive (or capital-intensive) processes for those that are more energy intensive. In other words, a manufacturer can shut down some of his energy-consuming machinery and hire a few workers to perform the same tasks manually.

The sensitivity of demand for good $x$ to price changes in good $y$ is called the *cross-elasticity* of demand and can be calculated simply:

$$E_{xy} = \frac{\text{percentage change in } Q_x}{\text{percentage change in } P_y}$$

The mathematical formula is:

$$E_{xy} = \frac{dQ_x/Q_x}{dP_y/P_y}$$

$$= \frac{dQ_x}{dP_y} \cdot \frac{P_y}{Q_x}$$

where $Q_x$ is the quantity of one good, and $P_y$ is the price of another good. Therefore, if the two goods are coal and natural gas, then the greater the number of switchable industrial furnaces, the closer will $E_{xy}$ be to unity.

It should be noted that the cross-elasticities of related goods are not always positive. In some cases, when the price of a good rises, demand for a related good falls. For instance, if the price of automobiles rises, demand for cars will fall

and so, consequently, will the demand for gasoline. Where the relationship is negative, the goods are called *complementary*.

Income is also an important nonprice factor that can affect demand. A person may spend a certain amount on energy to meet basic necessities whether his or her income rises, falls, or remains constant. But if an individual's income rises, he or she will probably spend more, especially on discretionary purchases, some of which may result in greater energy consumption. For instance, the same individual might elect to spend more on energy-consuming electric appliances, a second car, or more and longer automobile trips.

The responsiveness of demand for a good to changes in income($I$) – *income elasticity* of demand – can be calculated in much the same way as cross-elasticity, by the ratio:

$$E_I = \frac{\text{percentage change in } Q}{\text{percentage change in } I}$$

Here, as with all of the other measures, it is assumed that only income changes and all other things including price remain unchanged.

Such concepts of economic demand can be refined and embellished further, but the present discussion provides a sufficient framework for understanding energy demand. However, before we deal with the specifics of the demand picture, we might also consider the even larger context in which demand for energy takes place: the notion of the energy market itself. In considering our demand model, we assume a system where the allocation of energy resources is decided by the market, an international network of suppliers and consumers buying and selling according to demand and availability of supply – largely free of government intervention. In recent years, some observers have wondered whether the energy market works as the model would lead us to expect or whether it suffers from "defects." These critics have called on government to intervene in the market and determine allocations of energy resources. This course of action appeared to make sense in the late 1970s, when the market seemed unable to respond to rising prices; the price of oil in particular soared and demand hardly slackened. As we will see, the energy market proved more complex than critics thought at that time. But whether or not the market "works," there may be larger overriding policy issues that past experience has not and cannot definitely answer, that is, Does a free market lead to the most effective allocation of resources (i.e., does it work optimally)? Does it work equitably? Is it good policy to leave the future determination of energy consumption solely to the market?

## Energy use and economic activity

Before we consider these broader questions, we need to understand the relationship between energy consumption (demand) and economic growth – for it is the pursuit of growth that underlies or at least justifies most policies about resources. There is no doubt that energy consumption is a necessary component of industrial society, and we have already suggested that there probably is a positive relationship between income growth and growth in energy consumption. In general, let us continue to assert that as income rises, spending for energy consumption rises, when all else remains constant. But even with this assumption, we should consider that discretionary spending by definition may encompass a variety of options that are either energy-intensive or are not.

Consider especially how an industrial or commercial firm might react to increased income. We will assume that a business tries to maximize its profitability and will choose to invest some new income toward the increase in future output. This choice conceivably could involve lessening, not increasing, the burden of energy costs. Where the choice is specifically to increase output, the producer may invest in low-energy-intensity labor or capital – for instance, hire new workers or buy a low-energy-use capital good, like a computer, to speed up the production process. But what if there are few such substitution opportunities? What if the most profitable investment choices lead to

increasing energy consumption? Then firms will likely consume more energy as they expand production. And if this is generally the case throughout the economy, as income grows, energy consumption will inevitably grow as well. Indeed, if the opportunity to substitute labor or capital for energy is limited, then energy *must* grow as the economy expands. Presumably, significant output growth cannot be achieved any other way.

Until quite recently, many economists and energy analysts were convinced that increased energy demand was indeed a necessary outcome of economic growth. According to this view, the consequence of using less was inevitably diminished economic activity – recession – and ultimately a lower standard of living. Superficially, history seemed to support this contention. In the United States, from the mid-1930s on energy use in general rose not only in absolute terms, but on a per capita basis as well. At the same time, economic growth, as measured by aggregate national income or the total output of goods and services (gross national product, or GNP), grew as well on a per capita and an absolute basis. But remarkably, GNP appeared to grow at a fixed ratio to energy consumption. Even during the few dips in growth both seemed to follow in step.

As Figure 4.4 shows, prior to the 1930s there was little correlation between the growth in energy consumption per capita and the growth of GNP per capita. (In Fig. 4.4, note that GDP, or gross domestic product, equals GNP minus the net inflow of national income earned abroad; in other words, both are virtually the same.) But beginning just before World War II, per capita energy consumption and per capita GNP seemed to move in tandem. Indeed, looking at the graph it is hard to escape the conclusion that an exact correlation exists between GNP growth and the growth in energy use and has existed over the past half century. If true, it is essentially pessimistic because it implies that U.S. society has only two choices: either use more energy or live less well. According to this scenario, technical advances are unlikely to make it possible to grow economically with less energy consumption.

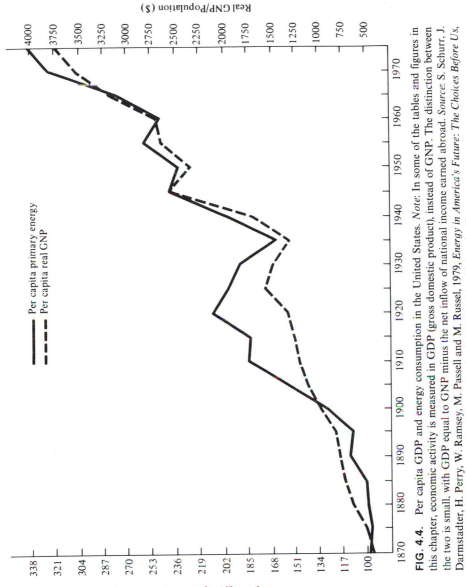

**FIG. 4.4.** Per capita GDP and energy consumption in the United States. *Note:* In some of the tables and figures in this chapter, economic activity is measured in GDP (gross domestic product), instead of GNP. The distinction between the two is small, with GDP equal to GNP minus the net inflow of national income earned abroad. *Source:* S. Schurr, J. Darmstadter, H. Perry, W. Ramsey, M. Passell and M. Russel, 1979, *Energy in America's Future: The Choices Before Us,* published for Resources for the Future, Inc., by Johns Hopkins University Press.

Also, this view questions the standard economic assumptions about the price mechanism in the energy market; since energy prices have not always grown at the same rate as GNP, does the price mechanism actually function or is energy consumption a prisoner of income elasticity alone? Again, if this is the case, we would have only two options: increased energy consumption or a lower standard of living.

Some free market economists, who might have been expected to have more faith in the price mechanism and to have envisioned more opportunities for substitution, supported this bleak scenario. In a 1972 study by the Chase Manhattan Bank, for example, forecasters could think of very few options – one was less television watching – for energy savings along with GNP growth.

However, such a view did seem rational in light of the pattern shown in Figure 4.4. The only evidence that ran counter to it was comparative data between U.S. consumption and that of most other industrialized countries. Though per capita GNP of Western European countries was comparable to per capita GNP of the United States, per capita energy consumption in Europe was typically lower, often one-third or more. Yet such comparative data, although suggestive, have limitations; no two countries are exactly alike. The United States, which is larger and had a greater amount of heavy industry (especially in the immediate postwar period), could not be compared exactly to France or the Netherlands or Great Britain.

But the conclusion that locked the GNP/energy relationship seemed to ignore what was actually taking place in the market. In real terms (that is, at a dollar amount with a fixed reference in time, adjusted for inflation – see definition in Chapter 9), prices for energy on the whole actually declined through much of the period from 1945 to 1972. Oil in particular exhibited real-price declines and was cheaper in 1970 (in deflated dollars) than it had been in 1945. Large hydroelectric projects lowered the real cost of electric power as well during much of this period, and also created vast power surpluses in parts of the United States. Utilities offered incentives to users to consume as much

electricity as possible: a pricing system that lowered the cost per kilowatt hour (KW-hr, see Appendix A) as customer use increased. What this suggests is that energy consumption actually may have been following the path classical economics would have predicted – but *based on price not income.* Indeed, it would imply that the lockstep growth path of GNP and energy was a mirage of sorts, induced by *lower* prices for energy; in fact, if real energy prices would have simply stayed constant, it is possible that demand would not have grown as rapidly and a gap between GNP growth and energy growth would have developed. As matters stood, falling real prices encouraged consumption and discouraged technical innovation for greater efficiency; there was no particular incentive to reduce energy intensiveness of production when the price of that factor of production[3] was declining. On the contrary, much growth was fueled by energy-intensive, labor-saving technological change.

Of course, if in fact demand growth was tied to price more than income, then according to the basic market model theory, if prices rose, it would (or should) lead to a decline in consumption without necessarily any fall in GNP. But when energy prices began to rise in the early 1970s – both through oil price jumps and rising electric rates – the energy market seemed notably defective,[4] and the pessimists, correct. Energy demand fell to be sure; per capita energy consumption in the United States as measured in British thermal units (BTUs, see Appendix A) plunged from 351 million BTUs/year in 1973 (the year in which the first oil shock took place) to 340 million in 1974 and 327 million in 1975, and total consumption fell from 74.3 quadrillion ($10^{15}$) BTUs, or 74.3 *Quads,* to 70.5 (see Table 4.1). But GNP fell too, and the U.S. economy experienced a severe recession. This suggested that GNP growth and energy use were not simply correlated, they were causally linked. The fall in energy use was not merely a sign of a slowing economy. Rather, the rise in energy prices led to a fall in demand, which in turn appeared to "cause" a recession – lasting until the energy price hikes were absorbed through increases in the general price

**Table 4.1.** U.S. gross national product, population, and energy consumption

| Year | Gross national product (billion $ 1982) | Population (millions) | Energy consumption[a] Total | Per 1982 dollar of GNP | Per person |
|------|------|------|------|------|------|
| 1973 | 2,744.1 | 211.4 | 74.3 | 27.07 | 351 |
| 1974 | 2,729.3 | 213.3 | 72.5 | 26.58 | 340 |
| 1975 | 2,695.0 | 215.5 | 70.5 | 26.18 | 327 |
| 1976 | 2,826.7 | 217.6 | 74.4 | 26.31 | 342 |
| 1977 | 2,958.6 | 219.8 | 76.3 | 25.79 | 347 |
| 1978 | 3,115.2 | 222.1 | 78.1 | 25.07 | 352 |
| 1979 | 3,192.4 | 224.6 | 78.9 | 24.71 | 351 |
| 1980 | 3,187.1 | 227.3 | 76.0 | 23.83 | 334 |
| 1981 | 3,248.8 | 229.6 | 74.0 | 22.77 | 322 |
| 1982 | 3,166.0 | 232.0 | 70.8 | 22.37 | 305 |
| 1983 | 3,279.1 | 234.3 | 70.5 | 21.50 | 301 |
| 1984 | 3,489.9 | 236.5 | 74.1 | 21.22 | 313 |
| 1985 | 3,585.2 | 238.7 | 74.0 | 20.63 | 310 |
| 1986P[b] | 3,675.5 | 241.1 | 73.9 | 20.11 | 307 |

[a]Total energy consumption is in quadrillion BTU, consumption per 1982 dollar of GNP is in thousand BTU, and consumption per person is in million BTU.
[b]P = Preliminary data.

*Note*: The total may not equal sum of components due to independent rounding.

*Source*: Population data from U.S. Department of Commerce, Bureau of the Census; all other data from Energy Information Administration, *Annual Energy Review 1986*.

level (inflation). Sure enough, by 1976 GNP growth resumed and so did per capita (and total) energy consumption. This only strengthened the belief in a causal connection between energy use and economic growth. With energy usage growing again, fears of future shortages and depletion of scarce resources prompted some to argue that Americans needed to accept voluntarily a decline in their standard of living to protect and preserve the world's resources.

When prices soared higher in the late 1970s than they had earlier in the decade, the market again did not provide any ameliorating influence. Indeed, price shocks did not even cut demand at first. Instead, consumption rose, GNP rose, and, because energy was a factor in virtually all modern production, the general price level rose. In other words, the energy market displayed a nearly inelastic demand and the energy price shocks were absorbed through higher output prices. By 1978, per capita energy consumption had risen to 352 million BTUs and total national demand rose 11 percent above 1975 levels to 78.1 quads (Table 4.1). Total consumption rose still further in 1979.

By this time, pessimism about future energy consumption was at its greatest. Studies in the mid- and late 1970s by various experts from a variety of public and private organizations (Table 4.2) forecasted ever-rising demand for the future. Beginning with the Department of the Interior study of 1972, most forecasters projected energy demand in the United States to reach between 80 and 100 quads by 1980 and as high as 116 quads by 1985. As the table shows, for the year 2000, with two exceptions, the estimates ranged from 93 quads to a massive 196 quads – the latter suggesting that annual energy demand would nearly triple during the last quarter of the century.

Economic recession in the United States in 1980 and 1982 made the higher end of the 1985 estimates unlikely well before the final tally was recorded. Energy demand fell, as people had grown to expect, with a falling GNP. But by the mid-1980s, it had become clear that most of the estimates were off not just because of recession, but because they were fundamentally flawed.

Forecasters had apparently taken as given the inelasticity of the demand curve for energy, largely, it seems, based on a rationale rooted more in psychology than economics. According to this view, the key to reducing consumption was a psychological change in our attitudes and our life-styles toward a willingness to make do with less. But it was assumed that we would not change our behavior merely because of the price of energy; we would not drive smaller cars, would not turn down the heat in winter or turn off the air conditioner in summer, or alter our life-styles in any other fashion. Indeed, there were few signs in the late 1970s that Americans were willing to make such changes.

But energy prices were having an impact that

**Table 4.2.** Forecasts of U.S. gross energy consumption, 1980, 1985, 2000, 2010 (in quadrillion BTUs)

| Study | 1980 | 1985 | 2000 | 2010 |
|---|---|---|---|---|
| Landsberg et al. (1963) | 79.19 | 89.23 | 135.16 | — |
| U.S. Department of the Interior (1972) | 96.00 | 116.60 | 191.90 | — |
| (1975) | — | 103.50 | 163.40 | — |
| Ford Foundation Energy Policy Project (1974) | | | | |
|   Historical growth | 100.00 | 115.00 | 187.00 | — |
|   Technical fix | 88.00 | 92.00 | 123.00 | — |
|   Zero energy growth | 85.13 | 88.00 | 100.00 | — |
| Project *Independence Blueprint* (1974) | | | | |
|   Base case | 86.30 | 102.92 | 147.00 | — |
|   Conversion | 82.20 | 94.16 | 120.00 | — |
| U.S. *National Energy Outlook* (1976) | | | | |
|   Low growth | 80.23 | 90.72 | — | — |
|   High growth | 85.40 | 105.64 | — | — |
| United Nations (1976) | | | | |
|   Reference case | 86.50 | 102.90 | — | — |
| Edison Electric Institute (1976) | | | | |
|   Moderate growth | — | — | 161.00 | — |
| Institute for Energy Analysis (1976) | | | | |
|   Low growth | — | — | 101.40 | — |
|   High growth | — | — | 125.90 | — |
| *National Energy Plan* (1977) | — | 97.00 | — | — |
| Stanford Research Institute (1977) | | | | |
|   Base case | — | — | 143.20 | — |
|   Low growth | — | — | 109.40 | — |
| Electric Power Research Institute (1977) | | | | |
|   High case | — | — | 196.00 | — |
|   Conservation case | — | — | 146.00 | — |
| Workshop on Alternative Energy Strategies (1977) | | | | |
|   Low growth | — | — | 115.10 | — |
|   High growth | — | — | 132.00 | — |
| Brookhaven National Laboratory/Dale Jorgenson Associates (1978) | | | | |
|   Base case | — | — | 138.50 | — |
| National Research Council, National Academy of Sciences (1979) | | | | |
|   Low growth | — | 69.25 | 66.73 | 65.10 |
|   High growth | — | 94.29 | 144.28 | 191.60 |
| Resources for the Future (1979) | | | | |
|   Low growth | — | 80.52 | 95.00 | — |
|   High growth | — | 89.50 | 145.00 | — |
| *The Global 2000 Report to the President* (1980) | | | | |
|   Low growth | — | 90.00 | 129.70 | — |
|   High growth | — | 102.00 | 141.30 | — |
| Ross and Williams (1981) | | | | |
|   Low growth | — | 73.04 | 67.83 | 64.00 |
|   Business and usual | — | 80.09 | 93.00 | 102.00 |
| Actual consumption | 76.20 | 74.00 | — | — |

*Source*: Adapted from Le Bel (1982).

became more apparent as time passed. In fact, there was some evidence of the impact of the price mechanism from the mid-1970s. Although overall use and per capita use were increasing, there was one sign that we were using energy more efficiently as well. From 1975 to 1978, though consumption was rising (by 11 percent), consumption per dollar of output had fallen by 4 percent to about 25,000 BTU/per $GNP. That is, we were producing more with less energy.

Actually, though this statistic suggested greater energy efficiency, it is understandable that forecasters did not necessarily find it a proof of a trend. In fact, greater improvement in the $/GNP ratio could have signaled – and in part did signal – a shift from a more energy-intensive manufacturing economy to one that was more service oriented. Also the historical evidence was at least ambiguous. Though energy inputs per dollar of GNP had fallen in some years of the post-World War II era (though not fast enough to change the lockstep GNP/energy demand ratio), by the 1960s we clearly were using more energy per dollar of output. Further, even if there was some new move toward efficiency in the 1970s, it was lagging behind the growth in demand and appeared at best a restraint, not a brake, on demand growth.

But in reality, there was a new trend toward energy efficiency in the U.S. economy that became increasingly evident (a clear sign was the increasing fuel efficiency of new automobiles) through the first half of the 1980s. The energy consumption rate per dollar of GNP decreased 23 percent from 1975 to 1986 and fell every year in that period. Total energy consumption did not approach its 1979 peak of 78.9 quads through the middle of the next decade; total energy consumption was still 6 percent lower in 1986 than it had been in 1979, and was even slightly lower than it had been in 1973 (Table 4.1). Per capita energy consumption was down from its peak year (1978) by almost 12 percent and was 6 percent below the 1975 level. At the same time, GNP (in real, or constant dollar, terms) had grown 37 percent from 1975 to 1986. The link had been broken between the growth of energy consumption and GNP.

Why did the forecasters err? One of the main reasons was an underestimation of the impact of prices on the energy market. A strong case can be made (and has been especially by economist William Hogan in Sawhill & Cotton 1986) that the energy market does respond in a fashion that is consistent with the classical market model. This has not always been apparent, because insufficient weight has been given to the importance of time for market adjustment. The crucial point is that in the energy market, adjustments – changes in demand – take place over the long term. They may lag so far behind price changes, in fact, that at first the market appears unresponsive to price shifts, contrary to the assumptions of the economic model.

This argument, in addition to fitting recent history, is logical. Energy inputs, as we have noted, are necessary for so much of life in an industrial society. But energy is consumed primarily by our society's machines – relatively expensive durable goods for the most part (appliances, factory equipment, automobiles, and so on). To gain greater efficiency of energy use in the economy, machines usually have to be replaced. But is there enough incentive to replace them even in the event of a large increase in energy costs? In the short term, the answer is no.

Let us consider the effect of a 10 percent real increase in energy costs. In most cases, there are few short-term measures that people can adopt – short of doing without – that also make economic sense. There may be a few no-cost housekeeping measures available to individuals and businesses, for example, individuals can make sure they turn off the lights when they leave the room or without hardship can replace 75-watt light bulbs with 60-watt bulbs. But the biggest gains in efficiency are investment based, requiring an outlay of money for new equipment or the refitting or repair of old. The financing cost of new equipment (and sometimes even of repairs) is usually far greater than the potential energy savings will be. In industry, for example, a study in the early 1980s noted that energy accounted for an average of 4 percent of the factor costs of production. Thus, a 10 percent increase in energy costs would mean an increase of only 0.4 percent per unit of output. A new factory or even a new machine, on the other hand, can cost millions of dollars, with the cost of borrowing being far more burdensome to a business than the increase in the cost of energy. In steel, for example, one economist said that the cost of a new mill was "staggering" (quoted by Peck & Begg in Sawhill & Cotton 1986). It is far easier and less expensive for the factory to raise prices a small amount in the short run to cover the increased cost. Hence, prices rise while energy demand remains constant – leaving the impression of an inelastic demand curve. Individuals, too, have little incentive to spend several hundred dollars per month to finance a new car in order to save a few dollars a month at the gasoline pump.[5]

Yet the longer term presents a different picture. If energy prices are rising (and are expected to rise further), then there is an incentive for businesses and individual consumers to buy more fuel efficient durable goods when they are replacing them anyway – as they will over time. They, in fact, may have no choice but to buy greater fuel efficiency since it is competitively sensible for the durable goods supplier to build it into new models. And the higher energy prices go, the greater the demand for efficiency in new durables is likely to be. As a result of this pattern, over time, energy demand per dollar of GNP falls. Indeed, this lagged effect should continue even if energy prices stabilize or decline.

In 1982, Hogan demonstrated how the adjustment process had worked by looking at data from 1960 through 1982. He found that by using rates of adjustment tied to depreciation rates of capital goods[6] of between 10 and 20 percent annually, the elasticity of demand for energy approached unity. In other words, if aggregate energy prices doubled and all other variables remained unchanged, after several years (Hogan used between 6 and 12 years) demand would drop by half.

If Hogan's analysis was correct, then at the time he proposed it, the price shocks of the late 1970s had yet to take full effect. He predicted as much, arguing that even stable energy prices

would not negate a continuation of efficiency gains. The prediction was borne out. From 1984 through 1986, total energy consumption, per capita consumption, and consumption per/$GNP all fell in the United States. Meanwhile, the GNP rose about 6 percent. By this time, real dollar prices for energy were, on balance, actually falling. Although electric rates rose two out of the three years (Table 4.3) because of expensive construction projects (especially nuclear power plants, see Chapter 9), aggregate energy costs declined because of steep drops in oil and natural gas prices. The price declines, of course, would be expected to encourage increased consumption and there were beginning to be signs of that as well; consumption of petroleum, which had declined steadily between 1979 and 1983, edged up again in 1984 and rose slightly in 1985 and 1986. In other words, by the mid-1980s the price mechanism appeared to be working in both directions at once; earlier price hikes were leading to greater efficiency, but price declines were beginning to encourage increased demand. Using Hogan's concept, this is what one would expect.

Hand in hand with the error in economic

**Table 4.3.** Average U.S. energy prices

| Year | Crude oil | Natural gas | Coal Bituminous | Coal Anthracite | Electricity average |
|------|-----------|-------------|------------|-----------|---------|
| 1973 | 67.1  | 20.1  | 36.5  | 61.7  | 1.96 |
| 1974 | 118.4 | 27.3  | 68.2  | 102.2 | 2.49 |
| 1975 | 132.2 | 41.1  | 83.9  | 149.5 | 2.92 |
| 1976 | 141.2 | 53.1  | 85.0  | 153.9 | 3.09 |
| 1977 | 147.8 | 72.3  | 87.7  | 153.8 | 3.42 |
| 1978 | 155.2 | 83.6  | 97.9  | 152.7 | 3.69 |
| 1979 | 217.9 | 108.1 | 105.3 | 177.2 | 3.99 |
| 1980 | 372.2 | 144.8 | 109.4 | 185.9 | 4.73 |
| 1981 | 547.8 | 179.5 | 117.9 | 190.1 | 5.46 |
| 1982 | 491.7 | 222.2 | 122.1 | 214.0 | 6.13 |
| 1983 | 451.6 | 232.3 | 117.2 | 230.0 | 6.30 |
| 1984 | 446.2 | 239.9 | 115.9 | 208.7 | 6.52 |
| 1985 | 415.3 | 225.5 | 114.8 | 204.2 | 6.71 |
| 1986 | 218.3 | 168.0 | 111.7 | 200.6 | 6.40 |

*Note*: Fossil fuel prices: Cents per million BTUs. Electricity: Cents per kilowatt-hour

*Source*: *Energy Facts 1986*, Energy Information Administration, U.S. Department of Energy.

analysis that regarded growth in GNP and energy consumption in a fixed ratio was at least a tacit technical supposition – equally wrong – that little cost-effective energy efficiency would result from technical innovation. As we discuss in Appendix A, there are theoretical limits to efficiency; a heat engine can never surpass the ideal efficiency of, for instance, the Carnot cycle. The closer technology gets to the Carnot limit of efficiency, the less likely it is to achieve cost-effective improvements. But the improvement in energy demand per $/GNP of 23 percent in the United States and similar improvements in other developed countries[7] during the 1980s strongly suggested that given the price incentives, appreciable improvement in efficiency was both desirable and probable. This suggests in turn that no ideal technical limitations had been at work, that is, the gains were due to people's economic behavior within the range of what was technically feasible.

But what of the future? Is there more efficiency to be gained and if so how much? And can the market be relied on to get us there?

## The mixed lesson of Sweden

In examining the question of how much energy an industrial society needs, critics of U.S. consumption patterns often point to the example of Sweden. In the 1970s, the Swedes were portrayed as the paradigm of an energy efficient people, while Americans were depicted as energy wastrels. In 1980, Sweden recorded per capita income *above* that of the United States, had an industrial economy, and had the resources to permit its people to enjoy a life-style not so different from that of Americans. Yet Swedes used, on a per capita basis, about one-third less energy than Americans. Looking at such figures it was hard not to agree fully with the critics.

For other reasons, too, Sweden seemed to provide a good comparison. Not only does it possess a heavy, energy-intensive, industrial sector, it faces a climate that requires considerable energy expenditure for at least one season, and although the Swedish people are

not so automobile oriented as their American counterparts, most Swedes own cars.

The wide disparity at least gave people in the United States reason to question old assumptions about the potential for energy efficiency gains in the context of a growing economy. However, as we noted before, such comparisons should be looked at with some caution and skepticism.

Indeed, the Swedish case raises some important questions. First, there is reason to believe that, like Americans, Swedes are only reacting as we would expect to their energy market. With automobile fuels, for example, Swedes use less in part because they pay much higher prices, the effect of interventionist government energy policies that place high taxes on gasoline. This fact can lead logically to two diametrically opposite conclusions: either that government intervention is good because high prices keep demand down, or government intervention is pointless in the long term since demand will fall anyway when prices rise as they likely will over time. There may be other factors, of course, motivating government energy policies, such as the desire to reduce pollution, but these two opposing positions have advocates, and certainly either conclusion is possible from looking at the effects of energy policies worldwide.

There are also practical and social differences between Sweden and the United States that should not be ignored. The United States is a large country; Sweden in area is only a bit bigger than California and all of the Scandinavian countries combined take up less land area than Alaska. This does not simply mean that Americans routinely drive longer distances. It can also have important implications for industry. For example, in 1976, the Swedish cement industry was 18 percent more energy efficient than the U.S. cement industry. Small countries can build large, efficient, centralized plants to meet the needs of the national economy. In the United States, the cost of transporting cement would outweigh efficiency gains from centralization. Small, local, inefficient plants are more cost effective in regional markets than centralized plants would be shipping over a distance of 1,000 miles or more. Ironically, a government policy like Sweden's that kept prices of transportation fuels high might actually increase the advantage of the small inefficient plants, since the cost of shipping would be that much greater.

Finally, there is the issue of taste. If Americans prefer a personal car to mass transit, it may be wasteful, as even a fuel efficient car is less so than a full bus. But how do you change taste in the context of a free society? The point is that there is a profound difference between identifying potential areas for energy savings and providing an answer as to how – short of coercion – those savings can be realized.

## Conservation potential

### An overview

Thus far we have discussed the issue of reducing energy demand from the standpoint of increasing *energy efficiency*: the same (or greater) output with a lower energy input. Greater efficiency is one form of energy *conservation*. It is probably the most desirable form to most individuals because it offers the potential for lower demand and a consistently growing economy.

Conservation can also be defined in three other ways. First, it can mean the substitution of cheaper fuels for more expensive ones – a reduction in factor costs, not in total BTUs per unit of output. More important, as a nation, we can conserve by substituting scarce finite resources with more abundant or renewable ones; the switch from natural gas to hot water heating to solar heat (see Chapter 12) may result in the same number or even a greater number of BTUs consumed, but it conserves finite supplies of natural gas. Finally, conservation may simply mean using less. This could imply a lower standard of living – as some critics fear – or it could merely mean a different life-style – a qualitative change rather than a quantitative one. As a society, we might begin to shift from energy intensive activities – both work and leisure – to more labor intensive ones, for instance, orienting the economy more to

specialized craft production (which our society evolved from) and away from mechanized mass production. Because we are a free society and such change requires alterations in behavior and attitudes, it may only occur with education and/or over a period of energy privation where people come to expect less and have less access to abundant sources of inexpensive energy.

However, such possibilities remain highly speculative. In the realm of public policy, it is much easier and more politically acceptable for a government such as ours to focus on affecting the energy market rather than on mandating changes in attitudes and life-styles. As a result, we will confine our discussion to the potential of conservation through efficiency improvement and substitution among resources.

Actually, the potential for the latter is hard to quantify, especially since substitution in the future could be greatly affected by the development of alternative technologies, such as solar or synthetic fuels (see Chapter 12). However, it should also be noted that gains have already been observed in the shift from oil to alternatives. In electric generation, for example, oil-fueled generators accounted for almost 17 percent of all U.S. electricity in 1977, but only 5.5 percent in 1986. Abundant coal, on the other hand, was generating about 56 percent of U.S. electrical supplies in 1986, up almost 10 percent in a decade.[8] In the late 1980s, there were still many cost effective substitution possibilities left for industry especially.

More research has been done on quantifying the potential in the United States for conservation through efficiency, which could represent an actual reduction in aggregate (or per capita) energy use. These analyses generally assume the continuation of current population growth trends and try to factor in the impact of a growing economy. We should keep in mind that population growth will shift the demand curve to the right even if per capita demand remains constant; also, income growth, though not as powerful an influence on demand as people once believed, probably does play some role and thus if GNP is increasing, there will be some income effect on energy demand.

Two of these analyses of conservation

potential are represented in Table 4.2: A low growth scenario of the National Academy of Sciences (1979) and analysis performed by Ross and Williams (1981). Both show total energy consumption *declining* by the year 2000 and still further by 2010.[9]

The Ross and Williams projection is particularly interesting, first, because their initial estimate for 1985, made four years earlier, proved remarkably close to the actual figure; their forecast was less than one quad off the total of 74 quads. Second, Ross and Williams believed that their projected energy consumption could be realized with economic growth close to the historic U.S. average of 2.7 percent annually. The National Research Council, on the other hand, envisioned reduced energy consumption in the context of a more slowly growing economy.

Such an achievement – growth with conservation – appears possible without significant

new technological breakthroughs. Indeed, some have argued that the potential for saving is even greater. A study by the World Resources Institute, for example, argues that throughout the industrialized world per capita consumption can be reduced by about 50 percent (see Goldemberg, Johansson, Reddy, & Williams 1987). In order to meet that goal, the United States, with one of the highest consumption rates in world, would probably have to lower its rate by an even greater percentage than Ross and Williams suggest.

In order to evaluate the prospects for conservation, it is useful to examine energy consumption in three sectors – residential/commercial, transportation, and industrial – and examine where savings may be achieved (see Table 4.4). We will consider here mainly the potential for investment in conservation and investigate what can be done with present technology.

Table 4.4. U.S. energy consumption by sector (quadrillion BTU)

| Year | Residential/ commercial | Industrial | Transportation | Electric utilities | Total |
|------|------|------|------|------|------|
| 1973 | 24.1 | 31.5 | 18.6 | 19.9 | 74.3 |
| 1974 | 23.7 | 30.7 | 18.1 | 20.0 | 72.5 |
| 1975 | 23.9 | 28.4 | 18.2 | 20.4 | 70.6 |
| 1976 | 25.0 | 30.2 | 19.1 | 21.6 | 74.4 |
| 1977 | 25.4 | 31.1 | 19.8 | 22.7 | 76.3 |
| 1978 | 26.1 | 31.4 | 20.6 | 23.7 | 78.1 |
| 1979 | 25.8 | 32.6 | 20.5 | 24.1 | 78.9 |
| 1980 | 25.7 | 30.6 | 19.7 | 24.5 | 76.0 |
| 1981 | 25.2 | 29.3 | 19.5 | 24.8 | 74.0 |
| 1982 | 25.6 | 26.1 | 19.1 | 24.3 | 70.8 |
| 1983 | 25.6 | 25.8 | 19.1 | 24.9 | 70.5 |
| 1984 | 26.4 | 27.8 | 19.9 | 25.9 | 74.1 |
| 1985 | 26.8 | 27.0 | 20.1 | 26.5 | 74.0 |
| 1986P | 27.3 | 26.0 | 20.7 | 26.8 | 73.9 |

*Note*: Total is the sum of Residential/commercial, Industrial, and Transportation. Consumption by Electric utilities has been allocated to each end-use sector according to electric utility sales to that sector. Total may not equal sum of components due to independent rounding. (P = Projected.)

*Source*: *Annual Energy Review 1986*, Energy Information Administration, U.S. Department of Energy.

## The residential/commercial sector

This sector includes both single and multifamily homes (over 100 million units in the United States), as well as the service sector of the economy: offices, hospitals, hotels, warehouses, and stores. Although there are some significant differences between the energy needs of the residential and commercial sectors, their consumption is aggregated together because of some important similarities. About 60 percent of total energy demand in the combined residential/commercial sector is for space heating. Together, they also consume a significant amount of energy for lighting, hot water heating, air conditioning, cooking, and refrigeration (the latter two in institutional and business food services as well as home kitchens). Lighting, refrigeration, and most small appliances operate on electricity produced at centrally located generating stations. Space and hot water heating may be electric, but are predominately supplied by direct inputs of fuels, typically oil, natural gas, butane, or propane. The residential/commercial sector as a whole used about 27 quads in 1986 (Table 4.4).

In the wake of the energy price shocks as well as natural gas and oil shortages of the 1970s, some efforts were made toward conservation in the residential/commercial sectors. Initially, these took the form of zero-investment options – such as shutting off lights – or mild forms of doing without – such as setting thermostats at 68°F instead of 72°F in winter.

The turmoil in the energy markets of the 1970s also provided impetus for the wide dissemination of information to the residential/commercial sector about investment alternatives. Throughout the sector, there was some movement toward conservation investment. Most especially there was an attempt to improve insulation in houses and buildings so that space heating could operate more efficiently. Consumers invested in a range of options from inexpensive duct tape and caulk to new triple-glazed windows. In some instances, people replaced heating systems to take advantage of improved efficiency. New oil and gas furnaces in the early 1980s required about one-third less energy to produce the same amount of heat as older models. Some businesses invested in more exotic equipment, such as computerized control and monitoring devices that could regulate heat and light throughout an office complex.

Efficiency gains were noteworthy in new housing stock. However, some new homes constructed in 1980 were on average 30 percent more efficient per square foot than the existing housing stock. However, some new homes designed for efficiency were 90 percent better than average. At the same time, while these gains were impressive in themselves, they did not produce a dramatic fall in energy consumption by the sector because the turnover in the building stock of the United States is very slow; there are (and always will be) far more older homes than new ones.

Still, these developments slowed demand growth in the residential/commercial sector – in spite of a rapid expansion of the service economy and continued population growth. From 1976 to 1986, consumption in this sector had increased less than 10 percent in total, rising to 27.3 quads. Measured against the pre-recession peak of 1978, however, consumption in the sector was up less than 5 percent in 1986, while overall real GNP was up 18 percent.

Despite the gains in the early 1980s, it has been argued that the easy conservation (presumably zero and low cost investment) had been made (see Hirsch 1987). Yet easy is a qualitative judgment that is not universally supported. Indeed, conservationist scenarios envision large savings in all sectors including residential/commercial. In their low growth scenario, Ross and Williams see this sector's share of total consumption falling from about 34 to 30 percent by the year 2000. As they see an overall decline in energy demand, clearly they believe that this sector can conserve considerably more than it does.

Tested prototypes and some production models of domestic and commercial technologies seem to bear out such analysis. The most important gains are potentially in space heating, particularly home heating, which comprises more than 50 percent of the space heating category. To date, much of the savings in this area has come from better insulation. But often insulation improvements have been unsystematic, for example, by caulking around a drafty window. Insulating has also not been universal. Experiments have shown that experts were able to produce an average energy savings of 30 percent per unit through shell improvements alone in older homes. Those gains averaged more than 50 percent when replacement of old furnaces was included.

Such systematic efforts in patching up old buildings would produce massive savings, but they are not necessary, according to some conservationists, because of new technologies. It has been argued that replacing all home windows with so-called "superwindows" would lead to a huge reduction in fuel demand. These double-glazed windows reflect infrared heat radiation back into the house and have insulating xenon gas between the panes. As a result, they are as much as eleven times better at retaining heat than standard single-pane windows (see Rosenfeld & Hafemeister 1988). By one estimate, replacing all windows with superwindows would save about *four quads* per year (see Lovins and Lovins 1987).

Built-in energy efficiency in new construction and eventual improvement in old (for instance, through the inevitable replacement of old furnaces) may help keep down the rate of growth in energy demand for residential/ commercial space heat for the rest of the century. At the same time, with the lower prices for energy in the late 1980s, the market was providing few incentives for consumers to take advantage of existing conservation technology.

There also is potential for considerable conservation in other parts of the residential/ commercial sector. However, because most of the remaining energy demand is for electricity (hot water heating being one partial exception) the savings of scarce energy resources is indirect. We can measure potential savings in quads – residential/commercial electric use required about 17 quads worth of fuel resources in 1986 – although in this instance a better measure may be in terms of the amount of new electric generating capacity that could be

avoided or old capacity that could be shut down. The point is that the vast cost of building central power plants (see Chapter 9) represents not only future demand on energy resources, but also represents a drain of financial resources that might be used more productively elsewhere in the economy.[10]

Some conservation of electric power in this sector cannot be easy to achieve because of the importance of noneconomic elements, particularly taste. Switching from incandescent light bulbs to fluorescent bulbs, for example, would save energy. Currently available 11-watt fluorescent bulbs that fit standard light sockets produce the same amount of light as 60-watt incandescent bulbs, and fluorescents also last considerably longer. But U.S. consumers have been reluctant to switch, partly because of taste; fluorescent light is harsher. The taste factor seems to override price considerations, although understanding the price issue is not so simple in this case. Fluorescent bulbs are initially more expensive than incandescents that provide comparable light, but fluorescents typically have a lower total cost because of lower energy demand as well as longer life. Consumers, who are accustomed to making consumption choices according to initial price, often cannot and do not appreciate the total cost consequences without deliberate efforts at education by manufacturers or government agencies. However, in this instance even during the early 1980s, when information about energy costs was more widely disseminated (including information on fluorescent lighting), there was no great flight from the incandescent bulb, thus suggesting again the importance of taste. A switch probably will not occur in the future unless electric rates rise considerably or fluorescent light becomes less distinguishable from incandescent light (which is under development, see Goldemberg et al. 1987) or both.

In other areas, taste is unlikely to pose a barrier to conservation. Home refrigeration, which nationwide consumes about a quad of electric energy annually, is a notable example. As Table 4.5 indicates, average annual energy usage for 1983-model home refrigerators was 2 and 2.5 KW-hr/liter capacity for one- and

**Table 4.5.** Energy performance of refrigerators and refrigerator-freezers

| Brand (origin) | Model | Type | Capacity (liters) | Specific electricity use (KW hour per liter per year) | Electricity use (KW hour per year) |
|---|---|---|---|---|---|
| Average, 45 U.S. Models | (1983) | 1 door | 363 | 1.94 | 703 |
| Hitachi (Japan) | 617A | 1 door | 169 | 1.36 | 230 |
| Gram (Europe) | K215 | 1 door | 215 | 1.26 | 270 |
| Kenmore (United States) | 564.86111 | 1 door | 311 | 1.19 | 370 |
| National (Japan) | 211 | 1 door | 207 | 0.99 | 205 |
| Laden (Europe) | 40.830 | 1 door | 305 | 0.95 | 290 |
| Bosch (Europe) | KS2680SR | 1 door | 255 | 0.86 | 220 |
| Gram (Europe) | K395 | 1 door | 395 | 0.80 | 315 |
| Gram (Europe) | prototype | 1 door | 200 | 0.52 | 104 |
| Average, 488 U.S. Models | (1983) | 2 door | 518 | 2.46 | 1275 |
| Bosch (Europe) | KS3180ZL | 2 door | 310 | 1.77 | 550 |
| Amana (United States) | TSC18E | 2 door | 510 | 1.71 | 870 |
| National (Japan) | 291(HV/T) | 2 door | 290 | 1.65 | 480 |
| Amana (United States) | ESR14E | 2 door | 402 | 1.58 | 635 |
| Whirlpool (United States) | | 2 door | 487 | 1.54 | 750 |
| Electrolux (Europe) | TR1120C | 2 door | 315 | 1.51 | 475 |
| Amana (United States) | prototype | 2 door | 510 | 1.43 | 730 |
| Kelvinator (United States) | prototype | 2 door | 510 | 1.39 | 710 |
| Toshiba (Japan) | GR411 | 2 door | 411 | 1.31 | 540 |
| Amana/Kelvinator | conceptual | 2 door | 510 | 1.14 | 580 |
| Pedersen | prototype | 2 door | 510 | 0.94 | 480 |
| Schlussler | prototype | 2 door | 368 | 0.68 | 252 |

*Source*: Goldemberg et al. (1987).

two-door units respectively. Prototypes had already been built that used 50 percent or less energy than the average, and in one case barely more than a quarter than average. But even new models on the market in 1983 showed significant efficiency improvements. It was estimated that replacing all home refrigerators by the most efficient 1983 production models would save the equivalent energy of the output of eighteen 1,000 $MW_e$ (megawatts of electric capacity) power plants (Goldemberg et al. 1987).

Although every refrigerator will not be replaced, it should be kept in mind that some of the efficiency gains will still likely take place. Efficiency gains may have been added to production models because of the energy consciousness of the late 1970s and early 1980s. With the lessening of fears about energy costs, there will be less incentive to make increased efficiency a high priority among manufacturers. But they will not make new models less efficient. The real question for the future is not whether there will be some gain, but how much is possible. And the answer is that overall national consumption may be cut by more than 0.5 quads of energy for refrigerators alone.

Refrigeration technology can save energy in many ways. Constant temperature is maintained through an electrically powered compressor, and the amount of electricity needed will depend on the design of the compressor and its placement, as well as on the insulation of the refrigeration unit itself. (In other words, if it is well insulated, the compressor need not run so often.) Other home and office appliances and machines may provide less of an opportunity for conservation gain, but in most instances, there is enough room for improved energy efficiency[11] to make the gains envisioned by conservationists possible.

## The transportation sector

The transportation sector includes all means of transport for people and freight – rail, air, truck, bus, and automobile. Altogether, this sector accounted for 20.7 quads, about 28 percent of total demand in 1986 (see Table 4.4). Energy demand in this sector had been growing steadily since the recession of the early 1980s.

Most transportation in the United States uses a form of internal combustion engine, generally Otto-cycle or diesel-cycle engines; jet aircraft use a form of combustion turbine (see Chapter 3). All three are internal power plants that use the energy of refined petroleum. Altogether, Otto-cycle and diesel-cycle vehicles, primarily trucks and private automobiles, use around three-quarters of all transportation energy. And the largest part of that category, indeed the most significant factor in the transportation sector, is the private automobile, which consumes 9.5 quads of energy annually, almost half the sector total.

Conservationists have focused on the automobile, partly because of its prominence in our energy consumption patterns. But our orientation toward and dependence on internal combustion travel also carries a special danger that has worried energy analysts and government policymakers. There is, at present, limited opportunity for cost-effective substitution of resources. This problem affects the transportation sector more than any other. In the electric sector, for instance, the industry will build more plants that burn abundant coal resources or use renewable hydropower if we demand more electric energy. In the industrial sector, manufacturers can generate process heat with gas or coal or oil or electricity. Increased usage of electric power may have various undesirable impacts, including price increases and pollution, but at least resources are available and can be substituted.

Some automobiles can run on ethanol or utilize synthetic gasoline or synfuels. (In 1988,

the U.S. Congress passed legislation to encourage the substitution of methanol and ethanol for gasoline fuels). But synfuels are far more expensive than oil and the technology is in its infancy. As a practical matter, if we increase automobile use, we consume more oil and deplete this resource globally, thus posing a risk of social dislocation from supply disruptions from abroad.

The energy crisis of the 1970s brought two appeals from conservationists with respect to automobile use. First, they wanted to persuade Americans to use public transportation, both rail transport, which is in many instances electrically powered, and bus transport, which is more fuel efficient on a consumption per passenger basis. This appeal had limited effect.

The oil supply and price shocks also brought calls for greater fuel efficiency in automobiles, and the U.S. government was prodded in 1977 to mandate an average fuel efficiency for new cars. According to the government plan, by 1986 car makers were required to have an average new fleet efficiency of 27.5 miles per gallon (mpg). However, Congress later lowered the standard to 26 mpg, after petitioning from the auto manufacturers. But 26 mpg still represented an important improvement; in 1973, the average efficiency for the entire fleet of old and new cars was 13.1 mpg.

The government effort only added to a move by consumers – reacting to the higher prices of gasoline – for more efficient cars. The energy shocks helped sales of smaller, more efficient, foreign cars over larger, less efficient, American models and led to a period of crisis in the U.S. automobile industry. There had been an assumption in the industry that taste would, as with light bulbs, outweigh energy price considerations. This was not correct. Although it seemed that Americans prefer larger automobiles, a dramatic upturn in energy prices turned out to be a greater concern. What's more, American car makers had even ignored the potential for efficiency gains in large cars. Size does not wholly determine efficiency (weight is a factor, but it alone is not decisive). As a result of government intervention and consumer concerns with fuel savings, current generations of large cars are more efficient than such cars were a decade ago. As a whole, the overall fleet of U.S. automobiles had improved gas mileage to 18 mpg by 1986.

It is important to recognize the magnitude of resource savings from even a small incremental improvement in fleet gas mileage. If the 1986 figure of 18 mpg had been improved to 20 mpg there would have been an additional savings of 200,000 barrels (bbl) of refined petroleum per day, that is, more than 0.4 quads per year. Raising the average to 26 mpg, which will not be accomplished until all old cars are replaced, would save more than 1.5 million bbl/day (assuming the number of miles driven remains constant). As we can see, the rollback from 27.5 to 26 mpg, though a small increment, will result in significantly greater consumption.

The fleet average could be raised not just to 27.5 mpg but, conservationists argue, to 50 mpg or more. Already a few production models, both diesel and gasoline, can maintain a $50+$ mpg average. The implications of improvements of this magnitude are extraordinary. For example, if by 1995 the number of miles driven increases by 50 percent while at the same time mileage improves to 50 mpg, then demand would fall to *less than half* of current levels, a savings of around 4 quads annually.

Prototypes suggest the possibility of even greater gains. A few have averaged close to 100 mpg. Of course, some of these gains have been produced by lowering weight and reducing size, which might produce tradeoffs in terms of safety or comfort. They also may utilize exotic technologies that will need to prove themselves reliable and inexpensive enough so that the price of the car will not outweigh its efficiency improvements. Still, even with present techno- logy, there is the potential to raise the new fleet average to 35 mpg or more. Even greater gains may result from technological developments, including the development of strong lightweight materials such as ceramics, plastics, and composites.

## The industrial sector

Judging by the figures in Table 4.4, it would appear that the industrial sector was the most successful in pursuing conservation policies. The figures show a reduction in consumption of some 20 percent (6.6 quads) between 1979 and 1986. In fact, there have been some real conservation gains in U.S. industry and there can and will likely be more. However, these numbers do not give the full picture of energy use in the industrial sector.

Industry consumes energy primarily for process steam (such as in paper making) and heat (such as in the smelting of metals). Also, industrial firms use electric energy for running mechanical devices and in direct application for chemical – electrolytic – processes.

An important percentage of the energy savings has not come from conservation of any kind, but rather from a structural change in the American economy. As we have noted, over the last decades, and particularly since the 1970s, the percentage of output from energy-intensive manufacturing has dropped and that of the less energy-intensive service sector has risen substantially. (This accounts of course for some of the increase in consumption on the residential/ commercial side; it is not from waste but from the expansion of the service economy.) In this evolutionary process, factories have not been retrofitted for more efficient machines, they have been closed and the financial resources diverted to other endeavors; indeed some factories have been converted into shopping malls or otherwise adapted to the service sector. Even within industry itself there have been shifts away from energy-intensive manufacturing, such as steel, to such capital-intensive industries as biotechnology. According to one study, 11.2 percent of energy savings in industry, per unit of output, was attributable to the changes in the economy itself, giving a different mix of energy requirements (see Peck & Beggs in Sawhill & Cotton 1986 and Table 4.6). If anything that trend, accelerated through the early 1980s, although in the latter part of the decade the fall in the value of the U.S. dollar somewhat revived heavy (energy-intensive) industry, probably slowing the shift.

The structural changes in the economy explain some of the energy savings, but not all. Indeed, as Table 4.6 shows, 28.3 percent of the improvement in consumption by industry

**Table 4.6.** Estimate of 1981 fuel consumption at 1972 fuel efficiency

| Sources of reduced 1981 fuel consumption | Percentage |
|---|---|
| Shift in the output mix[a] | 11.2 |
| Historic trend of fuel saving[b] | 10.3 |
| Accelerated fuel saving[c] | 18.0 |
| Total increase in 1981 energy with 1972 efficiency and output mix[d] | 39.5 |

[a] Estimated using factors developed by Robert Marlay to measure the structural shifts in energy use in industrial output toward or away from energy-intensive industries.

[b] The 0.8 annual rate of the reduction in fuel per unit of production from 1953 through 1972 applies the historical trend to the 1972 efficiency, and that saving is quoted as a percentage of 1981 actual energy consumption.

[c] Calculated as a residual.

[d] The difference in what 1981 energy consumption, stated in BTUs, would have been without the improvement in energy efficiency after 1972 and actual energy consumption, expressed as a percentage of 1981 consumption.

Source: M. J. Peck and J. J. Beggs in Sawhill & Cotton, eds., (1986).

reported in the same 1982 study came from efficiency advances; industry was producing the same output with less energy. A little over 10 percent of the improvement represented a continuation of a historical trend; there have been technical improvements in manufacturing processes over time that have lowered the energy requirements for a given amount of output. But the remainder, 18 percent, of the savings represented a response to the rising energy costs of the 1970s, an accelerated drive for efficiency improvement as a means of keeping a crucial factor cost down. Using Hogan's analysis, we would conclude that the impetus toward still greater conservation in industry remains possible.[12]

Actually some of the gains in efficiency of the early 1980s also resulted from the recession during that period. It was not simply a matter of reduced demand for manufactured goods leading in turn to lower factory output and, as a consequence, less energy consumption.

Reduction in demand brought on by recession squeezes out high-cost marginal producers altogether, leaving more efficient plants with a greater share of the market when growth in demand resumes. Although with greater demand, as the United States began experiencing in the late 1980s, some of the less efficient plants may return to production, the rest simply are not there to revive. To capture some of the new demand, the low-cost producers who survived add even more efficient new capital stock instead, making industrial output still more efficient.

Of course, there has also been important investment in conservation by industry and, as in every sector, there can and will be more. As in the other two sectors, new plants and equipment are more efficient and the old stock will eventually be replaced by the new, although at any given time, a company will require a profit-enhancing rationale for making such investments. How much (and how quickly) the various industries decide to invest in conservation of course probably will depend on how important a cost energy is in production. As we noted, on average, energy represents a factor cost of only 4 percent. But that is an average over all industries. Some industries are far more energy intensive than others, and in these, substitution of capital for energy became cost effective with the kind of price shocks the market experienced in the 1970s.

The cement industry provides an excellent example of the potential for, and impact of, conservation investments in an energy-intensive industry.[13] In 1974, according to the study by Peck and Beggs, 25 percent of the cost of production was from energy. In this instance, a 10 percent cut in energy consumption (or at least in energy costs) would mean a significant reduction of 2.5 percent; a greater cut would be that much more attractive.

In the 1970s, the industry, in seeking to reduce energy costs (and protect fuel supplies), began large-scale substitution of coal for oil and natural gas. Note in Table 4.7 the very rapid change in the fuel mix from 1972 to 1980. Switching fuels alone during this period would have lowered energy costs per unit of output by 9 percent even if there had been no other

**Table 4.7.** Fuel sources for the U.S. cement industry (percent of fuel types)

| Year | Coal | Oil | Natural gas | Electricity |
|------|------|-----|-------------|-------------|
| 1972 | 30.3 | 12.9 | 39.3 | 11.3 |
| 1980 | 60.8 | 5.2 | 12.8 | 21.2 |
| 1985[a] | 82.7 | 1.5 | 4.8 | 10.0 |

[a]In 1985, 1 percent of the industry's fuel was classified as waste fuel.

*Source*: Adapted from U.S. Bureau of Mines, *Minerals Year book*, Vol. 1, annual eds., and U.S. Portland Cement Association (1985).

change in plant and equipment. But considerable efficiency gains were also achieved. In 1972, the U.S. industry used 6.75 million BTUs/ton of cement output. By 1980, that had fallen below 5.9 million BTUs – a 12.6 percent improvement (Fig. 4.5). The gains resulted principally from the replacement of antiquated plants with new ones that were no less than 33 percent more efficient. Since, as of 1981, the average age of cement plants was still around 19 years, there was clearly the potential for a great deal more improvement.

Trends in fuel switching and in greater efficiency continued into the mid-1980s (as we would have expected using Hogan's analysis), despite the drop in the price of oil and natural gas. As of 1985 (Table 4.7), petroleum had ceased to be a major factor in production and accounted for only 1.5 percent of the energy consumption of the cement industry. Natural gas also had little impact; only 5 percent of all cement was produced by natural gas. Coal, on the other hand, had become the fuel of choice. As for efficiency, on average, a ton of cement required only 5.05 million BTUs of energy, a further efficiency gain of around 14 percent. It is worth noting that in 1985 cement makers produced 58.85 million tons with about 100 trillion BTUs (0.1 quad) less energy than they would have needed in 1972.

Conservationists believe that the potential throughout all of U.S. industry for efficiency gains is great. One study estimates that by using the most efficient technology, by the year 2020, U.S. industry can consume 38 percent less energy annually than it currently does, even while real GNP rises by 50 percent (see Goldemberg et al. 1987). It seems doubtful that the price mechanism alone will provide sufficient incentives for such gains. If prices rise for oil, fuel switching still may be more cost effective than efficiency improvement; it will often be more easily and cheaply accomplished and there is the assurance that at least one resource, coal, will still be abundant. (The present shares of supply from each of the energy resources is shown in Table 4.8.) If the prices of energy resources stabilize, there will be little or no market incentive for conservation.

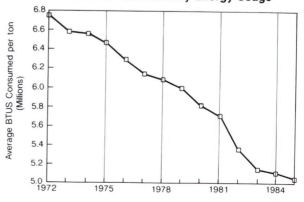

**FIG. 4.5.** U.S. cement industry energy usage. *Source*: Portland Cement Association, Market and Economic Research Department, July 1986.

**Table 4.8.** Supply sources for U.S. energy demand

|  | Supply (quads) | Share (%) |
|---|---|---|
| Petroleum | 32.7 | 43.3 |
| Natural gas | 16.5 | 21.8 |
| Coal | 17.6 | 23.2 |
| Hydropower and geothermal | 3.4 | 4.7 |
| Nuclear | 4.9 | 7.0 |
| Total[a] | 75.1 | 100 |

[a]1987 gross-energy supply.

*Source*: Adapted from *Oil & Gas Journal*, Pennwell Publishing, Tulsa, Ok, January 25, 1988.

## Conservation and policy

It is evident that the U.S. economy has the potential to conserve more energy, but important questions remain: What will it take to realize that potential? Will the solutions be equitable? Is a program of maximizing conservation even desirable? Policy choices extend across a vast range, from a decision by society to let the market decide to one that would require large-scale government intervention and control. At one end of the range, the government might only provide information; at the other end, it might be empowered to mandate conservation, control or ration supplies, and even use coercive measures to see that energy savings are realized.

The argument in favor of letting prices alone decide would appear to have a powerful precedent on its side in the United States. In the early 1980s, the market did adjust; the price mechanism did result in a search for efficiency and substitution. If left alone, we can expect to face rising prices and supply shortages, which in turn will lower demand – over time at an elasticity close to unity. Cost-effective technical improvements will be sought, thus lowering the energy requirements for economic growth. If there is any role for government in this scenario, it is at most educational, to help people make informed, rational market choices. For example, the government might help consumers understand how spending now for more efficient products will result ultimately in financial, as well as energy, savings.

Those who seek government intervention do so to help optimize the functioning of the allocation system or to further a philosophic end of preserving the world's resources or both. Although recent evidence has suggested that the market works better than some critics once had claimed, there are nevertheless two questions to raise about the market's behavior. The answers may support some kind of interventionist argument.

Both questions relate not so much to the demand for energy as to supply. First, the energy market, or at least its vital oil component, more than once changed its character in disturbing ways between the 1960s and the 1980s. In the 1960s, the oil market in the United States was essentially an oligopoly, defined by economists to be a market that is dominated by a small number of suppliers, a few large firms in this case. Supplies were largely domestic and abundant, and prices remained relatively low. But in the 1970s, as domestic demand soared far beyond supply, the center of control shifted to OPEC, which fixed prices and adjusted world production to suit its member nations. However, its power was broken in the 1980s, partly because demand in the industrialized world fell, but also because high prices encouraged exploration and new sources of supply by firms and countries outside the cartel, Norway and Great Britain, for instance. This diluted the power of OPEC by the mid-1980s to determine supply in the world market with the result that prices stabilized and then fell. Indeed, the resulting supply glut and falling prices allowed the market to stabilize much more quickly than if the only response would have been an adjustment in demand. Using Hogan's analysis again, we would have to say that on a demand basis alone – all else being equal – the market would have needed about a decade to adjust. In this instance, however, it took only a few years before market forces broke the cartel's power.

Thus, the first question is: What if the events that developed in the oil market of the 1970s recur in the 1990s or beyond? We must assume that by that time energy demand will have increased, which in turn will have led to oil shortages, price hikes, and the probable reestablishment of OPEC's control of this vital segment of the energy market. It must be remembered that despite its loss of market control in the 1980s, OPEC still has a large percentage of known and probable reserves of oil (see Chapter 2) and would surely regain its dominance in the event of another energy crisis. We can also assume a market that functions as before, that the elasticity of demand approaches unity over the course of a decade, or probably longer, since each price hike would have to be absorbed and it is unlikely that there would be just a single price hike. Which leads to the second question: What of new supply? It seems doubtful that there are vast new oil fields. Rather, by then we could be on the down slope of the bell curve of resource production seen in Chapter 2 for world oil. What's more, even if new supplies can be found, exploration and extraction costs are likely to be very high. If we compare the current cost of finding new fuels with the cost of fuel savings, the result favors conservation. As *Business Week* (March 16, 1987) has noted, "Saving oil is cheaper than finding it."

The uncertainty about new supplies is important to the policy debate. While the market was adjusting in the early 1980s, the world economy experienced a severe recession. Conservation and increased supply helped blunt the energy shock and end the hard economic times after about three years. Although we may assume that market forces will adjust to future price shocks, if there is no concurrent improvement in the supply picture, the time lag could mean economic hardship lasting two or three times as long – hardship that is likely to fall heaviest on the poor. Indeed, improvements in price in the mid-1980s have led to greater use and may result in faster rates of depletion and, consequently, to a production deficit that is more expensive and more painful for the economy to close.

The second point relates to the first: the longer the adjustment process, the longer that

vital energy supplies will carry political as well as economic consequences. The Arab states during the 1970s used oil as a weapon to pressure industrial countries into political stances. Whether the market works or not in regulating demand, policymakers cannot ignore the political consequences of waiting for it to make the necessary adjustments. In a free society such as ours, the free market may be preferred, but government officials and society as a whole could be faced with the decision to intervene in the market to preserve national political freedom of action. During the energy crisis years, the country and its leaders, particularly President Carter, discussed various interventionist steps and even passed some legislation to encourage conservation. However, even these steps were small and uncertain, and they were scaled back or reversed under President Reagan. Still, the debate continues whether to act to forestall another crisis before it comes.

An interventionist conservation policy could take many forms. Government could use powers of regulation, taxation, or coercion. In 1987, *Business Week* suggested that government should place the onus of conservation on industry, mandating a continuation and acceleration of the trend toward more energy-efficient products. The magazine noted the potential for a higher mpg average for automobiles and for improvement in energy efficiency of appliances.

Other experts have advocated different forms of intervention. One popular idea is for the government to offer tax credits for home and government conservation improvements or to place new taxes on imported oil or gasoline, keeping prices high to reduce demand. Policy could also be aimed at producing more efficient products or expanding substitution options through government-sponsored research. Critics of government policies have noted government subsidies through tax writeoffs[14] for oil and gas drilling and for large power plant construction. They have urged the government to give equivalent subsidies to conservation. At the most extreme level of intervention, the government could ration energy resources or seize energy distribution and transportation facilities.

Free market proponents view any market intervention with misgivings, and they are not without foundation. Not only has the energy market proven itself able in many respects to function over time, but more importantly governments are not always effective when they control markets. Centrally planned economies and government-controlled industries typically function less well than free counterparts. In the United States, government intervention in energy has at times helped, but it has also been disastrous. So government-mandated fuel efficiency in automobiles is given credit for declines in energy demand, but the government-sponsored research and development organiz-ation, the Synfuels Corporation, wasted large sums of money that could have gone for other purposes with few positive results (see Chapter 12). At the same time, although the market may be said to provide greater efficiency, it may not result – especially in the short run – in greater equity, and the social consequences of relying solely on market forces may be unacceptable.

Relying on the market may mean a policy choice of accepting economic and political problems, rather than trying to prevent them. A policy of prevention, however, may mean a loss of freedom and maybe governmental waste and mismanagement. Which course is prefer-able? This is the dilemma that our society faces and it will have to resolve in the decades ahead.

One further consideration is useful here. Americans may or may not be able to consume less energy while preserving economic growth. But underdeveloped countries really have no choice; they must increase consumption to develop advanced economies. If the industrial world does not conserve, it may find itself in competition for the world's resources. As we will see in the next chapter, this has political, economic, and ethical implications.

## Notes

1. There is a corresponding law of supply, which asserts that producers will supply an increasing amount of a good as the price rises.
2. *Capital* goods in economics refer to those goods

used in production, particularly plants and machinery. Commonly, capital is used to refer to money. In this book, we will use capital, alone, in the first sense; however, we will at times refer to financial resources specifically as *financial capital*.
3. *Factors of production* refer to those goods that are used in making other goods, including labor, capital, land, raw materials, and energy resources, all of which are inputs to production.
4. When a market for a good does not respond to price or other changes as the market model would predict, it is said to show *defects*. Indeed, markets where supply and demand developments do not lead to the *right* price and best allocation are deemed to have *failed*.
5. Sometimes, of course, the average consumer can realize an energy savings by buying a new capital good – a purchase, in fact, that would pay for itself in a few years through reduced energy costs. This often requires public education, since an individual will not make the purchase if he or she is unaware of the return and sees only the price of the capital good. But consumers have at times resisted purchases of new capital goods even when they were informed of the return. It has been argued that this represents a market failure, since people appear unwilling to accept what appears a good return on investment. At the same time, such a consumer may well be acting rationally, when we consider that the payback assumes an estimate of the future price of energy. Given the unpredictability of energy prices in the 1970s and early 1980s, the public has reason to be wary of accepting such estimates especially if they are for more than a year or two into the future (see also Chapter 11).
6. By *rate of adjustment*, we refer to the rate per year at which demand for energy changes with respect to percentage changes in price toward full adjustment at close to an elasticity of one. The *depreciation rate* refers to the annual percentage reduction in value of capital. In other words, it is assumed that wear and tear reduce the value of machines and buildings to the point where they no longer work or have a value of zero. Machines and buildings are assumed to have a steady (though different, depending on the machine or building in question) rate of deterioration. Although it is inevitable that one machine will deteriorate at a different rate than another, the depreciation rate is useful in deter-mining when capital generally will need to be replaced.

7. From 1980 through 1984, the GNP of all major industrial countries rose every year but one, 1982 – an average 2.2 percent per year. Energy demand in the aggregate fell, however, an average of 1 percent per year, and energy per unit of output fell every year by an average of 3 percent (Sources: The World Bank and International Monetary Fund, 1980–1985).

8. Substitution often reduced energy costs per unit of output. It did not necessarily reduce the amount of energy consumed. Burning coal may be cheaper than burning oil, but as many or perhaps more BTUs per unit of output could be consumed as a result.

9. The scenarios listed in Table 4.2 were developed between 1963 and 1981. During that period, estimates of future energy demand varied wildly, as can be seen in the table. Following the restabilization of the world oil market in the early 1980s, projections of demand have been less variable. These generally forecast a much lower rate of increase as a result of the conservation achieved from the crisis period. While the short-term indications in the late 1980s of increased consumption have caused concern, there appears to be no trend toward the precrisis rates of increase in energy demand.

10. There is an important economic concept called an *opportunity cost*. Essentially, whenever one good is produced, another good that might have been produced is not. Thus, there is a real cost to society however resources are expended.

11. In 1988, the U.S. Congress mandated a 25 percent improvement in home appliance efficiency. (The bill was pocket-vetoed by President Reagan, but was passed again by Congress and signed by the president.) This bill was designed to overcome a kind of market defect; inefficient appliances were still being sold because too often the buyers of the appliances were not the users. Home builders typically preferred to buy inexpensive but inefficient models since they had no concern about operating costs. As a result, several states mandated efficiency standards, but these differed with each other. Appliance manufacturers along with conservationists asked Congress to provide a uniform standard.

12. It is important to note that industry's response to the energy crisis of the 1970s was not always a move toward greater cost efficiency. That was the dominant trend, but some of the forces at work caused manufacturers to make choices that were not the most cost or energy efficient. For example, as Peck and Beggs (Sawhill & Cotton 1986) point out, it is often cheaper for a business to switch from oil to electricity rather than coal, even though the latter is the cheaper fuel per BTU. However, coal may require specialized equipment to protect the environment and becomes cost effective only as the scale of the operation (and the energy input) increases. Also, the shocks of the 1970s were twofold: price shocks and, with both oil and natural gas, supply shocks, during which energy resources were unavailable. Manufacturers opted for electricity in some cases to assure energy supplies, even though the choice may not have been the most cost effective.

13. Cement is made by combining limestone with small amounts of clay, iron ore, sand, and even oyster shells. This mixture is then fired in a kiln to produce small nuggets called clinker. The clinker is then ground with gypsum. The firing process consumes the most energy.

14. For many years, U.S. oil and gas drillers have been allowed a federal tax credit called the *oil depletion allowance*. This is presumed to account for depreciation of a natural resource property, and the owner is allowed to reduce his or her taxable income by a certain percentage each year. Drillers also have been allowed to deduct *intangible drilling costs* (that is, the cost of exploration for new resources) from their overall income. Makers and consumers of conservation and alternative energy equipment were given extra tax benefits in the late 1970s, but most of these were eliminated in the 1980s.

## Bibliography

Dorf, R. C. 1978. *Energy, Resources, and Policy.* Addison-Wesley, Reading, MA.

Freeman, S. D. 1974. *A Time to Choose.* The Energy Policy Project of the Ford Foundation, Ballinger, Cambridge, MA.

Goldemberg, J., T. B. Johansson, A. K. N. Reddy, and R. H. Williams. 1987. *Energy for a Sustainable World.* World Resources Institute, Washington, D.C.

Hirsch, R. L. 1987. "Impending United States Energy Crisis." *Science*, March 20, pp. 1460–70.

Kaplan, S. 1983. *Energy Economics: Quantitative Methods for Energy and Environmental Decisions.* McGraw-Hill, New York.

LeBel, P. G. 1982. *Energy Economics and Technology*. Johns Hopkins, Baltimore.

Lovins, A. B., and L. H. Lovins. 1987. "The Avoidable Oil Crisis." *The Atlantic*, December, pp. 22–30.

Reisner, M. 1987. "The Rise and Fall of Energy Conservation." *The Amicus Journal*, Spring, pp. 22–31.

Rosenfeld, A. H., and D. Hafemeister. 1988. "Energy-efficient Buildings." *Scientific American*, April, pp. 78–87.

Ross, M. H., and R. H. Williams. 1981. *Our Energy: Regaining Control*. McGraw-Hill, New York.

Sawhill, J. C., and R. Cotton (eds). 1986. *Energy Conservation: Successes and Failures*. The Brookings Institution, Washington, D.C.

Stobaugh, R., and D. Yergin (eds). 1983. *Energy Future*, 3rd edition. Vintage, New York.

# 5

# Global perspectives

## Introduction

For the United States and the industrial world, a key question is how to sustain economic growth in the face of a dwindling supply of conventional energy sources. But for the rest of the world energy-related questions are quite different. Other nations need energy resources and technology to emerge from what in many cases is great and chronic poverty. They need to create industrial economies, not just sustain them. Yet the cost of energy and the cost of growth itself often have been too great for them to pay – too great in environmental, financial, and even social terms. But the needs remain. How then do they acquire the energy resources and technologies for growth? How do they exploit what resources they do have? What policies should and can they adopt to achieve economic development?

And what policies should the industrial world adopt with respect to lesser-developed countries (LDCs)? Since the industrial world has most of the financial capital and the technical resources, they can play and, arguably, must play a major role in determining and furthering the path of economic development in the LDCs. Their plight has both moral and political implications. In fact, as we will see later in this chapter, the policies of the industrial world may be as important for the LDCs as their policies are for themselves.

Of course the LDCs differ widely, and their needs and problems differ in important ways. Some have abundant energy resources; others have none. Some are industrializing their economies rapidly and probably will join the ranks of the industrial world soon, regardless of what it does. Other countries have little hope of great advancement and, in some instances, are actually becoming increasingly impoverished. These developing nations, which encompass the bulk of the world's population, will seek to close the gap between themselves and the industrial world either with the help of rich nations or in conflict with them. The cost of aid to the industrial world might be great, but then the cost of conflict might be even greater. In the end, rich and poor will, perforce together, determine which path is chosen.

## World economic growth and energy demand

As we pointed out at the end of the last chapter, consumption of energy in some form will have to grow in the LDCs – if they are to grow economically. The statistics that bear this out speak for themselves.

Out of 119 countries reporting to the World Bank in 1985, 37 comprising more than half of the world's population, had per capita incomes

**Table 5.1.** Basic indicators

| | Population (millions) (mid-1985) | Area (thousands of square kilometers) | GNP per capita[a] | | Life expectancy at birth (years) 1985 |
| --- | --- | --- | --- | --- | --- |
| | | | Dollars (1985) | Average annual growth rate (percent) 1965–1985 | |
| Low-income economies | 2,439.4 | 32,547 | 270 | 2.9 | 60 |
| China and India | 1,805.5 | 12,849 | 290 | 3.5 | 63 |
| Other low-income | 633.9 | 19,698 | 200 | 0.4 | 52 |
| 1 Ethiopia | 42.3 | 1,222 | 110 | 0.2 | 45 |
| 2 Bangladesh | 100.6 | 144 | 150 | 0.4 | 51 |
| 3 Burkina Faso | 7.9 | 274 | 150 | 1.3 | 45 |
| 4 Mali | 7.5 | 1,240 | 150 | 1.4 | 46 |
| 5 Bhutan | 1.2 | 47 | 160 | — | 44 |
| 6 Mozambique | 13.8 | 802 | 160 | — | 47 |
| 7 Nepal | 16.5 | 141 | 160 | 0.1 | 47 |
| 8 Malawi | 7.0 | 118 | 170 | 1.5 | 45 |
| 9 Zaire | 30.6 | 2,345 | 170 | −2.1 | 51 |
| 10 Burma | 36.9 | 677 | 190 | 2.4 | 59 |
| 11 Burundi | 4.7 | 28 | 230 | 1.9 | 48 |
| 12 Togo | 3.0 | 57 | 230 | 0.3 | 51 |
| 13 Madagascar | 10.2 | 587 | 240 | −1.9 | 52 |
| 14 Niger | 6.4 | 1,267 | 250 | −2.1 | 44 |
| 15 Benin | 4.0 | 113 | 260 | 0.2 | 49 |
| 16 Central African Republic | 2.6 | 623 | 260 | −0.2 | 49 |
| 17 India | 765.1 | 3,288 | 270 | 1.7 | 56 |
| 18 Rwanda | 6.0 | 26 | 280 | 1.8 | 48 |
| 19 Somalia | 5.4 | 638 | 280 | −0.7 | 46 |
| 20 Kenya | 20.4 | 583 | 290 | 1.9 | 54 |
| 21 Tanzania | 22.2 | 945 | 290 | (.) | 52 |
| 22 Sudan | 21.9 | 2,506 | 300 | (.) | 48 |
| 23 China | 1,040.3 | 9,561 | 310 | 4.8 | 69 |
| 24 Haiti | 5.9 | 28 | 310 | 0.7 | 54 |
| 25 Guinea | 6.2 | 246 | 320 | 0.8 | 40 |
| 26 Sierra Leone | 3.7 | 72 | 350 | 1.1 | 40 |
| 27 Senegal | 6.6 | 196 | 370 | −0.6 | 47 |
| 28 Ghana | 12.7 | 239 | 380 | −2.2 | 53 |
| 29 Pakistan | 96.2 | 804 | 380 | 2.6 | 51 |
| 30 Sri Lanka | 15.8 | 66 | 380 | 2.9 | 70 |
| 31 Zambia | 6.7 | 753 | 390 | −1.6 | 52 |
| 32 Afghanistan | — | 648 | — | — | — |
| 33 Chad | 5.0 | 1,284 | — | −2.3 | 45 |
| 34 Kampuchea, Dem. | — | 181 | — | — | — |
| 35 Lao PDR | 3.6 | 237 | — | — | 45 |
| 36 Uganda | 14.7 | 236 | — | −2.6 | 49 |
| 37 Vietnam | 61.7 | 330 | — | — | 65 |
| Middle-income economies | 1,242.1 | 38,071 | 1,290 | 3.0 | 62 |
| Lower middle-income | 674.6 | 16,090 | 820 | 2.6 | 58 |
| 38 Mauritania | 1.7 | 1,031 | 420 | 0.1 | 47 |
| 39 Bolivia | 6.4 | 1,099 | 470 | −0.2 | 53 |
| 40 Lesotho | 1.5 | 30 | 470 | 6.5 | 54 |

**Table 5.1.** (*Cont.*)

| | Population (millions) (mid-1985) | Area (thousands of square kilometers) | GNP per capita[a] | | Life expectancy at birth (years) 1985 |
|---|---|---|---|---|---|
| | | | Dollars (1985) | Average annual growth rate (percent) 1965–1985 | |
| 41 Liberia | 2.2 | 111 | 470 | −1.4 | 50 |
| 42 Indonesia | 162.2 | 1,919 | 530 | 4.8 | 55 |
| 43 Yemen, PDR | 2.1 | 333 | 530 | — | 46 |
| 44 Yemen, Arab Rep. | 8.0 | 195 | 550 | 5.3 | 45 |
| 45 Morocco | 21.9 | 447 | 560 | 2.2 | 59 |
| 46 Philippines | 54.7 | 300 | 580 | 2.3 | 63 |
| 47 Egypt, Arab Rep. | 48.5 | 1,001 | 610 | 3.1 | 61 |
| 48 Côte d'Ivoire | 10.1 | 322 | 660 | 0.9 | 53 |
| 49 Papua New Guinea | 3.5 | 462 | 680 | 0.4 | 52 |
| 50 Zimbabwe | 8.4 | 391 | 680 | 1.6 | 57 |
| 51 Honduras | 4.4 | 112 | 720 | 0.4 | 62 |
| 52 Nicaragua | 3.3 | 130 | 770 | −2.1 | 59 |
| 53 Dominican Rep. | 6.4 | 49 | 790 | 2.9 | 64 |
| 54 Nigeria | 99.7 | 924 | 800 | 2.2 | 50 |
| 55 Thailand | 51.7 | 514 | 800 | 4.0 | 64 |
| 56 Cameroon | 10.2 | 475 | 810 | 3.6 | 55 |
| 57 El Salvador | 4.8 | 21 | 820 | −0.2 | 64 |
| 58 Botswana | 1.1 | 600 | 840 | 8.3 | 57 |
| 59 Paraguay | 3.7 | 407 | 860 | 3.9 | 66 |
| 60 Jamaica | 2.2 | 11 | 940 | −0.7 | 73 |
| 61 Peru | 18.6 | 1,285 | 1,010 | 0.2 | 59 |
| 62 Turkey | 50.2 | 781 | 1,080 | 2.6 | 64 |
| 63 Mauritius | 1.0 | 2 | 1,090 | 2.7 | 66 |
| 64 Congo, People's Republic | 1.9 | 342 | 1,110 | 3.8 | 58 |
| 65 Ecuador | 9.4 | 284 | 1,160 | 3.5 | 66 |
| 66 Tunisia | 7.1 | 164 | 1,190 | 4.0 | 63 |
| 67 Guatemala | 8.0 | 109 | 1,250 | 1.7 | 60 |
| 68 Costa Rica | 2.6 | 51 | 1,300 | 1.4 | 74 |
| 69 Colombia | 28.4 | 1,139 | 1,320 | 2.9 | 65 |
| 70 Chile | 12.1 | 757 | 1,430 | −0.2 | 70 |
| 71 Jordan | 3.5 | 98 | 1,560 | 5.8 | 65 |
| 72 Syrian Arab Republic | 10.5 | 185 | 1,570 | 4.0 | 64 |
| 73 Lebanon | — | 10 | — | — | — |
| *Upper middle-income* | 567.4 | 21,981 | 1,850 | 3.3 | 66 |
| 74 Brazil | 135.6 | 8,512 | 1,640 | 4.3 | 65 |
| 75 Uruguay | 3.0 | 176 | 1,650 | 1.4 | 72 |
| 76 Hungary | 10.6 | 93 | 1,950 | 5.8 | 71 |
| 77 Portugal | 10.2 | 92 | 1,970 | 3.3 | 74 |
| 78 Malaysia | 15.6 | 330 | 2,000 | 4.4 | 68 |
| 79 South Africa | 32.4 | 1,221 | 2,010 | 1.1 | 55 |
| 80 Poland | 37.2 | 313 | 2,050 | — | 72 |
| 81 Yugoslavia | 23.1 | 256 | 2,070 | 4.1 | 72 |
| 82 Mexico | 78.8 | 1,973 | 2,080 | 2.7 | 67 |
| 83 Panama | 2.2 | 77 | 2,100 | 2.5 | 72 |
| 84 Argentina | 30.5 | 2,767 | 2,130 | 0.2 | 70 |

**Table 5.1.** (*Cont.*)

| | Population (millions) (mid-1985) | Area (thousands of square kilometers) | GNP per capita[a] | | Life expectancy at birth (years) 1985 |
| | | | Dollars (1985) | Average annual growth rate (percent) 1965–1985 | |
|---|---|---|---|---|---|
| 85 Korea, Republic of | 41.1 | 98 | 2,150 | 6.6 | 69 |
| 86 Algeria | 21.9 | 2,382 | 2,550 | 3.6 | 61 |
| 87 Venezuela | 17.3 | 912 | 3,080 | 0.5 | 70 |
| 88 Greece | 9.9 | 132 | 3,550 | 3.6 | 68 |
| 89 Israel | 4.2 | 21 | 4,990 | 2.5 | 75 |
| 90 Trinidad and Tobago | 1.2 | 5 | 6,020 | 2.3 | 69 |
| 91 Hong Kong | 5.4 | 1 | 6,230 | 6.1 | 76 |
| 92 Oman | 1.2 | 300 | 6,730 | 5.7 | 54 |
| 93 Singapore | 2.6 | 1 | 7,420 | 7.6 | 73 |
| 94 Iran, Islamic Republic | 44.6 | 1,648 | — | — | 60 |
| 95 Iraq | 15.9 | 435 | — | — | 61 |
| 96 Romania | 22.7 | 238 | — | — | 72 |
| *Developing economies* | 3,681.5 | 70,618 | 610 | 3.0 | 61 |
| *Oil exporters* | 523.3 | 12,785 | 1,060 | 3.1 | 58 |
| *Exporters of manufactures* | 2,098.3 | 22,473 | 520 | 4.0 | 64 |
| *Highly indebted countries* | 554.5 | 21,213 | 1,410 | 2.5 | 62 |
| *Sub-Saharan Africa* | 418.0 | 21,874 | 400 | 1.0 | 50 |
| *High-income oil exporters* | 18.4 | 4,012 | 9,800 | 2.7 | 63 |
| 97 Libya | 3.8 | 1,760 | 7,170 | −1.3 | 60 |
| 98 Saudi Arabia | 11.5 | 2,150 | 8,850 | 5.3 | 62 |
| 99 Kuwait | 1.7 | 18 | 14,480 | −0.3 | 72 |
| 100 United Arab Emirates | 1.4 | 84 | 19,270 | — | 70 |
| *Industrial market economies* | 737.3 | 30,935 | 11,810 | 2.4 | 76 |
| 101 Spain | 38.6 | 505 | 4,290 | 2.6 | 77 |
| 102 Ireland | 3.6 | 70 | 4,850 | 2.2 | 74 |
| 103 Italy | 57.1 | 301 | 6,520 | 2.6 | 77 |
| 104 New Zealand | 3.3 | 269 | 7,010 | 1.4 | 74 |
| 105 Belgium | 9.9 | 31 | 8,280 | 2.8 | 75 |
| 106 United Kingdom | 56.5 | 245 | 8,460 | 1.6 | 75 |
| 107 Austria | 7.6 | 84 | 9,120 | 3.5 | 74 |
| 108 Netherlands | 14.5 | 41 | 9,290 | 2.0 | 77 |
| 109 France | 55.2 | 547 | 9,540 | 2.8 | 78 |
| 110 Australia | 15.8 | 7,687 | 10,830 | 2.0 | 78 |
| 111 Finland | 4.9 | 337 | 10,890 | 3.3 | 76 |
| 112 Germany, Federal Republic | 61.0 | 249 | 10,940 | 2.7 | 75 |
| 113 Denmark | 5.1 | 43 | 11,200 | 1.8 | 75 |
| 114 Japan | 120.8 | 372 | 11,300 | 4.7 | 77 |
| 115 Sweden | 8.4 | 450 | 11,890 | 1.8 | 77 |
| 116 Canada | 25.4 | 9,976 | 13,680 | 2.4 | 76 |
| 117 Norway | 4.2 | 324 | 14,370 | 3.3 | 77 |
| 118 Switzerland | 6.5 | 41 | 16,370 | 1.4 | 77 |
| 119 United States | 239.3 | 9,363 | 16,690 | 1.7 | 76 |

**Table 5.1.** (*Cont.*)

| | Population (millions) (mid-1985) | Area (thousands of square kilometers) | GNP per capita[a] | | Life expectancy at birth (years) 1985 |
| | | | Dollars (1985) | Average annual growth rate (percent) 1965–1985 | |
|---|---|---|---|---|---|
| *Nonreporting nonmember economies* | 362.6 | 25,826 | — | — | 69 |
| 120 Albania | 3.0 | 29 | — | — | 70 |
| 121 Angola | 8.8 | 1,247 | — | — | 44 |
| 122 Bulgaria | 9.0 | 111 | — | — | 71 |
| 123 Cuba | 10.1 | 115 | — | — | 77 |
| 124 Czechoslovakia | 15.5 | 128 | — | — | 70 |
| 125 German Democratic Republic | 16.6 | 108 | — | — | 59 |
| 126 Korea, Democratic Republic | 20.4 | 121 | — | — | 68 |
| 127 Mongolia | 1.9 | 1,565 | — | — | 63 |
| 128 USSR | 277.4 | 22,402 | — | — | 70 |

[a]GNP = gross national product.

*Source*: Adapted from the World Bank (1987).

of under $400, less than *one-fortieth* the income of the average American (Table 5.1). Another 23 countries had per capita incomes of under $1,000. In other words, over half the world's nations, with more than three-quarters of the population, have less than one-sixteenth the per capita income of the United States. In fact, only 32 countries had incomes over $3,000, 19 of which were fully industrialized. The rest of this group included oil exporters, like Kuwait and Oman, as well as emerging industrial countries, such as Hong Kong and Singapore. Still, the picture is clear; most of the people of the world are very poor, in some instances desperately so. Most of the world's wealth is in the hands of a few.

Not surprisingly, the disparity of energy consumption between industrial and poor countries is similarly great (Table 5.2). Of the poorest 37 nations – sometimes called the least-developed countries – China has the highest per capita consumption, which, as of 1985 was still 93 percent lower than per capita consumption in the United States. However, China, which

has an expanding industrial capacity, is atypical of the LDCs. More common are nations like Mali or Ethiopia or Nepal, where per capita consumption of commercial energy is less than *one-three hundredth* that of the United States. On average, the poorest countries consume only about 4 million BTUs per person per year, just over 1 percent of U.S. demand.

Actually, these figures must be qualified to a certain extent. They represent consumption of *commercial* energy – fossil fuels, electricity, and so on. In many LDCs, a considerable amount of energy consumption is from noncommercial sources. For heating and cooking, many people use wood fuel acquired by gathering, not by payment. Others burn such combustible substances as dried animal dung, again obtained without payment and thus outside of the energy marketplace. Consequently, it is more difficult to state with certainty the actual energy consumption in many nations. Studies have indicated, however, that in some countries noncommercial sources provide as much as 98 percent of total energy consumption. This

**Table 5.2.** Commercial energy use

| | Average annual energy growth rate (percent) | | Energy consumption per capita (kilograms of oil equivalent) | Energy imports as a percentage of merchandise exports |
| --- | --- | --- | --- | --- |
| | Energy production 1980–1985 | Energy consumption 1980–1985 | 1985 | 1985 |
| *Low-income economies* | 6.7 | 5.7 | 306 | 32 |
| *China and India* | 6.8 | 5.9 | 382 | — |
| *Other low-income* | 3.8 | 3.9 | 86 | 33 |
| 1 Ethiopia | 5.5 | −1.8 | 17 | 43 |
| 2 Bangladesh | 18.8 | 7.6 | 43 | 41 |
| 3 Burkina Faso | — | −0.4 | 20 | — |
| 4 Mali | 25.0 | 4.1 | 25 | 55 |
| 5 Bhutan | — | — | — | — |
| 6 Mozambique | −22.1 | 1.4 | 86 | 37 |
| 7 Nepal | 16.7 | 8.6 | 17 | 49 |
| 8 Malawi | 5.1 | −2.6 | 39 | 23 |
| 9 Zaire | 3.8 | 0.7 | 73 | 12 |
| 10 Burma | 6.7 | 6.3 | 74 | 3 |
| 11 Burundi | 23.7 | 14.8 | 26 | 18 |
| 12 Togo | −2.1 | −5.4 | 47 | — |
| 13 Madagascar | 12.5 | −10.5 | 33 | 34 |
| 14 Niger | 21.0 | 7.0 | 48 | 3 |
| 15 Benin | — | −0.7 | 35 | 23 |
| 16 Central African Republic | 1.9 | 2.1 | 33 | 1 |
| 17 India | 9.6 | 6.4 | 201 | 30 |
| 18 Rwanda | 9.0 | 6.1 | 43 | 25 |
| 19 Somalia | — | 2.0 | 82 | 43 |
| 20 Kenya | 12.2 | −5.7 | 103 | — |
| 21 Tanzania | 2.9 | 2.8 | 39 | — |
| 22 Sudan | 0.8 | 0.3 | 61 | 51 |
| 23 China | 6.2 | 5.7 | 515 | — |
| 24 Haiti | 4.1 | 1.9 | 55 | — |
| 25 Guinea | 0.2 | 0.7 | 53 | — |
| 26 Sierra Leone | — | −1.3 | 82 | 63 |
| 27 Senegal | — | −2.3 | 110 | 17 |
| 28 Ghana | −15.6 | −7.4 | 131 | 9 |
| 29 Pakistan | 8.8 | 9.4 | 218 | 52 |
| 30 Sri Lanka | 9.2 | 3.4 | 139 | 33 |
| 31 Zambia | 1.4 | (.) | 412 | 29 |
| 32 Afghanistan | 2.3 | 11.7 | 73 | 2 |
| 33 Chad | — | — | — | 1 |
| 34 Kampuchea, Democratic | −2.1 | 1.6 | 58 | — |
| 35 Lao PDR | 3.5 | 8.0 | 58 | — |
| 36 Uganda | 3.3 | 5.2 | 24 | — |
| 37 Vietnam | −1.2 | (.) | 76 | — |
| *Middle-income economies* | 2.9 | 2.7 | 886 | 16 |
| *Lower middle-income* | 2.0 | 4.4 | 358 | 21 |
| 38 Mauritania | — | 0.4 | 127 | 23 |
| 39 Bolivia | −0.7 | −1.5 | 263 | 1 |

**Table 5.2.** (*Cont.*)

| | Average annual energy growth rate (percent) | | Energy consumption per capita (kilograms of oil equivalent) | Energy imports as a percentage of merchandise exports |
|---|---|---|---|---|
| | Energy production 1980–1985 | Energy consumption 1980–1985 | 1985 | 1985 |
| 40 Lesotho | — | — | — | — |
| 41 Liberia | 1.2 | 0.7 | 345 | 16 |
| 42 Indonesia | 0.2 | 4.4 | 219 | 12 |
| 43 Yemen, PDR | — | 17.7 | 750 | — |
| 44 Yemen, Arab Republic | — | 20.0 | 117 | — |
| 45 Morocco | −4.5 | (.) | 237 | 50 |
| 46 Philippines | 19.6 | 1.8 | 255 | 44 |
| 47 Egypt, Arab Republic | 9.3 | 7.9 | 588 | 10 |
| 48 Côte d'Ivoire | 28.9 | 0.6 | 166 | 14 |
| 49 Papua New Guinea | 7.7 | 3.2 | 235 | 25 |
| 50 Zimbabwe | −4.4 | −2.4 | 427 | 1 |
| 51 Honduras | 2.5 | 1.7 | 201 | 28 |
| 52 Nicaragua | 1.0 | 0.3 | 259 | 21 |
| 53 Dominican Republic | −5.0 | 3.3 | 372 | 71 |
| 54 Nigeria | −4.6 | 9.0 | 165 | 3 |
| 55 Thailand | 56.1 | 6.6 | 343 | 33 |
| 56 Cameroon | 17.2 | 7.7 | 145 | 1 |
| 57 El Salvador | 3.1 | 0.9 | 186 | — |
| 58 Botswana | 0.1 | 1.2 | 380 | — |
| 59 Paraguay | 15.1 | 6.1 | 281 | 57 |
| 60 Jamaica | 5.4 | −5.0 | 954 | 59 |
| 61 Peru | −0.3 | 0.7 | 543 | 4 |
| 62 Turkey | 7.2 | 6.8 | 712 | 53 |
| 63 Mauritius | 2.8 | −0.1 | 311 | 23 |
| 64 Congo, People's Republic | 12.4 | 5.7 | 232 | 1 |
| 65 Ecuador | 7.8 | 11.1 | 720 | 1 |
| 66 Tunisia | −0.1 | 4.4 | 546 | 19 |
| 67 Guatemala | 7.2 | −2.7 | 176 | 17 |
| 68 Costa Rica | 7.1 | 0.6 | 534 | 14 |
| 69 Colombia | 5.4 | 2.4 | 755 | 14 |
| 70 Chile | 3.0 | −1.2 | 726 | 16 |
| 71 Jordan | — | 9.8 | 771 | 73 |
| 72 Syrian Arab Republic | 2.9 | 4.0 | 838 | 76 |
| 73 Lebanon | −8.4 | −2.1 | 777 | 33 |
| *Upper middle-income* | 3.2 | 2.2 | 1,510 | 14 |
| 74 Brazil | 12.6 | 3.2 | 781 | 37 |
| 75 Uruguay | 20.8 | −3.1 | 745 | 30 |
| 76 Hungary | 2.7 | 0.7 | 2,974 | 21 |
| 77 Portugal | 9.3 | 4.3 | 1,312 | 36 |
| 78 Malaysia | 21.0 | 7.7 | 826 | 9 |
| 79 South Africa | 3.3 | 0.8 | 2,184 | 1 |
| 80 Poland | 1.9 | 0.8 | 3,438 | — |
| 81 Yugoslavia | 3.9 | 2.6 | 1,926 | 31 |
| 82 Mexico | 4.8 | 1.2 | 1,290 | 1 |

**Table 5.2.**  (*Cont.*)

| | Average annual energy growth rate (percent) | | Energy consumption per capita (kilograms of oil equivalent) | Energy imports as a percentage of merchandise exports |
|---|---|---|---|---|
| | Energy production 1980–1985 | Energy consumption 1980–1985 | 1985 | 1985 |
| 83  Panama | 11.1 | 0.5 | 634 | — |
| 84  Argentina | 3.6 | 2.2 | 1,468 | 6 |
| 85  Korea, Republic of | 9.3 | 5.0 | 1,241 | 24 |
| 86  Algeria | 5.7 | 11.8 | 1,123 | 2 |
| 87  Venezuela | − 3.6 | 1.7 | 2,409 | 1 |
| 88  Greece | 12.2 | 2.3 | 1,841 | 66 |
| 89  Israel | − 21.5 | 2.4 | 1,949 | 21 |
| 90  Trinidad and Tobago | − 3.8 | − 5.3 | 3,641 | 4 |
| 91  Hong Kong | — | 6.6 | 1,264 | 5 |
| 92  Oman | 14.3 | 14.3 | 2,683 | — |
| 93  Singapore | — | − 1.1 | 2,165 | 34 |
| 94  Iran, Islamic Republic | 9.3 | 3.8 | 1,026 | 4 |
| 95  Iraq | − 6.1 | 2.1 | 662 | 1 |
| 96  Romania | 1.5 | 1.2 | 3,453 | — |
| *Developing economies* | 4.1 | 3.8 | 502 | 17 |
| *Oil exporters* | 1.6 | 3.7 | 629 | 5 |
| *Exporters of manufactures* | 6.0 | 4.3 | 555 | 22 |
| *Highly indebted countries* | 2.1 | 2.3 | 776 | 10 |
| *Sub-Saharan Africa* | − 2.6 | 2.3 | 107 | 10 |
| *High-income oil exporters* | − 14.5 | 8.4 | 3,699 | 2 |
| 97  Libya | − 7.9 | 11.1 | 3,042 | 6 |
| 98  Saudi Arabia | − 19.6 | 8.1 | 3,653 | 1 |
| 99  Kuwait | − 4.8 | 7.7 | 4,569 | 4 |
| 100  United Arab Emirates | − 5.5 | 7.4 | 5,102 | — |
| *Industrial market economies* | 1.8 | 0.1 | 4,958 | 21 |
| 101  Spain | 8.2 | 0.2 | 1,932 | 45 |
| 102  Ireland | 14.2 | 1.8 | 2,627 | 11 |
| 103  Italy | 1.5 | 0.4 | 2,606 | 30 |
| 104  New Zealand | 9.2 | 3.5 | 3,823 | 13 |
| 105  Belgium | 14.5 | − 1.3 | 4,666 | 17 |
| 106  United Kingdom | 2.6 | (.) | 3,603 | 14 |
| 107  Austria | − 1.0 | − 0.7 | 3,217 | 18 |
| 108  Netherlands | − 0.3 | − 0.1 | 5,138 | 21 |
| 109  France | 9.8 | − 0.4 | 3,673 | 25 |
| 110  Australia | 5.9 | 0.9 | 5,116 | 7 |
| 111  Finland | 11.4 | − 0.6 | 4,589 | 24 |
| 112  Germany, Federal Republic | 0.9 | − 0.1 | 4,451 | 17 |
| 113  Denmark | 63.1 | 0.3 | 4,001 | 19 |
| 114  Japan | 5.0 | 1.4 | 3,116 | 32 |
| 115  Sweden | 8.6 | 2.2 | 6,482 | 18 |
| 116  Canada | 3.4 | 0.4 | 9,224 | 5 |

**Table 5.2.** (*Cont.*)

| | Average annual energy growth rate (percent) | | Energy consumption per capita (kilograms of oil equivalent) | Energy imports as a percentage of merchandise exports |
| | Energy production 1980–1985 | Energy consumption 1980–1985 | 1985 | 1985 |
|---|---|---|---|---|
| 117 Norway | 5.9 | 2.8 | 8,920 | 7 |
| 118 Switzerland | 1.8 | 1.7 | 3,952 | 11 |
| 119 United States | 0.2 | −0.4 | 7,278 | 26 |
| *Nonreporting nonmember* | | | | |
| *economies* | 2.9 | 2.9 | 4,487 | — |
| 120 Albania | 8.2 | 7.5 | 1,267 | — |
| 121 Angola | 10.3 | 2.4 | 207 | 1 |
| 122 Bulgaria | 3.3 | 1.5 | 4,332 | — |
| 123 Cuba | 28.1 | 0.9 | 1,075 | 13 |
| 124 Czechoslovakia | 1.0 | 0.6 | 4,853 | — |
| 125 German Democratic Republic | 3.9 | 1.2 | 5,680 | — |
| 126 Korea, Democratic Republic | 2.7 | 2.9 | 2,118 | — |
| 127 Mongolia | 7.4 | 5.5 | 1,313 | — |
| 128 USSR | 2.9 | 3.3 | 4,885 | — |

*Source*: Adapted from the World Bank (1987).

narrows the gap in one respect – total energy demand – between rich and poor, although the poor countries still do not approach the industrial nations in consumption. For example, Bangladesh was estimated to derive about 70 percent of its per capita energy from noncommercial sources, but that nation still consumed less than one-fortieth as much energy per person as the average industrial country.

Yet in a real sense, the reliance on noncommercial energy sources, so far from closing the gap between rich and poor, underscores it and indicates why demand for conventional commercial energy in the LDCs must grow. In most instances, noncommercial sources are consumed inefficiently using the most rudimentary technology (see Goldemberg et al., 1987). Wood and other vegetative fuels (all called *biomass fuels*, see Appendix C) can run an automobile or power plant. But to do the former, for example, wood must be converted to alcohol and the internal combustion engine

needs to be modified to use it. The resulting technology is at present far more costly than a gasoline-powered automobile.

Conventional technologies for power plants or industrial processing also are overwhelmingly geared toward fossil fuels. Therefore, wood can be used, but it would require whole new industries with enormous investments in new technology and new capital equipment – investments poor nations cannot make. It can be argued that not every nation needs to have cars or large-scale industrialization. But today technology for agricultural production or for raw material extraction depends on commercial energy resources as much as manufacturing industries do. Without it, a country cannot compete in world markets for any commodities and services with countries that employ modern technology. And without entry into world markets, economic growth in these poorer countries is likely to be slow or nonexistent. Although income and energy use are not absolutely linked, there is little doubt that income growth in the poor nations will require significant increases in energy consumption.

Yet the prospect of growth in commercial energy demand in the LDCs has disquieting implications. Consider the following: From 1965 to 1985, the rate of growth of demand in the 73 poorest countries was about 4 percent annually. If we project a rate of about 3.5 percent growth per year of energy demand (which was typical of the 1970s) for all of the LDCs, by the year 2020, given present population trends, per capita consumption would still average only about one-third that of the industrial world in the early 1980s. But this would entail about a seven-and-a-half-fold increase in aggregate consumption for the Third World. The *additional* amount of energy consumption for the LDCs alone would be more than what the entire world consumed in 1980. Meeting this new level of demand would certainly strain world fuel resources. However, even assuming that the resources would be available, the financial implications alone are staggering. Even at a scaled-down 2.5 percent annual growth rate, the World Bank estimated in 1983 the investment would have to be huge. Not only will an LDC have to spend billions of dollars per year on energy equipment – for example, new power plants – it will have to spend millions more each year on energy resources. The World Bank estimated that each year the average LDC would need to spend 15 percent more (over current levels) of its export earnings on energy resources (see Goldemberg et al. 1987).

Export earnings, it must be emphasized, represent income that really matters to an LDC. In international markets, all trade is paid for in the currencies of industrial nations, so-called hard currencies. Currencies in the LDCs often have little value outside of their own national borders. If these countries want new technology, they must pay for it in dollars or Japanese yen or the currency of another industrial nation. If they want a major development project that requires a construction company from the United States or Japan or Western Europe, they must pay in industrial currencies. If they want

to borrow money, they must borrow dollars or yen and pay back interest and principal in the same currency. And if they want to buy oil or gas or coal or even electric power from another country, they have to pay in hard currency. To get hard currency, however, the LDCs have to sell products in world markets, particularly to the industrial nations. Except for direct foreign aid from industrial powers or from smaller sources such as tourism, exports are the only means to hard currency, and hard currency is essential for development. But if demand for energy resources requires 15 percent more than is earned from export income (and it could be more if energy prices increase as a result of heightened demand), there will be less hard currency for building factories and housing and for development generally.

In many cases, the percentage of export income already required for energy resources is burdensome. On average, the poorest nations already spend the equivalent of about 33 percent of export income on energy resources. But this figure does not reflect the problem fully, because some LDCs, like China, have enough of their own resources so that they rarely need to import any. At the same time, eight of the poorest countries, and several slightly better off LDCs, already spend over 40 percent of export earnings on commercial energy resources, and a few spend over 50 percent. An additional 15 percent burden would be intolerable and 15 percent is only an average. Several nations might need to spend nearly all of their export earnings on energy resources.

Of course, energy demand may not grow at a rate of 2.5 percent. Technical efficiency improvements may permit the same level of development with a vastly lower energy requirement, if LDCs can gain access to the most efficient technology. It is also possible that more poor nations – like China – may be able to exploit indigenous resources, so that while their energy use rises, they will spend less of their export earnings on it. Supplies of some fossil fuels may be dwindling, but as we will see that does not mean that all of the energy resources of the LDCs are being fully exploited.

## World energy supply

The more an LDC can meet its own energy needs with indigenous resources, the better off it is; hard currency can be spent for capital goods instead of oil and gas. The energy crisis of the 1970s stimulated a number of countries, along with international energy corporations, to search for more fossil fuels. Indigenous supplies were discovered where no one had previously thought to look. But in considering the possibilities for local supply, we should not limit indigenous resources to fossil fuels. It is true that such fuels are crucial given contemporary technology, and those countries without fossil resources probably will have to rely on the international fossil fuel markets. At the same time, other resources of the underdeveloped world could contribute significantly to local energy self-sufficiency.

## World fossil-fuel markets and supplies

### World oil

Oil is the dominant fuel in worldwide use today; about 41 percent of all commercial fuel is oil based. After World War II, increasing supplies and low prices helped oil gain ascendancy in world energy markets. But it achieved and has maintained dominance for other reasons as well. It can be easily transported and stored, and it has been the only fuel that is technically appropriate for automotive transportation. As the automobile became more important worldwide, so did oil. By the start of the energy crisis, oil composed 48 percent of the commercial fuel market. By then, oil and oil-based products had become basic staples of modern life. No country in the world existed without some demand for oil, and virtually the whole world was affected by the vicissitudes of the oil market.

In 1973, world oil production was about 58 MBD (million barrels of oil per day, which is equivalent to 21 BB$\ell$/year, see Chapter 2). Despite rising prices, production jumped to over 65 MBD by the late 1970s. As we might expect, a disproportionately large share of this was consumed by industrial nations – almost 65 percent. Eastern bloc countries – the USSR and other Eastern European countries – accounted for over 16 percent while the LDCs (minus China) also consumed about 16 percent. After 1982, of course, oil prices fell and so did the share of oil in world energy demand. Still, the share remained at 40 percent into the mid-1980s. Demand continued to rise in the LDCs while it fell in the industrial nations, although the latter still consumed more oil than the rest of the world combined.

Although demand is ubiquitous, supply is not. The distribution of oil production is shown in Table 5.3 for the 1970s and early 1980s. OPEC members comprise the largest group. Though only a handful of non-industrialized countries, OPEC produced about 30 MBD (11 BB$\ell$/year), or close to 50 percent of the world's daily production by the late 1970s. By 1984, however, with demand in decline and production outside of OPEC growing, OPEC's ascendancy over the market ended and its percentage of daily output fell steeply, while the oil production of the rest of the world slowly rose. This situation continued through the remainder of the 1980s (Cambridge Energy Research Associates 1988).

The OPEC countries are concentrated largely in the Middle East, as can be seen from the listings of oil production in Table 5.3. The concentration of resources in the OPEC nations, which these production statistics reflect, may be seen in Table 5.4. The Middle East has the largest quantity of reserves and estimated recoverable resources of any of the regional groupings shown. We will discuss the philosophical implications of such a distribution of wealth later in this chapter, but now we consider the profound impacts it has on the world oil market.

While it dominated world production, OPEC effectively set the prices for oil and, of course, raised them several times over. The impact of OPEC's control of the market on industrial nations was profound, contributing heavily to inflation and recession in the industrial countries and stimulating conservation. But the effect of OPEC's domination of the oil market on developing

**Table 5.3.** World crude oil production, 1973–1984 (In millions of barrels a day)

|                              | 1973 | 1977 | 1978 | 1979 | 1980 | 1981 | 1982 | 1983 | 1984 |
|------------------------------|------|------|------|------|------|------|------|------|------|
| *Oil exporting countries*    | 31.1 | 31.7 | 30.6 | 31.5 | 27.8 | 23.8 | 19.8 | 18.5 | 19.5 |
| Saudi Arabia                 | 7.6  | 9.2  | 8.3  | 9.5  | 9.9  | 9.8  | 6.5  | 4.9  | —    |
| Iran, Islamic Republic of    | 5.9  | 5.7  | 5.4  | 3.1  | 1.7  | 1.4  | 2.0  | 2.4  | —    |
| Venezuela                    | 3.4  | 2.2  | 2.2  | 2.4  | 2.2  | 2.1  | 1.9  | 1.8  | —    |
| Indonesia                    | 1.3  | 1.7  | 1.6  | 1.6  | 1.6  | 1.6  | 1.3  | 1.4  | —    |
| Nigeria                      | 2.0  | 2.1  | 1.9  | 2.3  | 2.0  | 1.4  | 1.3  | 1.3  | —    |
| United Arab Emirates         | 1.5  | 2.0  | 1.8  | 1.8  | 1.7  | 1.5  | 1.3  | 1.1  | —    |
| Iraq                         | 2.0  | 2.5  | 2.9  | 3.6  | 2.8  | 1.2  | 1.2  | 1.1  | —    |
| Libyan Arab Jamahiriya       | 2.2  | 2.1  | 2.0  | 2.1  | 1.8  | 1.2  | 1.2  | 1.1  | —    |
| Algeria                      | 1.1  | 1.2  | 1.2  | 1.2  | 1.1  | 1.0  | 1.0  | 1.0  | —    |
| Kuwait                       | 3.0  | 2.0  | 2.1  | 2.5  | 1.7  | 1.1  | 0.8  | 1.1  | —    |
| Oman                         | 0.3  | 0.3  | 0.3  | 0.3  | 0.3  | 0.3  | 0.3  | 0.4  | —    |
| Qatar                        | 0.6  | 0.4  | 0.5  | 0.5  | 0.5  | 0.4  | 0.3  | 0.3  | —    |
|                              |      |      |      |      |      |      |      |      |      |
| Natural gas liquids output   | 0.3  | 0.4  | 0.4  | 0.5  | 0.6  | 0.7  | 0.7  | 0.6  | —    |
|                              |      |      |      |      |      |      |      |      |      |
| *Other developing countries* |      |      |      |      |      |      |      |      |      |
| Net oil exporters            | 2.8  | 4.6  | 5.3  | 5.8  | 6.4  | 6.7  | 7.3  | 7.5  | 7.7  |
| Mexico                       | 0.5  | 1.1  | 1.3  | 1.6  | 2.1  | 2.6  | 3.0  | 3.0  | 3.0  |
| China                        | 1.1  | 1.9  | 2.1  | 2.1  | 2.1  | 2.0  | 2.0  | 2.1  | 2.1  |
| Egypt                        | 0.2  | 0.4  | 0.5  | 0.5  | 0.6  | 0.6  | 0.7  | 0.7  | 0.8  |
| Malaysia                     | 0.1  | 0.2  | 0.2  | 0.3  | 0.3  | 0.3  | 0.3  | 0.4  | 0.4  |
| Other                        | 0.9  | 1.1  | 1.2  | 1.3  | 1.3  | 1.3  | 1.3  | 1.3  | 1.4  |
|                              |      |      |      |      |      |      |      |      |      |
| Net oil importers            | 1.5  | 1.5  | 1.5  | 1.5  | 1.5  | 1.7  | 1.9  | 2.1  | 2.3  |
| Argentina                    | 0.4  | 0.4  | 0.5  | 0.5  | 0.5  | 0.5  | 0.5  | 0.5  | 0.5  |
| Brazil                       | 0.2  | 0.2  | 0.2  | 0.2  | 0.2  | 0.2  | 0.3  | 0.3  | 0.4  |
| India                        | 0.1  | 0.2  | 0.2  | 0.2  | 0.2  | 0.3  | 0.4  | 0.5  | 0.6  |
| Romania                      | 0.3  | 0.3  | 0.3  | 0.3  | 0.2  | 0.2  | 0.2  | 0.2  | 0.3  |
| Other                        | 0.5  | 0.4  | 0.4  | 0.4  | 0.4  | 0.5  | 0.5  | 0.6  | 0.6  |
|                              |      |      |      |      |      |      |      |      |      |
| *Industrial countries*       | 13.8 | 13.3 | 14.1 | 14.7 | 14.8 | 14.8 | 15.2 | 15.6 | 15.9 |
| United States                | 11.0 | 9.9  | 10.3 | 10.2 | 10.2 | 10.2 | 10.3 | 10.3 | 10.2 |
| Canada                       | 2.1  | 1.6  | 1.6  | 1.8  | 1.8  | 1.6  | 1.6  | 1.6  | 1.6  |
| United Kingdom               | —    | 0.8  | 1.1  | 1.6  | 1.7  | 1.8  | 2.1  | 2.3  | 2.6  |
| Norway                       | —    | 0.3  | 0.4  | 0.4  | 0.5  | 0.5  | 0.5  | 0.6  | 0.7  |
| Australia                    | 0.4  | 0.4  | 0.4  | 0.4  | 0.4  | 0.4  | 0.4  | 0.4  | 0.4  |
| Other                        | 0.3  | 0.3  | 0.3  | 0.3  | 0.3  | 0.3  | 0.3  | 0.4  | 0.4  |
|                              |      |      |      |      |      |      |      |      |      |
| *Other countries*            | 9.1  | 11.4 | 11.9 | 12.2 | 12.5 | 12.5 | 12.6 | 12.7 | 12.7 |
| USSR                         | 8.6  | 11.0 | 11.5 | 11.8 | 12.1 | 12.2 | 12.3 | 12.4 | 12.4 |
| Other                        | 0.5  | 0.4  | 0.4  | 0.4  | 0.4  | 0.3  | 0.3  | 0.3  | 0.3  |
|                              |      |      |      |      |      |      |      |      |      |
| Total                        | 58.3 | 62.5 | 63.4 | 65.7 | 63.0 | 59.5 | 56.8 | 56.4 | 58.1 |

*Source*: The World Bank (1985).

**Table 5.4.** World crude oil reserves and resources in BB$\ell = 10^9$ bb$\ell$ as of 1987

|                              | $Q_d$ | $R$ | $Q_\infty$ |
| ---------------------------- | ----- | --- | ---------- |
| North America                | 230   | 34  | 247        |
| Central and South America    | 124   | 89  | 220        |
| Africa                       | 28    | 55  | 162        |
| Middle East                  | 148   | 402 | 598        |
| Asia-Pacific                 | 38    | 37  | 96         |
| Western Europe               | 23    | 22  | 68         |
| Eastern Europe               | 119   | 61  | 472        |
| Total                        | 710   | 700 | 1863       |

*Note*: North America: Continental United States, Canada, Alaska. Central and South America: Mexico, Central and South America. Africa: Nigeria, Algeria, Libya. Middle East: Saudi Arabia, Kuwait, United Arab Emirates, Qatar, Iraq, Egypt, Iran. Asia-Pacific: China, Indonesia, Malaysia, India. Western Europe: Great Britain, Norway. Eastern Europe: USSR, Romania.

*Sources*: Exxon (1984), World Bank (1983), WAES (1977), IMF (1984), and Cambridge Energy Research Associates (1988).

lands – both on oil exporters and importers – was even more dramatic.

First, there was a tremendous transfer of wealth to the exporting countries. Suddenly, some poor LDCs became extremely wealthy, accumulating vast sums of money termed *petrodollars*. In some countries, the dollars were immediately earmarked for development projects to make the transition from poor to rich permanent. To speed that process along, some exporting nations not only spent their petrodollars but borrowed heavily on the expectation of continually rising oil prices and thus billions of more petrodollars to come. Banks in the industrial countries were happy to loan billions based on that expectation.

In the meantime, the poor oil-importing countries were devastated as their energy bills soared to a huge portion of their export earnings. In many cases, they borrowed, too, either to develop indigenous energy resources or to develop their own exports to pay for increasing energy costs. Although they were hurt by rising oil prices, they depended on the

persistence of high inflation that rising oil prices helped to generate. Inflation had led to increases in the prices of all commodities and because many of the non-OPEC LDCs exported other commodities – minerals as well as agricultural products – they hoped that ever-increasing prices would increase their incomes and help them pay their debts.

Both importing and exporting LDCs, then, were counting on a continuation of rising prices and oil market control by OPEC. But OPEC's loss of control of the market and economic policies in industrial countries that curbed inflation combined to hurt oil exporters and importers alike. Of course those that had not borrowed substantially (such as Kuwait) were affected only marginally. But borrowers, including such OPEC members as Nigeria and Venezuela as well as many non-OPEC nations, suffered. Outstanding debts drained so many national financial resources that growth was hampered. Countries instituted austerity measures that lowered people's already low living standards and often were still insufficient to produce enough revenue to pay interest on the debt.[1]

Countries could not pay their debts at all, much less under the terms accepted when they had taken the loans. Consequently, a world financial crisis persisted throughout the 1980s. The crisis also affected the industrial powers. Even though they enjoyed general prosperity during the period, they held most of this debt and the debt problem continued to threaten the stability of the world's banking system. By the late 1980s, industrial governments were still concerned about finding solutions and were considering forgiving more than $100 billion in debt owed by the poorest African nations, many of whose debts to foreign banks and governments were significantly higher than their gross national products.

While the industrial powers clearly gained the upper hand over OPEC in the 1980s, their triumph may be short-lived. As we saw in Chapter 2, OPEC remains the largest group of potential producers with the largest natural endowment of oil. Even though it does not control the market, OPEC is the linchpin of the oil trade. Now and in the future that trade is likely to be focused especially in the Persian Gulf region, where supertankers pick up crude oil for delivery to markets mostly in the industrial world. (Indeed, many LDCs and industrial nations do not buy from OPEC directly, but rather purchase oil on the *spot* market in Rotterdam, Holland.) As long as the world has a major need for oil, OPEC may be in the position to reassert its control over supply and price, again with profound impacts throughout the world.

For nonexporting LDCs, that prospect has to be daunting. It is unlikely that there will be significant new finds of oil in nonproducing countries. To the extent they depend on oil, the nonexporting LDCs will be subject to OPEC along with the industrial nations. In the past, OPEC members considered reducing the impact of price hikes on the poorest nations – either through aid grants or by selling oil through special contracts at discounted prices. Of course, prices fell anyway and OPEC was not put to the test. Without either alternative energy sources or special price considerations, a renewal of the price hikes of the 1970s could cripple the importing nations of the underdeveloped world.

## World coal

While oil is the most utilized fossil fuel resource, coal is the largest. In Chapter 2, we noted total world coal resources in terms of billions of tons of coal. But we can better compare the size of coal and oil resources by looking at them in terms of the amount of energy – for example, number of Btus – extractable from both. On that basis, we can consider a ton of coal as equivalent to a certain number of barrels of oil. In Table 5.5, world coal has been converted into billions of barrels of oil equivalent (BBOE), and we can plainly see that on this basis as well coal is the largest fossil fuel resource. It has an energy equivalent almost five times greater than oil as measured by reserves ($R$) and is nearly 40 times larger in terms of the (estimated) *remaining undiscovered resources* ($Q_\infty - Q_d$). Yet coal is utilized far less than oil, with an $R/P$ ratio almost nine times that of oil.

As with oil, the distribution of world coal

**Table 5.5.** World coal resources compared with oil and natural gas

| Resource measures[a] | World coal resources | | World oil resources BB$\ell = 10^9$ bbl | World natural gas resources BBOE |
|---|---|---|---|---|
| | Bmt units | BBOE | | |
| Production (P) | 2.45 Bmt/year | 11.7 BBOE/year | 21 BB$\ell$/year | 8.3 BBOE/year |
| Reserves (R) | 663 Bmt | 3,160 BBOE | 700 BB$\ell$ | 425 BBOE |
| R/P ratio | 270 years | 270 years | 33 years | 52 years |
| Recoverable resources ($Q_\infty$) | 5,380 Bmt | 25,600 BBOE | 1,863 BB$\ell$ | 1,200 BBOE |
| Remaining resources ($Q_\infty - Q_d$) | 4,610 Bmt | 22,000 BBOE | 1,153 BB$\ell$ | 775 BBOE |

[a]See Chapter 2 for precise definitions.

*Sources*: Wilson 1980, U.S. Geological Survey 1983, and Cambridge Energy Research Associates (1988).

resources is as important a question as the total mass amounts. Only this time, the largest supplies of the resource are not in LDCs; they are here in the United States and in East bloc countries, particularly the USSR and Poland (Fig. 5.1). South Africa, Australia, and Germany also have significant supplies. But there is not much coal in the developing countries. Only China has a very large supply; India has some reserves but they are relatively small compared with any of the other major coal producers.

A more detailed breakdown of coal resources

## Economically Recoverable Coal Reserves of the World (10⁹ tce)[a]

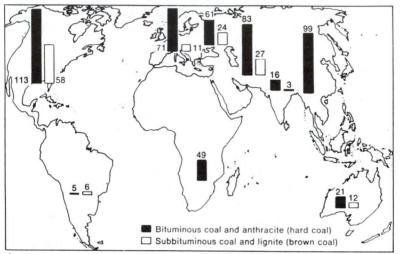

[a]tce = metric tons coal equivalent

## World Coal Resources and Reserves, by Major Coal-Producing Countries (mtce)[b]

|  | Geological Resources | Technically and Economically Recoverable Reserves |
| --- | --- | --- |
| Australia[a] | 600,000 | 32,800 |
| Canada[a] | 323,036 | 4,242 |
| People's Republic of China[a] | 1,438,045 | 98,883 |
| Federal Republic of Germany[a] | 246,800 | 34,419 |
| India[a] | 81,019 | 12,427 |
| Poland[a] | 139,750 | 59,600 |
| Republic of South Africa | 72,000 | 43,000 |
| United Kingdom[a] | 190,000 | 45,000 |
| United States[a] | 2,570,398 | 166,950 |
| Soviet Union | 4,860,000 | 109,900 |
| Other Countries | 229,164 | 55,711 |
| Total World | 10,750,212 | 662,932 |

[b]mtce = million metric tons coal equivalent

**FIG. 5.1.**   World coal resources distribution. Adapted from Wilson (1980).

**Table 5.6.** Coal resources in the developing world

| Country/region | Recoverable reserves $R$ (mtc)[a] | Production [mtc/yr] 1977 | Production [mtc/yr] 2000 | Import requirements 2000 (mtc/yr) | Export potential (mtc/yr) |
|---|---|---|---|---|---|
| India and Indonesia | 12,000 | 72 | 305 | 0 | 5 |
| China | 99,000 | 373 | 1,465 | 0 | 30 |
| East/Southeast Asia | — | 1 | 20 | 60–180 | 0 |
| Africa ((LDCs)[c] | 6,000 | 25[b] | 180[b] | 6–10[b] | 25 |
| Latin America | 11,000 | | | | 25 |
| LDC Total | >128,000 | 471 | 1,970 | 66–190 | 85 |

[a] 1 mtc = million metric tons of coal
   = $12$–$24 \times 10^{12}$ BTU (thermal equivalent)
   = $1,000$–$2,500$ GW-hr (electric energy equivalent of $1 \times 10^6$ KW-hr)
[b] These figures are combined for Africa and Latin America.
[c] The developing nations of Africa. The Republic of South Africa is not included in this definition.

*Sources*: Wilson (1980) and International Energy Agency (1982). After Cassedy & Meier (1988).

for the developing world is shown in Table 5.6. Here we can see that several countries have at least small indigenous supplies. In Latin America, for example, there is some coal, although most of it is in just two countries: Colombia and Brazil. In Asia, there are small, proven reserves in Indonesia and Bangladesh.

Table 5.6 also shows the production estimates and import and export potential for LDCs with coal resources. Outside of China and India, production in the developing world has equaled only 1 percent of total world output. However, projections of potential production are striking, not only for the increases in China and India, but also in Africa and Latin America. All of these areas have the potential to increase exploitation of coal resources significantly. With these increases, production in LDCs could reach levels that are significant on a world scale for the first time in history. Since coal can be substituted for oil and gas in electric power and industrial applications, it promises to become a major source of energy if these production goals are reached and would reduce the reliance of these countries on the oil market. Indeed, coal could provide a major portion of the fuels for development of countries with indigenous supplies. And international

trade in coal between LDCs (see last column – Export potential – of Table 5.6) has even been suggested (see Cassedy & Meier 1988).

Coal also has the potential to supply more of the energy needs of countries without resources. There has been a growing world coal trade that eventually could rival the oil trade. A coal trade did not exist in the post-World War II era prior to the energy crisis, and it would take many years to develop. But a study (WOCOL 1980) declared that such a trade was practical and would offer importing nations fuel at costs far below those of oil. Furthermore, there is little in the structure of a prospective world coal market that would lend itself to the formation of a cartel, like OPEC, if for no other reason than the dominance in reserves by the United States and the USSR, who would have little to gain by cartelization. However, the LDCs might balk at too great a reliance on coal because the resource is primarily controlled by the industrial world; reliance on imported coal would mean dependence on rich nations, which might be unacceptable to countries that in the not-so-distant past were exploited by these same (then colonial) industrial powers.

However, coal imports have been on the rise. The coal trade from the United States alone had increased by about one-third during the 1980s over the 1970s and is projected to rise still more by the year 2000. In fact, according to the World Coal Study (Wilson 1980), exports from just four countries – the United States, Canada, South Africa, and Australia – may top 400 million metric tons of coal per year (mtc/yr) by the year 2000. With some effort, these four countries could expand their combined exports to over 700 mtc/yr by the end of the century. Most of the trade would go to the industrialized countries in Western Europe and Japan; some, however, is projected for the developing countries. Asian nations, especially the rapidly growing economies of the Pacific Rim such as Taiwan and South Korea, are likely to be the major importers (as much as 180 mtc/yr). The WOCOL study also projects that Central and South American nations will increase the fraction of coal in their energy mix.

But for a large-scale coal trade to develop, there must be more than demand and exploitable resources. A potential coal exporter will require a large investment in railroads, ports, and ships, which, by and large, does not now exist and would require many years to develop. Even the United States, the largest potential exporter of coal, is developing its export capability slowly and tentatively in the face of ever-shifting world energy markets. This means that unless some steps are taken to accelerate the process, developing nations cannot depend on the coal trade soon to meet the bulk of their energy needs regardless of the demand for, or the price of, oil.

## World natural gas

Natural gas resources comprise about one-third of the world's conventional fossil fuels. Even more than coal, it is well suited for use in the country or region where it is produced. While it can be transmitted economically through overland pipelines, overseas export requires liquefaction (creating liquefied natural gas or LNG). This process is not only expensive, it requires special ships and creates new safety concerns. LNG is so flammable and explosive that an accident can literally level a harbor area.

Yet natural gas is probably the most desirable fossil fuel. It is cleaner burning than oil or coal and has wide applications – from space heating to industrial heat to electric generation. It also can be used in the petrochemical industries for the creation of organic compounds, including fertilizers.[2]

The world's natural gas resources are much smaller in terms of equivalent thermal energy values than coal (see Table 5.5). Still, at present rates of consumption, the remaining resources are likely to last well into the twenty-first century, but may start to become scarce later in the century. As we saw in Chapter 2, it seems likely that the logistics curve for conventional world gas production will follow a time path similar to that of oil.

Despite the probable worldwide scarcity of natural gas, some significant new reserves are being found. Beginning in the early 1980s, a number of new gas deposits were identified, many of them in developing countries. At present, natural gas deposits exist in 50 developing countries, although the exact size of these reserves has yet to be determined.

**Table 5.7.** World natural gas resources in 1978 (trillions of cubic feet)

| | Reserves | | Undiscovered | Total |
|---|---|---|---|---|
| | 1978 | 1987 | 1978 | 1978 |
| United States | 211 | 187 | 80.0 | 291.0 |
| Canada | 95 | 98 | 350.0 | 445.0 |
| Other Western Hemisphere | 121 | 227 | 93.4 | 214.0 |
| Western Europe | 140 | 219 | 204.8 | 345.0 |
| Eastern Europe | – | 29 | – | – |
| Iran | 500 | 489 | 400.0 | 900.0 |
| Other Middle East | 198 | 595 | 402.3 | 600.0 |
| Africa | 183 | 249 | 167.4 | 350.0 |
| Asia/Pacific | 123 | 225 | 326.7 | 450.0 |
| USSR | 875 | 1,450 | 2,180.0 | 3,055.0 |
| China | 28 | 31 | 272.0 | 300.0 |
| Totals | 2,473 | 3,799 | 4,476.6 | 6,950.0 |

*Source*: Adapted from *Petroleum Economist*, 1979, and Cambridge Energy Research Associates, 1988.

**Table 5.8.** Natural gas resources in the developing world

| Region | LDC | Recoverable Reserves R (tcf)[a] 1982 | 1987 | Production 1980 P (tcf/yr) | 1980 R/P ratio (yrs) |
|--------|-----|------|------|------|------|
| *Asia/* | Brunei | 7.3 | 7.0 | 0.3 | 24 |
| *Pacific* | India | 12.0 | 17.6 | 0.1 | 133 |
| | Indonesia | 23.5 | 73.0 | 1.5 | 15 |
| | Malaysia | 15.0 | 52.2 | — | — |
| | Pakistan | 15.7 | 22.4 | 0.2 | 69 |
| | China | 28 | 30.7 | — | — |
| | Thailand | 11 | — | — | — |
| | Bangladesh | — | 12.6 | — | — |
| | Regional estimate | 112.5 | 215.5 | 2.5 | — |
| *Middle* | Abu-Dhabi | 20.0 | 183.5 | 0.2 | 95 |
| *East* | Bahrain | 9.0 | 6.9 | — | — |
| | Iran | 485.0 | 489.4 | 1.2 | 415 |
| | Iraq | 27.0 | 26.3 | 0.5 | 55 |
| | Kuwait | 30.8 | 36.4 | 0.4 | 77 |
| | Qatar | 60.0 | 156.7 | 0.2 | 331 |
| | Saudi Arabia | 110.0 | 139.9 | 1.2 | 96 |
| | Turkey | 0.5 | — | — | — |
| | Regional estimate | 741.5 | 1,039.1 | 4.3 | — |
| *Africa* | Algeria | 131.5 | 105.2 | 1.5 | 82 |
| | Egypt | — | 10.2 | — | — |
| | Ivory Coast | 3 | — | — | — |
| | Libya | 23.8 | 25.7 | 0.8 | 29 |
| | Nigeria | 42.0 | 84.0 | 0.9 | 43 |
| | Cameroon | — | — | — | — |
| | Morocco | — | — | — | — |
| | Tunisia | — | — | — | — |
| | Tanzania | — | — | — | — |
| | Zaire | — | — | — | — |
| | Regional estimate | 200.3 | 225.1 | 2.5 | — |
| *Central and* | Argentina | 22.0 | 23.6 | 0.5 | 47 |
| *South America/* | Ecuador | — | 4.0 | — | — |
| *the Caribbean* | Mexico | 64.5 | 76.5 | 0.6 | 116 |
| | Venezuela | 42.0 | 95.0 | 1.3 | 31 |
| | Trinidad | 12.0 | 10.4 | 0.1 | 108 |
| | Bolivia | — | — | — | — |
| | Brazil | — | — | — | — |
| | Chile | — | — | — | — |
| | Regional estimate | 140.5 | 209.5 | 2.5 | — |
| Total | | 1,194.3 | 1,689.2 | 11.8 | 101 |

[a]Trillion cubic feet of natural gas, which equals $10^{15}$ BTU (1 quad) or 100,000 GW-hr (electric equivalent).

*Sources*: International Energy Agency (1982), World Bank (1983), and Cambridge Energy Research Associates 1988.

Nevertheless, these resources may be critical for economic development in several countries.

The world distribution of natural gas is depicted in Table 5.7 and a breakdown for the developing world is shown in Table 5.8. As we can see, along with the most oil, the Middle East region has the most natural gas. In fact, this gas is *associated* with oil, that is, it exists in the same geological deposits as petroleum and escapes when the oil is extracted. Natural gas reserves that exist independently of oil are called *nonassociated* deposits. Nonassociated gas deposits have been discovered in such diverse locations as Bolivia and Bangladesh (Cassedy & Meier 1988).

Newly discovered natural gas deposits hold out the hope of energy self-sufficiency and a path to economic development for a number of LDCs. Besides having a variety of uses, natural gas is also easy and cost effective to exploit and develop. The World Bank noted that natural gas requires the lowest capital outlay and of all commercial energy sources, gas can be most reliably utilized and distributed to domestic markets in LDCs that have the resource.[3]

## Conventional and nonconventional fossil fuels

Oil, coal, and natural gas resources exist in deposits scattered throughout the world. But there are more countries with little or no energy resource endowments than those with abundant resources. Some analysts hope that the have-not nations may be able to exploit what are termed *nonconventional fossil fuels*. This category includes nonconventional natural gas, tar sands, and shale oil, all of which are described in Appendix C. But they are all nonconventional because the techniques of exploration and the technologies of extraction have not yet been developed sufficiently to allow large-scale production at a cost approaching that for conventional sources (see Chapter 2). If it were possible to exploit these resources fully, the world's supply of available fossil fuels would increase immensely and the distribution would be far more widespread (see Table 5.9). But for now this scenario is hypothetical. It is important to recognize that although resources may exist,

**Table 5.9.** Comparison of conventional and nonconventional fossil-fuel reserves

|  | World known recoverable reserves (1981) | | | |
|---|---|---|---|---|
|  | Standard units | Quads | BTU distribution (%) | |
| Petroleum | 566.0 BB$\ell$ | 3,225.0 | 9.47 ⎫ | |
| Natural gas | 2,230.0 TCF | 2,299.0 | 6.75 ⎪ | Conventional |
| Coal | 651.7 Bmt | 14,858.0 | 43.65 ⎬ | |
| Natural gas liquids | 363.0 TCF | 1,456.0 | 4.28 ⎭ | |
| Bitumens | 26.7 BB$\ell$ | 107.9 | 0.30 ⎫ | |
| Shale oil (estimated) | 8,612.0 BB$\ell$ | 10,352.0 | 30.41 ⎬ | Nonconventional |
| Tar sands | 300.0 BBOE | 1,740.0 | 5.11 ⎭ | |
| Total |  | 34,036.9 | 100.00 | |

*Source*: Adapted from P. G. Le Bel (1982). *Energy Economics and Technology.* Johns Hopkins, Baltimore.

they are not always exploitable in a cost-effective way and might not be for years even if an immediate effort were made to develop new technologies (see Chapter 11). Indeed, if an LDC were to try to exploit its nonconventional resources at present, the project would probably retard economic growth, not improve it. The financial resources would be better spent on building a manufacturing or agricultural capacity or improving a national infrastructure. Whether desirable or not, advanced technologies are more likely to come from industrial nations, which are in a better position to spend more on research and development. Thus, the direction of technological development probably will be affected by the way future need is perceived in the industrial world.

# World renewable energy resources

Lack of fossil fuels does not alone determine a country's indigenous supply of energy resources. There are also renewable energy resources, a diverse category that includes hydropower, wind and solar energy, and harvested biomass, particularly wood.

In the industrialized world, most interest in renewable energy arises out of a desire to displace fossil fuels for economic, environmental, or even aesthetic reasons. In much of the underdeveloped world, renewable resources play a fundamental role. Wood offers the best example. It is an extremely important fuel in many nations, providing the basic energy requirements of living: cooking and home heating. In many rural LDC villages, wood fuel is the primary, if not the only, available energy resource. The demand for wood was unaffected by the oil scarcity that hit the industrial world, although its availability is affected by different kinds of forces – such as population growth and migration. As we will see in considering several of the world's renewable resources, there may be both opportunities and serious problems in expanding the use of indigenous renewable resources in developing countries.

## Hydropower

Hydropower is presently the largest and best developed renewable energy resource. A conventional technology for generating electricity, (as we saw in Chapter 3), it provides reliable and economic service throughout the world. In the developed world, hydroelectricity utilization grew during the energy crisis. In the developing world, hydropower projects have taken on an almost symbolic character. They are viewed as a symbol of progress, a representation of modern technological development. Huge dams with their human-made lakes have a special grandeur, although they also require significant amounts of money, which the LDCs often cannot obtain.

Because of the wealth of information on regional watersheds throughout the world, hydroresources are the best defined renewable resource. The resources are defined in terms of *potential hydroelectric capacity* (see Chapter 3) measured in kilowatts, megawatts, or gigawatts ($GW_e$ where $1 \ GW_e = 1,000 \ MW_e$) of electric output capacity. Table 5.10 lists global hydroresources, showing the potential generating capacity by continent and the share already exploited. The table also shows subtotals for both the developed and underdeveloped nations.

As we can readily see, the world potential hydrocapacity is very large: $2,200 \ GW_e$. A typical large thermal power plant has an output capacity of about $1,000 \ MW_e$ or $1 \ GW_e$; therefore, the world's hydrocapacity is the equivalent of 2,200 large thermal plants, about *double the present installed world generating capacity* for power plants of all types and sizes. Because of the distribution of hydroresources, not all thermal plants could be replaced. Some could be replaced, but more importantly a great deal of new hydroelectric power that is renewable and thus immune to the problems of fossil fuel prices and supplies can be generated in many countries.

As of the late 1980s, only about 17 percent – $365 \ GW_e$ – of the world's hydropotential had been exploited (Table 5.10). This includes some

**Table 5.10.** World hydroelectric generation capacities

| Region | Potential capacity (GW$_e$) | Installed capacity (GW$_e$) | Potential installed (%) |
|---|---|---|---|
| North America | 356 | 129 | 36 |
| Europe | 163 | 96 | 59 |
| USSR | 250 | 30 | 12 |
| Industrial world subtotals | 769 | 255 | 33 |
| Latin America | 432 | 34 | 8 |
| Africa | 358 | 17 | 5 |
| Asia | 610 | 53 | 9 |
| Oceania | 45 | 7 | 15 |
| LDC subtotals | 1,445 | 111 | 7 |
| World | 2,214 | 366 | 17 |

*Source*: Adapted from World Energy Conference (1980) and Worldwatch (1981).

spectacularly large projects such as the Itaipa on the border of Brazil and Paraguay. But it also encompasses small hydroprojects; for example, there are 92,000 small hydrostations in China (aggregate capacity of 6.6 GW$_e$). Only 7 percent of the potential hydrocapacity in the developing world has been utilized. This accounts for almost 45 percent of the existing electric generation in the LDCs, but clearly expansion of hydrogeneration is a feasible option in the developing world.

At the same time, the distribution of hydro-endowment is anything but uniform, particularly with regard to need. Asia, taken as a whole, has half the world average of water runoff per capita. The vast arid strip across Africa known as the Sahel is nearly devoid of any water resources. Rainfall in many regions is subject to wide variations; thus any hydroelectric plants depending on rain runoff will suffer the same variations and will require thermal-generating equipment as a backup source to avoid the sporadic curtailment of electricity. Even with significant rainfall, there is no guarantee that a country will have significant hydroelectric resources. It also needs a hilly or mountainous terrain, and that requirement excludes many nations of the world.[4]

In most potential sites, new hydrocapacity is cost competitive with thermal power plants.

Large hydroprojects cost no more to build and often less to maintain than any thermal plant of comparable size, and there is no fuel cost (see Chapter 9). But there can be a major expense that will reduce hydropower's cost competitive-ness: long-distance transmission lines. Fre-quently, hydrosites are far from population centers, thus requiring the expense of high-voltage electric transmission lines. In some cases, it has paid to transmit hydrogenerated electricity long distances. In Zaire, large hydroresources (potentially 40 GW$_e$) exist on the Zaire River 1,700 kilometers from the country's rich copper resources. By 1980 over 2 GW$_e$ of Zaire River hydropower had been developed and about 0.5 GW$_e$ was being trans-mitted to copper-smelting operations at the mines. The electric generation-transmission combination proved far cheaper than transpor-ting the raw copper ore 1,700 kilometers to the hydroelectric source. But in other cases, the expense of setting up transmission facilities has been too great. As a result, either the industrial project has been brought closer to the hydro-power source (as in the case of aluminum smelting in the Pacific Northwest) or the power itself has been forgone for an alternative means of electric generation.

Still, there are many potential hydrosites in the developing world that can be exploited.

As the need grows, some will be developed, although countries will have to find the financial resources to do so.

## Geothermal power

The interior heat of the earth is an energy resource available to a few countries. Called *geothermal energy*, this resource in the form of steam can generate electricity and provide heat for purely thermal applications, such as space heating and industrial processing. Potential geothermal power sites are located in the western United States, Italy, and Iceland in the developed world and Mexico, the Philippines, and Indonesia in the LDCs. The present world use of geothermal energy is about $2\,GW_e$ in generating plants and about $7\,GW$ for thermal applications. The potential for further development is limited under the present technology of exploration and use (see Appendix C). In any case, geothermal energy is a limited resource worldwide.

## Wind and solar power

The principles of wind and solar energy conversion are reviewed in Chapter 12 and Appendix C. But there are several points we should make here. Although the technology exists to convert wind or sunlight into electricity or to convert sunlight into useful thermal energy, wind and solar energy are not widely applicable at present. First, given the current state of solar and wind technologies, the wind does not blow strongly enough or long enough nor is the sunlight strong or regular enough to make either a useful energy resource in many countries. Second, even where these resources are available and sufficiently strong, they are always intermittent, that is, they vary over time and are totally unavailable at certain times of the day or for certain weather conditions. For some applications, such as electric power,[5] an intermittent supply of energy is not acceptable and therefore these technologies must be used together with conventional energy sources to assure the continuity of supply. For other applications, however, an intermittent supply can indeed by tolerated in a working technology.

Windmills have been used since ancient times to drive pumps for irrigation and land reclamation (in the Netherlands especially). For these functions, the continuous availability of the wind energy is not as important as the average supply over time; it must be sufficient to pump the amount of water needed over the given time period. Modern irrigation pumps that are electrically driven from wind or solar (photovoltaic) generators are proving useful in the Third World, especially in agricultural areas that are too remote from power lines. Solar and windpower are also useful to industries such as in oil refining or water desalinization, where an intermittent supply can be tolerated as long as annual production is sufficient and where the product itself (refined oil or desalted water) can be stored if a continuous flow of output is needed.

For applications where interruptions in supply cannot be tolerated, technological developments will be needed to make energy storage inexpensive and efficient (see Chapter 12 and Appendix C). Currently, storage is very expensive and wind and solar technologies thus have limited use, especially in poorer countries. To pay for all of the additional equipment required for sufficient storage[6] or to provide backup sources run by conventional fuels, a poor country would place itself at a disadvantage by adopting wind or solar technologies. Large-scale solar- or wind-based energy would increase the costs of industrial production and take more from hard currency earnings than conventional systems. A heavy reliance on solar or wind energy using the present technology would slow a poorer nation's economic development.

Environmentalists who wish to limit the expanded use of fossil fuels have advocated the large-scale introduction of solar and wind technologies at dispersed sites in developing nations. These advocates point out that they are attractive energy sources – environmentally benign, and completely renewable – and any country that adopted them would avoid polluting the environment. But if an LDC must pay for these energy sources, it will remain poor

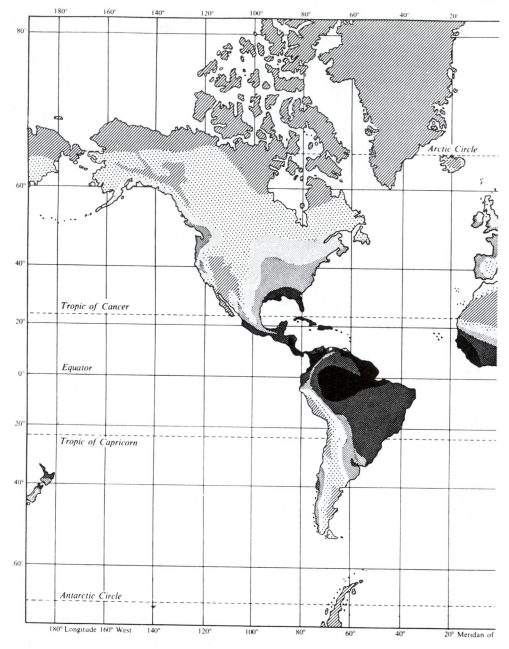

**FIG. 5.2.** Geographical distribution of fuel-production potential of world forest lands (tonnes C/ha/an = tons of carbon per hectare, annually). *Source*: Earl (1975).

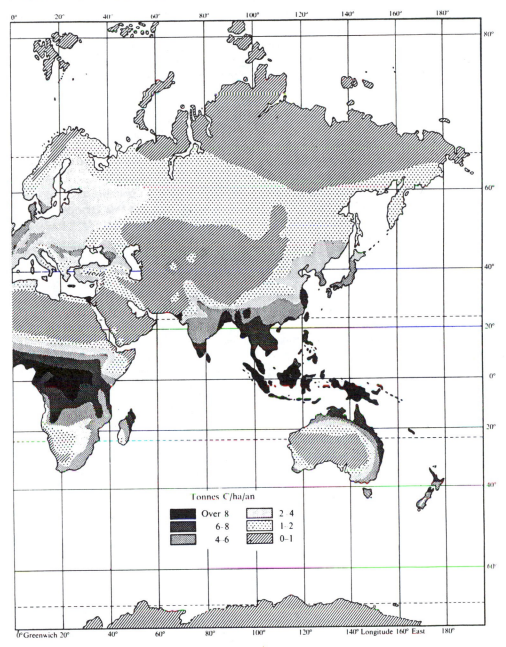

for the immediate future. Conventional technology is cheaper and will help an LDC develop.
Although the environment is important, it may
be hard to convince an LDC to be concerned
with air pollution when its people cannot get
enough to eat.

The potential for worldwide use of these two
renewable energy resources will depend on
geography, the cost of conventional energy
sources, and the future of the technology. Solar
devices will always depend on significant
amounts of sunlight and wind machines on
strong, regular wind. These requirements from
nature limit the usefulness of these technologies.
But as each technology advances, it will work
more efficiently and cost less, increasing its
economic viability as an energy source in the
regions where it is feasible, especially if the cost
of fossil fuels rises. As we will see in later
chapters, there is some prospect for real change
a decade or two into the future both in terms
of relative costs and of technological advance.
But this time horizon, short as it may seem, is
an important period for the developing world,
a time in which nations may either grow or fall
further and more hopelessly into poverty.

### Forest and biomass energy

The final major category of world renewable
resources is called *biomass*. Biomass resources
are primarily various types of vegetation, either
cultivated (like sugar cane) or not (like most
wood). Wood – timber – is by far the largest
resource. The world's total forest area is about
$4 \times 10^9$ hectares (1 hectare = $10^4$ square meters),
out of a total land area of about $15 \times 10^9$
hectares. The distribution of these forest lands
is shown in Figure 5.2. It shows the production
potential of wood based on the rate of annual
carbon fixation (photosynthesis) that is possible in a given region. How productive an area
is determines how much wood can be harvested
annually for fuel (see Earl 1974 and Jhirad 1987).
Not surprisingly, the most productive forest
areas are in the tropics and the developing
world.

The world's forests produce an estimated
*annual increment* (that is, growth) of about

$13 \times 10^9$ tons of wood that has a fuel-heating
value equivalent to about $8 \times 10^9$ tons of coal.
However, only about one-eighth of this annual
growth is presently being harvested. About half
of the annual wood harvest is consumed as fuel,
with the rest being used for lumber or the
production of industrial products such as resins.
Altogether, wood provides only about 7 percent
of total world energy production, but an
estimated three-quarters of it is consumed in
LDCs. In fact, while the world as a whole
harvests less wood than it grows yearly, LDCs
that depend on wood fuel are in many cases
severely depleting their forest resources.

Wood is the prime fuel for many of the
world's poorest countries, for example, 90
percent of the sub-Saharan African population
depends on wood as a fuel. Wood is either
burned directly or is turned into a derivative
fuel, like charcoal or alcohol (see Appendix C).
Charcoal is often used in industrial applications.
But it is not adequate for the industrialization
needs of the LDCs. (Charcoal-burning kilns
like those depicted in Figure 5.3 have a low
efficiency.) In many cases, the potential energy
production as measured by annual forest-growth
increments is not sufficient to supply the growing fuel needs of those countries; harvesting
is already exceeding annual growth. Forests are
being stripped for miles around population
centers (Figure 5.4). In all, 11 million hectares
per year are being deforested and although
logging and overgrazing by livestock contribute

**FIG. 5.3.** Wood kilns for charcoal production in
Uganda. *Source*: Earl (1975).

**FIG. 5.4.** A deforested area in Nepal caused by excessive fuel gathering and cattle grazing. *Source*: Earl (1975).

to the problem, most deforestation is due to fuel gathering. Attempts have been made to reforest millions of hectares annually, but estimates still show net losses. According to some projections (Repetto 1985), most of the rain forests will be gone by the turn of the century if present trends continue.

Hopes for continued and even expanded use of wood biomass have focused both on changing policies and technology. It has been proposed that people in poor countries be given (or have the opportunity to buy very inexpensively) efficient wood cooking stoves (see Goldemberg et al. 1987). The use of these stoves, which would embody existing technology, could reduce deforestation, especially if an effort was made to manage wood supplies. Increased rural electrification in the underdeveloped world would probably also reduce deforestation. However, building and providing the stoves would alone require an estimated investment of $1 billion annually (Goldemberg et al. 1987). The distribution of these stoves and the ongoing distribution of wood are expensive and could create logistical problems. Indeed, in some countries, the lack of infrastructure and internal political strife would make such distribution spotty, if not impossible.

There is hope for expanded use of forest resources through the use of forestry management or forest plantations (see Earl 1974 and Anderson & Fishwick 1985). Such operations,

which may be called *energy-crop plantations*, balance harvesting with replanting and land fertility to achieve a sustainable production. Whereas some efforts are presently being made to make forest and energy-crop production truly renewable, at present they have had a limited effect on the deforestation problem.

Finally, it seems that more general policies and planning will be needed to deal not just with biomass energy management, but with related issues of population pressure, land use, and alternative fuel development. Without such planning and the financial investment to carry it out, economic growth may be severely hindered in many LDCs. Development of other areas of the energy sector may also be hampered. For example, hydroelectric dams are affected by silting, as rain carries away soil from barren deforested areas. Agriculture will be similarly set back by soil erosion. In other words, deforestation threatens not only energy supplies, but really the entire life of several LDCs. Already, a country such as Mali is on the verge of economic collapse; each day many of its people need to sweep the encroaching desert sands from their front door. The potential for catastrophe from deforestation in other nations looms very large.

## Nuclear energy worldwide

Nuclear energy (see Chapter 7) has also been introduced in some countries of the developing world. Its status and prospects for the future depend not on fuel resources but on the status of nuclear technologies themselves – which will be determined in the industrial countries. The production of uranium-based fuel does not at present threaten to deplete available reserves, and newer nuclear technologies such as breeder reactors (Chapter 8) or fusion (Chapter 12) are not in commercial use even in the industrial world. Their ultimate impact on resources and on energy development worldwide is uncertain.

Worldwide, there are over 300 nuclear power plants in operation (Figure 5.5). While construction of nuclear plants has virtually stopped in the United States, it continues in a few

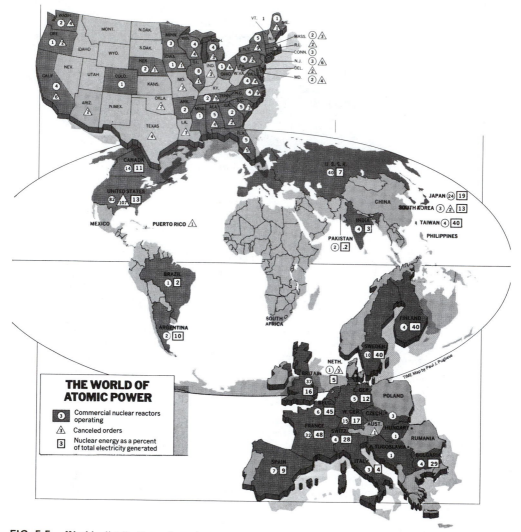

**FIG. 5.5.** World distribution of nuclear power plants. *Source*: *Time*, February 13, 1984.

Eastern and Western European countries as well as in South Korea and Japan. Concerns over the safety and costs of such plants, however, have dampened development even in the countries with ongoing nuclear construction (see Chapters 7 to 10).

There is one central point in considering the opportunity for development of nuclear energy outside the industrial or Eastern bloc countries; the technology of nuclear power at present is too complex for most countries to develop their own capability, and even those LDCs with uranium resources often can do no more with their natural wealth than sell them to the industrial world. Nuclear technology is monopolized by the Western industrial powers, the USSR, and China, and since there is at least some convergence of the technology of nuclear power and nuclear weapons, the great powers are not willing to transfer their nuclear capability without controls. To the extent that the LDCs can obtain nuclear energy, they do so only under restrictive conditions designed to prevent the proliferation of nuclear weapons. Although questions have been raised about the effectiveness of those conditions, they nonetheless limit the availability of nuclear power to the LDCs. At the same time, given the extremely high costs of nuclear power plants, the need for expertly trained personnel to run the plants, and the difficulties connected with nuclear safety and waste disposal, nuclear power seems a reasonable option only for those LDCs (such as Korea) that have limited indigenous conventional resources and have already made a great deal of progress toward industrialization.

## Resources and development: the North–South question

Having some energy resources, such as uranium, clearly does not assure either energy supplies or economic development. For many countries the issue is moot anyway; they have no resources. In either case, the distribution of resources puts them at a disadvantage. Since development in poor nations is critically dependent on the availability of energy, somehow they need to maintain supplies of energy resources that will not drain all of their financial resources.

Clearly some countries are not able to meet their resource needs and grow economically – or at least they are unable to do so without help from the rich nations. The rich have provided aid, but some policymakers and philosophers have wondered just how far that aid should go, and even if it should continue at all. Although the experts prescribe aid if it fulfills a moral obligation or satisfies a practical necessity, there is not universal agreement that transfers of wealth accomplish either purpose.

From an ethical standpoint, a sense of charity, if nothing else, would require the rich nations of the world to provide substantial assistance to the poor. Traditional ethics demands help for those in need, and poor nations are definitely in great need. But acceptance of the obvious remedy – transfers of resources and technology from rich to poor (or North to South, reflecting the geographic distribution of rich and poor) – is not universal. Indeed, there are two limitationist viewpoints that advocate diametrically opposed policies on resource and wealth transfers. Opponents of aid are not, it should be emphasized, simply callous, but rather they believe they pursue a greater good: the survival of human civilization, if not the planet itself (see Hardin 1968, in Aiken & La Follette 1977; Ehrlich & Ehrlich 1968; Ophuls 1977; and discussions by Shrader-Frechette 1981 and Cawley in Welch & Meiwald 1983).

The opposition to wealth transfers has been based on a particular concept of ethics, called *lifeboat ethics*. According to this view, every nation is like a lifeboat tossed in a hostile sea, threatened by pollution, resource depletion, and population explosion. The rich boats, the industrial countries, have a moderate load with respect to their carrying capacities; the poor boats are overcrowded and ill-provisioned, in imminent danger of being swamped. The rich boats could take some of the people off the poor boats (immigration) or transfer provisions (supply resources).

In the view of lifeboat ethics, sharing would actually be the greater evil because the poor countries have such deep and severe problems that sharing resources only will lead to greater misery for them in the long run. Helping these poor countries only prevents them from dealing with the important issues like population; aid prevents them from learning the hard lessons of experience, and thus prevents them from developing the policies needed to survive on the carrying capacities of their own lifeboats. Sharing will in fact be worse for everyone because it will lead to a more rapid depletion of resources worldwide. What the industrial countries give up in aid they often cannot replace, leading to great suffering everywhere – if not now then later in future generations – as resources dwindle. As Garret Hardin put it, "Complete justice [for all the people of the world will lead to] complete catastrophe."

In Hardin's view, the problem lies in human

nature, which leads to what he calls "the tragedy of the commons." In theory, we believe that resources of the world are held in common by all, and all are entitled to them. But as individuals, we seek to maximize our well-being and use resources to that end. In the process, however, resources – the air and the water as well as nonrenewable resources – become over-used, depleted, or destroyed. The tragedy then is that in the pursuit of what is good for ourselves, we collectively bring the world to the brink of ruin. People of the underdeveloped world are not any different from people in the industrial nations in this regard. Given easy access to resources and technology, they would increase population, pollution, and would consume more and more resources, speeding the global disaster that lifeboat advocates see in progress. (There are, it should be noted, more than a few examples of waste, corruption, and mismanagement of LDC aid programs, meaning that wealth transfers often fail to have their intended effects anyway.)

This may seem cruel to the underdeveloped world, but if one assumes that a catastrophe is looming that threatens civilization, then an extreme remedy appears necessary. Also, life-boat advocates do not let industrialized countries off easily, either. In adopting a lifeboat approach, a society would be committed to curtailing the use of fossil fuels, to limiting population growth, and indeed to limiting or eliminating economic growth and further indus-trialization. These policies would have to be enforced – by coercion if necessary or as Hardin (1968) put it, "mutual coercion, mutually agreed upon."

The legitimacy of the lifeboat argument has, not surprisingly, been challenged on several grounds. From some ethical perspectives, there are absolute standards and if helping the poor is ethically right, it would be mandatory to help them even if the end of the world results (see Watson in Aiken & LaFollete 1977). Still, even among those who envision catastrophe, this policy prescription seems morally untenable and unnecessary. One view, in particular, advocates an opposite policy approach.

This view, called *spaceship ethics*,[7] sees every person on the planet as a passenger on spaceship earth who not only can be supported, but must be (see Fuller 1969); the ship cannot make its voyage while some of the passengers ride first class and others suffer. This has a practical as well as a moral rationale. The suffering passengers will revolt, throwing the ship into turmoil. The advocates of spaceship ethics also see the transfer of wealth as a matter of justice, since industrialization of the North depended to an extent on exploitation of resources from the South. Even today, industrial nations depend on developing lands for natural resources, markets for goods, or sources of inexpensive labor for manufacturing. The transfer of wealth back to the poor nations would, according to this view, recognize our mutual interdependence and redress historic wrongs (see Fuller 1969 and a discussion by Shrader-Frechette 1981).

However, the two arguments underlying spaceship ethics – global peace and historical justice – make a better case against a lifeboat position than they do for the spaceship approach. The concept, at least as it appears in a number of important versions (especially Fuller 1969), seems impossibly utopian and appears at best impractical and potentially dangerous. To equalize wealth would require a complete change of life-style in the North. Fuller argued for a ban on fossil fuels, elimination of automobile ownership (cars would only be rented), and (in agreement with lifeboat theorists) the effective end of economic growth. Such changes would be extremely difficult to effect and could result in social upheaval.

But there is also what seems an inconsistency in the spaceship approach. Theorists imply that the changes would be, and should be, carried out democratically and willingly by industrial nations. This seems unlikely. Yet without the willingness on the part of the people, it is difficult to imagine how such massive social changes could take place without an authoritarianism comparable to, or much greater than, the type lifeboat ethics would require. Indeed, if lifeboat ethics is at all correct about human nature, spaceship ethics without political authoritarianism is doomed. People are not likely to agree to change their habits and life-styles and forcing them to do so impinges on human freedom. Experience seems to give some support to the lifeboat view that people maximize their own comfort and well-being – at least until a crisis is present and tangible to society. As a result, to accept either spaceship or lifeboat theories, one has to decide that the good of the world requires a restriction of human freedom and a limitation on what we would consider democratic rights. As long as society remains democratic and committed to freedom of choice about most personal matters, neither system is likely to gain ascendancy without a major world disaster.

Still, deprivation of the poor nations does touch the conscience of wealthy countries and our ethics demands a response to their continued poverty. The practical point of spaceship theorists also is worth considering; great suffering will in time threaten the world's political and social stability. People who are extremely impoverished and whose situation continues to deteriorate become understand-ably desperate and adopt terrorism, violent revolution, or any other means to change their condition. To prevent it, many in the North see little choice but to continue to transfer resources and technology to the South. At the same time, policymakers agree that there are important questions to consider in aiding the LDCs. For example, how can wealth and technology be transferred most effectively? What should the North require of the South to see that aid is utilized toward economic advancement? Should there be conditions placed on domestic policies? For example, some have suggested that aid be conditional on the adoption by the LDCs of more market-oriented economic policies to ensure growth (see Pachauri 1985).

Transfers of energy resources and technology raise their own set of questions. If the North transfers conventional energy technologies, it probably will increase pollution and perhaps lead to a dependence by LDCs on increasingly scarce resources. An alternative would be to transfer the most energy-efficient technologies or new technologies, either free of charge or at extremely low cost. This might include efficient biomass burners or solar and wind energy systems, including backup sources. Technolo-gies such as solar or wind are not cost effective if they have to be purchased at market prices, but at subsidized prices they may be. However, such technology transfers, even as outright grants, will have limited applications. First, as we have indicated, a technology must fit a specific site; a wind generator, for example, serves no useful purpose where winds are light and irregular. More important, in some countries, advanced technologies cannot be effectively employed because neither the infra-structure nor the institutions exist to support and service them. The maintenance of solar-collection systems has been a particular problem in West Africa for this reason (Cassedy & Meier 1988). Still, if subsidized and sustained by industrial nations, solar, wind and other advanced technologies can contribute to thermal and electric energy supplies in a number of LDCs.

Industrial nations can also benefit from the transfer. Expanded use of renewable resources would lessen the reliance of LDCs on scare fossil fuels and would protect the world's environ-ment. What's more, such a program would require that industrial nations expand and improve the capacity to manufacture techno-logies that use renewable resources. This could in fact lower the cost of production sufficiently to make alternative technologies more cost competitive with conventional energy techno-logies.

Even with renewable resources, countries will need some fossil fuel resources for the foresee-able future, and poorer nations need to be assured of both the reliability of supply and price. It has also been argued that the North cannot alone carry the burden because the North does not control the world's fossil fuels. Rich Southern nations, OPEC countries in particular, would make special arrangements to keep poorer LDCs from facing future price and supply shocks that could disrupt economic development (see Pachauri 1985).

Of course, there has been an ongoing transfer of resources from rich to poor for many years. In 1986, industrial nations provided $37 billion in aid to poorer countries and wealthy OPEC countries added more than $3.5 billion.[8] What's more, in recent years, countries have been meeting, especially through the United Nations Conference on Trade and Development (UNCTAD), to develop a global strategy to make such aid programs more effective. Though it has been difficult to arrive at specific policies for countries to adopt, there has been agreement on the basic principles of how the world economy can function more equitably and efficiently. Essentially, the goal is to have industrial nations coordinate policies to keep the world economy stable and growing and in that context to alleviate the debt problem and make it possible for more aid and investment to flow to the LDCs. The LDCs, on the other hand, have agreed generally on the need to make their economies more open to foreign investment as well as to improve the way they mobilize indigenous resources.[9]

In the end, however, the relationship of North to South will depend on how much recognition there is of mutual interest. If the self-interest of nations, rich and poor, requires greater interdependence, then there will be more of an effort to transfer wealth and technology to enhance the world economic system. In that context, the more a nation's standard of living increases, the greater the opportunity for its people to participate in the world markets and that in turn will boost the wealth of all nations. This approach would satisfy a sense of distributive justice, but it raises other questions. Most important, can there be improvement in all of the world's living standards without increasing pollution, population pressure, and resource depletion? This question cannot be answered with certainty. But if the answer is no, then transfers from North to South would raise world living standards at the expense of the world itself and would pay off for no one in the end.

## Notes

1. Often these austerity measures were required by the International Monetary Fund (IMF), which tried in the 1980s to impose strategies that would solve the debt crisis. As of the late 1980s, many of the austerity plans had disintegrated as a result of domestic pressures within LDCs, and the debt crisis was still far from solved.

2. Nitrogen fertilizers are manufactured using natural gas as an input. Natural gas provides an inexpensive source of hydrogen in reactions with steam that artificially fix nitrogen from the air, first forming ammonia ($NH_3$) and then other nitrogen compounds.

3. This view is discussed in *The Energy Transition in Developing Countries*, World Bank, 1983.

4. Mountainous terrain can also be a deterrent to development. The site can be too remote and too rugged for hydroelectric development.

5. We refer, of course, to continuous electric power generation. As we will see in later chapters, solar energy may provide a substitute for fossil fuel generation during periods of peak use – called peak load – especially when this period occurs during the time of strongest sunlight.

6. In considering the cost of capital equipment, we must include the cost of financing – that is, interest on borrowed money (see Chapter 9 for a discussion of financing costs).

7. The metaphor of spaceship earth is credited to Adlai Stevenson, who as the U.S. ambassador to the United Nations invoked the image in a 1965 speech (noted in Shrader-Frechette 1981).

8. It should be noted that by 1985 several OPEC nations were net receivers of development assistance. Three – Iran, Iraq, and Qatar – had as recently as 1978 been net givers, and Iraq and Qatar together provided over $1 billion in assistance in 1980. The fall in oil prices and the Iran–Iraq war changed their fortunes. OPEC countries increased assistance to poorer LDCs until 1980, when the amount of aid peaked at $9.6 billion. OPEC countries were giving less than half that amount four years later. In any case, the amounts given did little to alleviate the financial crisis in the non-OPEC LDCs.

9. The agreement on principles is related in a summary of UNCTAD VII (July 9 – August 3, 1987), which appeared in *Finance & Development*, a publication of the IMF and the World Bank. The article by C. B. Boucher and W. E. Siebeck is entitled "UNCTAD VII: New Spirit in North-South Relations?"

## Bibliography

Aiken, W., and H. LaFollette (eds.). 1977. *World Hunger and Moral Obligation*. Prentice-Hall, Englewood Cliffs, NJ.

Anderson, D., and R. Fishwick. 1985. *Fuelwood Consumption and Deforestation in African Countries: A Review*. The World Bank, Washington, DC.

Blair, J. 1978. *The Control of Oil*. Random House, New York.

Brown, L. R. (project director). 1986. *State of the World 1986*. Norton, New York.

Cambridge Energy Research Associates. 1988. *World Oil Trends*. Arthur Anderson, Cambridge, MA.

Cassedy, E. S., and P. M. Meier. 1988. "Planning for Electric Power in Developing Countries in the Face of Change." *Planning for Changing Energy Conditions*. Energy Policy Studies, Vol. 4. University of Delaware. Transaction Books, New Brunswick, NJ.

Earl, D. E. 1974. *Forest Energy and Economic Development*. Clarendon Press, Oxford.

Eckholm, E. 1979. *Planting for the Future: Forestry for Human Needs*, Worldwatch Paper No. 26. Worldwatch Institute, Washington, DC.

Ehrlich, P. R. 1968. *The Population Bomb*. Ballantine, New York.

Ehrlich, P. R., and A. H. Ehrlich. 1970. *Population, Resources and Environment Issues in Human Ecology*. Freeman, San Francisco.

Fuller, B. 1969. *Operating Manual for Spaceship Earth*. Pocket Books, New York.

Goldemberg, J., T. B. Johansson, K. K. N. Reddy, and R. H. Williams. 1987. *Energy for a Sustainable World*. World Resources Institute, Washington, DC.

Gorse, J., and D. Steeds. 1987. *Desertification in the Sahelian and Sudanian Zones of Western Africa*. World Bank Technical Paper No. 61. Washington , DC.

Hardin, G. 1968. "The Tragedy of the Commons." *Science 162*, December 13, pp. 1243–8.

Jhirad, D. 1987. "Renewable Energy in Developing Countries: Priorities and Prospects." *The Energy Journal*, Vol. 8, special LDC issue.

Ophuls, W. 1977. *Ecology and the Politics of Scarcity*. Freeman, San Francisco.

Pachauri, R. K. 1985. *The Political Economy of Global Energy*. Johns Hopkins University Press, Baltimore.

Perrine, R. L., and W. G. Ernst (eds.). 1985. *Energy for Ourselves and Our Posterity*. Prentice-Hall, Englewood Cliffs, NJ.

Repetto, R. (ed.). 1985. *The Global Possible: Resources, Development and the New Century*. Yale University Press, New Haven.

   1986. *World Enough and Time – Successful Strategies for Resource Management*. Yale University Press, New Haven.

Simon, J. 1981. *The Ultimate Resource*. Princeton University Press, Princeton, NJ.

Shrader-Frechette, K. S. 1981. *Environmental Ethics*. Boxwood Press, Pacific Grove, CA.

Welch, S., and R. Miewald. (eds.). 1983. *Scarce Natural Resources*. Sage, Beverly Hills, CA.

Wilson, C. L. 1980. *Coal-Bridge to the Future: Report of the World Coal Study*. Ballinger, Cambridge, MA.

Wilson, C. L. (project director). 1977. *Energy: Global Prospects 1985–2000*. McGraw-Hill, New York.

The World Bank. 1983. *The Energy Transition in Developing Countries*. Washington D.C.

   1987. *World Development Report 1987*. Oxford University Press, New York.

World Energy Conference. 1980. *Survey of Energy Resources 1980*. Prepared by Federal Institute for Geosciences and Natural Resources, Hannover, Federal Republic of Germany for the 11th World Energy Conference, Munich, September. World Energy Conference, London.

Worldwatch Institute. 1981. *Rivers of Energy: The Hydropower Potential*. Worldwatch Paper 44, Washington, DC.

# Power generation: the technology and its effects

# 6

# Fossil fuels – impacts and technology

## Coal

Fossil fuels – oil, natural gas, and coal – power most of the conventional technologies we discussed in Chapter 3. Yet when we think of fossil fuels we generally think first of oil. More than any other primary energy source, it is a concern and presence in our daily lives. Oil pricing and oil embargoes and gasoline lines make the news. In contrast, coal appears to be of secondary concern. We do not line up for coal, nor do we hear much about its price. Nevertheless, in this chapter on fossil fuels, we will focus more on the technology and impact of coal than on either oil or natural gas. The reason should be apparent by now. As we have seen, coal is by far the largest fossil fuel resource available. It has the physical capacity to supply the world's energy needs for centuries. While superabundant energy sources may be developed, such as nuclear fusion or large-scale solar technologies (see Chapter 12), at this time coal is the most abundant conventional energy source. As oil and gas supplies dwindle in the decades ahead, coal probably will be the fuel to replace them.

Coal is not a perfect substitute for oil and gas. Actually, there are processes that can turn coal into liquid or gaseous forms, although these processes have not been perfected (see Chapter 12). Coal cannot be burned in its natural state in, for example, internal combus-

tion engines. As we noted in Chapter 4, these engines account for a significant portion of U.S. energy demand. Still, coal's importance should not be underestimated. It is used primarily as a fuel for making steam, especially for turbines that generate electricity and for industrial heat. Industrial energy use and electric power generation account for more than 50 percent of all the fuel consumed in the United States. Coal does not supply all of this need, but it could and, given the size of the resource, it may supply most of it in the decades ahead.

We have chosen to focus on coal for another reason as well. Not only is it the largest fossil resource, it also has the severest impacts during routine operation. These problems relate mostly to what are called *spillovers* or *externalities*, the unintended costs to society that result from using a technology. Of course, any large technological operation carries with it some risk of unintended effects. But with coal, the spillovers can be severe enough that, if they are not carefully controlled, they can do serious harm to both people and the environment. Many of these impacts have parallels in petroleum and natural gas technologies, and we will point out some instances where coal impacts coincide with those of oil and gas. But coal provides us with the best example of the impacts of fossil fuel technology – impacts that raise questions beyond those of technical workability and encompass basic human values as well.

## Coal – technological use, present and future

The use of coal as a fuel has been inextricably linked with the rise of the industrial world. But following World War II, coal use was largely displaced by oil and natural gas.[1] Oil and gas not only became readily available, they had proven easier and cheaper to transport and store, and in most cases, were cleaner to burn than coal. Despite large annual increases in energy consumption in the United States, coal demand actually declined by about one-third from the 1940s through the 1960s (see Fig. 6.1). In fact, the level of demand in 1943 (651 million tons) was not surpassed until 1975. During the energy crisis, coal supplied less than one-fifth of the total U.S. energy demand, including less than half of the fuel for electric generation and only about one-sixth of the fuel for industrial heating. These small shares of total energy supplies by coal were typical of other industrialized countries as well during that period.

With the soaring price of oil, coal use, especially in electric power generation, increased. This switch has been slow because of constraints by the same disadvantages in transportation and handling that had contributed to the decline in coal demand in the first place. Nevertheless, by 1986 coal supplied about 56 percent of the energy for electric generation, over 15 quads. Indeed, the United States's electric utilities alone were consuming more coal annually than the country as a whole did in any single year before 1980. This trend toward increasing coal consumption is expected to continue. Total usage is expected to rise to over one-third of the U.S. energy demand. By the year 2000, U.S. consumption probably will exceed 1 billion tons per year, up by one-third from its level in the late 1980s of over 800 million tons.[2]

Just how quickly coal demand rises will depend on several factors: growth in overall energy demand, the cost of oil and gas, environmental restrictions and pollution control requirements, limitations on the use of land and water, and transportation costs and problems. As coal gradually becomes a more important component of our energy mix, the impacts on our environment, our economy, and perhaps on our society will be significant.

## The impacts of coal

Coal may affect water, air, land, and wildlife, as well as human health and safety. These can occur in any aspect of its use from *extraction, transportation, preparation,* and *combustion* to

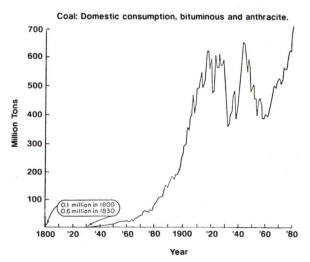

**Coal: Domestic consumption, bituminous and anthracite.**

0.1 million in 1800
0.6 million in 1830

**FIG. 6.1.** Historic consumption of coal in the United States. *Source*: U.S. Department of Energy. (1980)

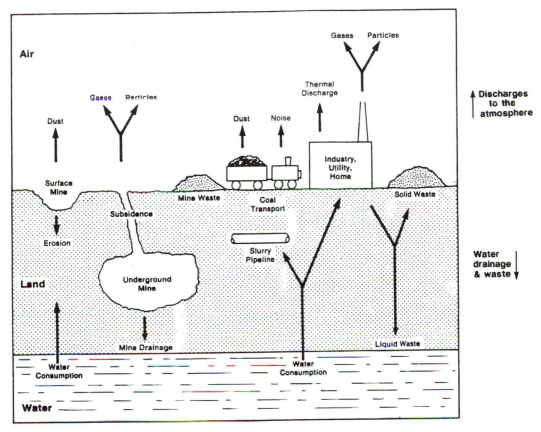

**FIG. 6.2.** The cycle of coal use. Adapted from Wilson (1980) and Office of Technology Assessment, U.S. Congress (1979).

the final *disposal* of its by-products (see Fig. 6.2). At every step of the process, the impacts, if uncontrolled, can lead to serious consequences.

Before we discuss the specifics of the impacts of coal, we should make two points. First, all coal is not the same and its impacts vary according to the kind of coal in question. And second, it matters where coal is found – in terms of its depth at a particular site and in terms of its geographic location.

Essentially there are four types of coal: lignite, subbituminous, bituminous, and anthracite. Table 6.1, which gives data on representative samples of each type, shows that the most energy per pound is found in anthracite coal.

But anthracite reserves are small and it is used in few electric generating plants. Clearly, bituminous coal, which is second in energy content, has a tremendous advantage in size of reserves (Table 6.2). Indeed, not only are the reserves enormous but they are widespread, as Figure 6.3 indicates. But bituminous coal also has by far the highest sulphur content and that increases some of the negative impacts, as we will see.

Lignite and subbituminous coal present other problems. Lignite has the lowest heat content and crumbles easily, which makes it difficult to transport and store. Subbituminous coal, which is low in sulphur and has a higher energy content than lignite, would seem ideal. But here

**Table 6.1.** Representative types of coal

| Type of coal rank | Lignite | Subbituminous | Bituminous | Anthracite |
|---|---|---|---|---|
| | Power plant data (%) | | | |
| Representative location | McLean, North Dakota | Sheridan, Wyoming | Muhlenberg, Kentucky | Lackawanna, Pennsylvania |
| *Physical composition* | | | | |
| Moisture | 37 | 22 | 9 | 4 |
| Volatiles | 28 | 33 | 36 | 5 |
| Fixed carbon | 30 | 40 | 44 | 81 |
| Ash | 6 | 4 | 11 | 10 |
| Total | 101 | 99 | 100 | 100 |
| *Chemical composition* | | | | |
| Sulphur | 0.9 | 0.5 | 2.8 | 0.8 |
| Carbon | 41 | 54 | 65 | 80 |
| Other | 58 | 45 | 32 | 19 |
| Total | 100 | 100 | 100 | 100 |
| *Energy, Btu/lb* | 7,000 | 9,610 | 11,680 | 12,880 |

**Table 6.2.** National quantities of coal (1979)

| | | | | |
|---|---|---|---|---|
| Sulphur content (%) | 0.7 | 0.5 | 2.2 | 0.6 |
| Billion tons estimated reserves | 42.9 | 182.4 | 241.9 | 7.3 |

*Sources*: U.S. Department of Energy, *Coal Data; Cost and Quality of Fuels: 1979*; and *Demonstrated Reserve Base of Coal in the United States on January 1, 1979*, May 1981.

is where the issue of location enters; more than 90 percent of the reserves are in Montana and Wyoming. To distribute it around the country requires transporting it as much as 2,000 miles. As with any commodity, the further it must be shipped the greater are the costs for its use. It also matters where in the ground coal is found. Is it near enough to the surface so that top soil can be removed and the coal simply scooped up? Or must shafts be dug and the coal extracted by underground mining techniques? It should be noted that both surface and underground mining have environmental impacts, but different mining techniques will affect both the costs of extraction and the extent of the spillovers.

## Coal extraction

The extraction of coal is a dirty and potentially dangerous process. Underground mining, principally in the Appalachian regions of the Eastern United States (Fig. 6.3), has harmed local water supplies. As water seeps into coal seams, particularly in abandoned mines where there is no attempt to deal with drainage, it reacts with compounds in the coal. Acids, such as sulphuric acid, form and then leach into underground aquifers and drain into rivers and streams. As a result, thousands of miles of streams in Appalachia are heavily polluted and cannot support aquatic life.

Water-borne chemicals also seep out of waste

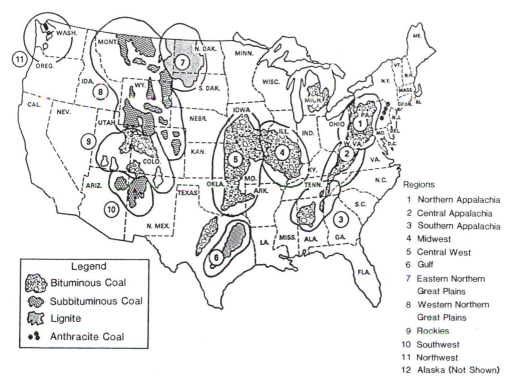

**Legend**
- Bituminous Coal
- Subbituminous Coal
- Lignite
- Anthracite Coal

**Regions**

1 Northern Appalachia
2 Central Appalachia
3 Southern Appalachia
4 Midwest
5 Central West
6 Gulf
7 Eastern Northern
  Great Plains
8 Western Northern
  Great Plains
9 Rockies
10 Southwest
11 Northwest
12 Alaska (Not Shown)

**FIG. 6.3.** Regional distribution of U.S. coal resources. *Source*: Energy Information Administration, U.S. Department of Energy (1981).

that is separated from the coal after mining; over the decades, in Appalachia, tons and tons of waste have been piled up without regard to its potential for environmental degradation. The waste piles are of course also unaesthetic and they can become sources of air pollution as well. Wastes contain flammable components, and compaction and oxidation can lead to spontaneous combustion within the piles (see Bethell & McAteer in Ridgeway 1982). The fires can smoulder for years, spewing pollutants into the air.

Underground extraction can also threaten the stability of the ground. Years of digging can result in land subsidence – an actual caving in of the earth. Buildings have been damaged or destroyed by land collapses, often many years after a mine has been exploited and closed. According to government estimates, some 2 million acres of land have subsided and several

times that amount is endangered. Cumulative damage to property by land subsidence is in the hundreds of millions of dollars (see Penner & Icerman 1976).

However damaging underground mining has been to the environment and to the general public, these dangers have been slight compared to those historically faced by underground coal miners. Underground mining is one of the most dangerous jobs in the world. Since 1900, mining accidents – mostly cave-ins and explosions – have killed over 100,000 coal miners in the United States and injured four million more. Early in the century, mining accidents resulted in an average of more than 2,000 deaths per year. Even in the 1960s, casualty rates were high. The year before Congress passed the Coal Mine Health and Safety Act of 1969 there were 9,450 injuries and 311 fatalities among U.S. coal miners. With such difficult working conditions, it is not surprising that the coal mines have historically produced perhaps the most contentious labor–management relations of any industry in the United States.

Although the human toll from accidents has been significant, it paled beside the toll from disease directly associated with mining. By 1969, there were over 50,000 miners or ex-miners who showed the adverse consequences of years of mine employment. They suffered from *black lung disease* (pneumoconiosis), a condition that results from prolonged inhalation of coal dust. Miners were said at the time to have a 50–50 chance of coming down with black lung.

Surface mining may be less hostile to workers, but produces more noticeable environmental damage. Prior to congressional passage of the Surface Mining Control and Reclamation Act of 1977, surface coal mining operations had scarred thousands of acres in Appalachia and the Midwest – marring them to the point that they could not be reclaimed for other uses. This was especially true where the technique of *highwall mining* was used.[3] In this method, which was frequently employed in Appalachia, the coal was dug out of the side of a mountain beginning where the coal seam reached the surface. Typically, in the past, the top soil and rock and vegetation covering the seam – called the overburden – were simply scooped up and dumped down the side of the hill, leaving a steep slope. At best, this method left unsightly mounds called *spoils piles*. But in fact, rocks from the overburden sometimes tumbled down hills into fields and even backyards.[4] Meanwhile, the digging continued in toward the center of the mountain until there would be a sheer wall of rocky rubble, as much as 100 feet high and miles in length, where soil and vegetation had been – a wall that would be left, sheer and barren, once the seam had been exploited (see Dorf 1978).

Besides destroying the land, surface mining, like underground mining, also impaired water resources. In fact, acid drainage from surface operations and spoils banks has been worse from surface mining than from underground operations.

Some have pointed to the environmental and human costs of coal mining as the main reason to slow the expansion of coal use. As we will see later in this chapter, however, many of these problems can be alleviated, if not solved, by technology. But it is important to note that coal extraction may also present social and economic issues that have no easy solutions.

Perhaps the most important social question connected to the expanded extraction of coal is land use. In the United States, some farmers and Native Americans who own land under which lie vast fields of coal have opposed mining these fields regardless of the economic incentives offered. Personal values are given a higher priority. Farmers prefer to keep their land undisturbed, if for no other reason than because the land has been farmed by their families for generations. A number of these farmers have refused outright to. sell or lease their land for surface mining, because to them the traditional tie to the land takes precedence over profit.

Native Americans may place an even higher value on certain tribal lands, not only because they have had them longer but because they have special cultural meaning for the entire tribe. As one Northern Cheyenne said, "this is our last piece of land, and if we lose it, we'll be Indians without lands in the future" (quoted in Ridgeway 1982.) Indeed, in some instances, the land is said to have meaning even beyond cultural value. Some have ascribed a religious or moral value to their land as well. In any case, the loss of such land cannot be compensated for in any way. Of course, the position of both the farmers and the Native American tribes would conflict with the rest of society if it adopts a policy goal to increase the exploitation of coal. But short of enacting coercive measures, there is little that a free society can do to force anyone to give up his or her land.

## Coal transportation and preparation

Coal must be transported from the mine to the point of use. Actually about 10 percent of all coal is burned at the mine mouth to generate electric power. But the rest must be shipped, and transportation becomes an increasing concern with the expansion of coal use. As we have noted, the cleanest coal is in the West, far from present points of major use. If we are to expand coal usage, we are presented with the problem of either burning more dirty coal or transporting coal for thousands of miles.

In the United States, coal is transported in river barges or by truck or railroad. More than half is shipped by rail. Western coal is frequently placed in what are called *unit trains* – trains of 100 coal cars or more – carrying a total of 10,000 tons on each trip from the mine. These trains travel on established rail lines that typically pass through towns and cities. It is believed that increased coal use would lead to a greater number of railroad accidents and, as train after train rumbles through populated areas, would create disruptions to daily life (see Horwitch in Stobaugh & Yergin 1983). The laying of new track could allow trains to avoid some population centers or trains could ride on elevated tracks over city traffic. But such measures would be costly. If the railroads bore the cost, it would undoubtedly increase the cost of shipping and, consequently, the cost of coal fuel. Already, there have been instances where the cost of rail transport made it cheaper for buyers in one U.S. city to import coal from abroad rather than transport it from the West (see Horwitch).

An alternative to the rails is a *coal slurry*, where coal is pulverized, mixed with water, and sent through a pipeline. This has advantages, not the least of which is that coal-fired boilers often use pulverized coal. After the water is removed and the coal is dried, the slurried coal is ready for use. Slurry pipelines are a cheaper means of transport than rails. One slurry pipeline can take the place of a tremendous amount of rail traffic. The Black Mesa coal slurry pipeline is 273 miles long and transports 5 million tons of coal annually from the coal deposits at the Black Mesa in Arizona to a power plant in Nevada. This would require 150 rail cars per day if it were done by railroad.

But the Black Mesa line is unusual, and attempts to build more like it probably will meet resistance – particularly over issues of land use and water rights. Slurry pipelines, like railroads and highways, require uninterrupted strips of land, termed *rights of way*. In the nineteenth century, railroads acquired rights of way over vast areas. These acquisitions were not popular and railroad owners often were seen as despoilers of the land. Some of these proposed slurry lines would require new rights of way, with consequent disputes over land use. Attempts to build slurry pipelines would certainly meet additional opposition on environmental grounds, since disruptive impacts can be expected along the route of a pipeline. Natural habitats might be adversely affected by the construction process alone. Also, slurry lines might spring leaks, posing a threat of environmental damage because of the highly acidic nature of the slurry water.

Yet the greatest area of dispute is likely to be over water rights. Coal slurry lines require massive amounts of water; the Black Mesa pipeline uses 2,700 gallons per minute. Water rights are an especially critical issue in the Western coal lands because water is relatively scarce there. Ranchers, farmers, and others who rely on local sources have opposed use of water for slurries. In 1978, Western congressmen led the fight against a bill that would have permitted construction of slurry pipelines.[5] Any large-scale slurry development would surely spark further opposition.

As for the preparation process of coal for use, it consists of crushing and cleaning coal for improved and cleaner combustion. A potential for environmental damage is present because preparation facilities produce large volumes of waste and often consume considerable quantities of water. On the whole, however, in normal operation, the environmental impact of the preparation process itself is low.

## Coal combustion

Combustion converts coal into useful heat energy (as we saw in Chapter 3), but it is also the part of the process that engenders the greatest environmental and health concerns. Coal combustion releases a variety of air pollutants, including suspended particulates, sulphur dioxide, nitrogen oxides, carbon monoxide, and certain fine particles. Carbon dioxide is also produced in combustion. While perhaps not strictly a pollutant (being a natural product of all combustion), it nonetheless is of great concern, as we will discuss later on.

Sulphur dioxide ($SO_2$) is the major air pollutant today from coal combustion; electric generation and industrial heat account for about 90 percent of the $SO_2$ emissions, almost 70 percent of which comes from power plants. Sulphur dioxide causes widespread material damage to buildings (the Taj Mahal, for example) and can destroy works of sculpture made of limestone, marble, or similar materials. Sulphur dioxide also aggravates metal corrosion.

But sulphur dioxide is a far greater concern to public health because it produces respiratory irritation. In high concentrations, it can increase acute respiratory ailments and, where local concentrations are very high, it can lead to a significant number of deaths. It should be noted that the evidence for most of these fatalities is *epidemiological*, that is, derived from public health statistics. As a consequence, the victims usually cannot be identified as having died directly from pollution. However, it has been observed that the death rate, especially among vulnerable groups like the elderly and the chronically ill, suddenly rises during periods of high levels of sulphur dioxide. Indeed, during this century, air pollution levels in a few locales have been so high that hundreds of deaths have

**Table 6.3.** Health effects, threshold concentrations, air quality standards, and margins of safety for fossil fuel pollutants

| Pollutant | Lowest best judgment estimate for an effects threshold | Adverse effect | Standard | Percent safety margin[a] | |
|---|---|---|---|---|---|
| | | | | Lowest best judgment estimate for an effects threshold | For all best judgments estimates |
| Sulphur dioxide | 300 to 400 $\mu g/m^3$ (short-term) | Mortality harvest | 365 $\mu g/m^3$ | None | None to 37 |
| | 91 $\mu g/m^3$ (long-term) | Increased frequency of acute respiratory disease | 80 $\mu g/m^3$ | 14 | 14 to 212 |
| Suspended sulfates | 8 $\mu g/m^3$ (short-term) 8 $\mu g/m^3$ (long-term) | Increased infections in asthmatics | None None | None None | None None |
| Total suspended particulates | 70 to 250 $\mu g/m^3$ (short-term) | Aggravation of respiratory diseases | 260 $\mu g/m^3$ | None | None to 15 |
| | 100 $\mu g/m^3$ (long-term) | Increased prevalence of chronic bronchitis | 75 $\mu g/m^3$ | 33 | 33 to 133 |
| Nitrogen dioxide | 141 $\mu g/m^3$ (long-term) | Increased severity of acute respiratory illness | 100 $\mu g/m^3$ | 41 | 41 to 370 |
| Carbon monoxide | 23 $mg/m^3$ (8 hour) 73 $mg/m^3$ (1 hour) | Diminished exercise tolerance in heart patients | 10 $mg/m^3$ 40 $mg/m^3$ | 130[b] 82[b] | 130 to 610[b] 82 to 788[b] |
| Photochemical oxidants | 200 $\mu g/m^3$ (short-term) | Increased susceptibility to infection | 160 $\mu g/m^3$ | 25 | None to 363 |

[a]Safety margin = effects threshold minus standard divided by standard × 100.
[b]Safety margins based upon carboxyhemoglobin levels would be 100 percent for the 8-hour standard with a range of 100 to 40 judgment estimates. For the 1-hour standard the lowest safety factor would be 67 percent with a range of 67 to 400 percent.

*Source*: NAS, NRC (1977).

resulted. In London in 1952, for example, 3,900 more deaths than were expected occurred during four days of very high levels of pollution; sulphur dioxide levels were especially high (see Wilson et al. 1980).[6]

Present air quality standards as established by the U.S. government limit pollutant densities within specific localities to what are called *threshold values*. At these levels, according to the concept, health effects begin to become significant for each pollutant (see Table 6.3 and the discussion in Chapter 10). Other industrialized countries have similar air quality standards. Differences of opinion persist, however, as to the adequacy of these standards for public health protection.

Other pollutants whose effects are less well understood include the nitrogen oxides ($NO_x$, where $x$ is an integer denoting the particular oxide compound) and the related photochemical oxidants, particularly ozone ($O_3$). (The photochemical oxidants result from the reaction of nitrogen oxides with hydrocarbons.) About half of these come from stationary sources such as

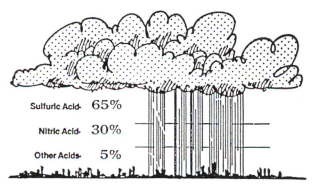

Sulfuric Acid- 65%

Nitric Acid- 30%

Other Acids- 5%

TYPICAL NORTHEASTERN U.S. ACID RAIN COMPONENTS

**(a)**

**FIG. 6.4.** Acid rain. (a) Acid rain components. *Pure rain* has a pH of between 5.6 and 5.7. These pH values take into consideration the amount of acidity created by the reaction of rainwater with normal levels of atmospheric carbon dioxide. Acid precipitation (rain, snow, sleet, or hail) has a pH 5.6 or below. (b) The pH scale. The pH (*potential hydrogen*) scale is a measure of hydrogen ion concentration. Hydrogen ions have a positive electrical charge and are called *cations*; ions with a negative electrical charge are known as *anions*. A substance containing equal concentrations of cations and anions, so that the electric charges balance is neutral, has a pH of 7. However, a substance with more hydrogen ions than anions is acidic and has a pH less than 7; substances with more anions than cations are alkaline and have pH measures above 7. Thus, as the concentration of hydrogen ions increases, the pH decreases with increasing acidity. *Source*: U.S. Environmental Protection Administration (1980).

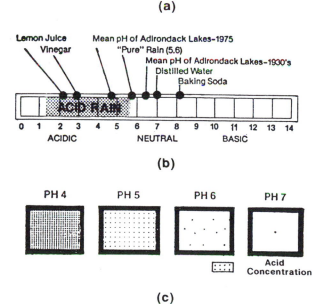

Lemon Juice
Vinegar

Mean pH of Adirondack Lakes-1975
"Pure" Rain (5.6)
Mean pH of Adirondack Lakes-1930's
Distilled Water
Baking Soda

ACID RAIN

0   1   2   3   4   5   6   7   8   9   10   11   12   13   14
ACIDIC            NEUTRAL            BASIC

**(b)**

PH 4            PH 5            PH 6            PH 7

Acid Concentration

**(c)**

coal-burning furnaces and boilers, whereas the other half is emitted by mobile sources such as cars and trucks. Heavy concentrations of the two in a locality cause *smog*, as has been experienced in the Los Angeles air basin especially. Smog produces respiratory distress and eye irritations as well as aesthetic degradation and visibility limitations.

So-called *fine particles* and carbon monoxide are the remaining air pollutants. Particles in the stack gases of coal combustion vary in size from 0.01 micrometers ($\mu m = 10^{-6}$ meters) to $10 \mu m$ in diameter. The smallest particles, in the range of 0.01 to $1 \mu m$, are the most dangerous to health. They are thought to be deposited in the respiratory system and add to pollution-related respiratory distress. Carbon monoxide (CO) is a natural product of combustion of the carbon in coal is well as petroleum that can, in high concentrations, affect the heart.[7]

Despite attempts to reduce carbon monoxide, nitrogen oxides, and ozone, the U.S. government reported in 1987 that in many metropolitan areas carbon monoxide and ozone levels were above standards. Los Angeles, where most air pollutants are generated by gasoline-burning automobiles, recorded 144 days where carbon monoxide and ozone levels were above the threshold levels.

Although air pollution is the most apparent local effect of coal combustion at the point of use, coal combustion also affects the environment in *remote* regions. Many scientists believe that two pollutants of all fossil fuel combustion (coal burning in particular) cause *acid rain*. Although evidence at this time is not absolutely conclusive, indications point strongly to sulphur oxide and nitrogen oxide pollutants as primary sources of acid rain. The acid in acid rain is about one-half to two-thirds sulphuric acid, and most of the rest is nitric acid (see Fig. 6.4a). According to most theories, the acids form high in the atmosphere where the oxides are transformed into sulphates and nitrates that then react with moisture to form acids. The acids then fall with the rain.

The heaviest concentrations of acid rain have been recorded in the Northeastern United States and Eastern Canada. But many scientists feel that most of the constituents of acid rain are produced in the U.S. Midwest and that prevailing wind patterns carry the chemicals several hundred miles eastward. In any case, regions of the Northeast have noted unusually acidic rain. Normally, rain has an acid balance, measured on a scale called the *potential hydrogen* (or pH) value of 5.6. But rain with pH values ranging down to 5.0 or lower (increasingly acidic; see Fig. 6.4 b,c) are regularly recorded in some places. Acid rain has had devastating impacts on particular lakes and streams in the Northeast United States and Canada. As they have become progressively more acidic, these bodies of water have gradually lost their stocks of fish and other aquatic life. In recent years, naturalists have observed harmful effects on other wildlife, certain tree species, and agriculture, and they expect the effects will grow more marked unless greater pollution controls are exercised.

## Coal disposal

Every coal burning plant produces solid waste products (such as fly ash from the flue and bottom ash from the furnace), all of which must be disposed of. Even though such products represent only a small percentage (by volume or weight) of the coal input, the absolute amounts are huge because of the massive quantities of coal used. For example, a typical 1,000 MW power plant operating for a day will burn close to 10,000 tons of coal. If solid wastes weigh only 10 percent, then 1,000 tons of waste must be disposed of for each day of operation.

The problem of coal waste disposal revolves around the issue of land use. Large landfill areas must be provided, either adjacent to each generating site or at sites available for low cost transportation (rail or water) for the waste. This can be difficult to find, especially when a plant is located in a densely populated region where landfill sites are becoming overutilized. Care must be taken in selection of these waste sites for both aesthetic and environmental reasons. The particular environmental concern, here as with much of coal technology, is the risk of seepage of harmful chemicals into ground water.

## Control technologies and their costs

It is possible to reduce the environmental impacts from the use of coal. But how much of a reduction is necessary? How much is desirable? The amount of abatement and control required has been an issue literally for centuries. As we noted in Chapter 1, in the fourteenth century, ordinances in London forbade coal burning entirely because of the dirty smoke it produced. Today, there is general agreement on the need for some control, and there is technology to provide a good deal of it. But reducing the impacts of coal often has steep financial costs.

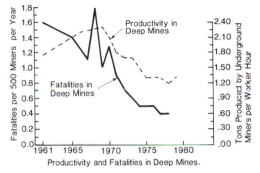

**FIG. 6.5.** Coal mine safety and productivity in deep coal mines. *Source*: U.S. General Accounting Office, March 1981, *Low Production in American Coal Mining*. Adapted from Chapman, 1983.

### Extraction controls

In underground mining, property damage and environmental degradation (acid drainage, uncontrolled fires, and land subsidence) can be reduced through more careful procedures while the mine is worked. For example, after extraction, mine tunnels can be collapsed in a controlled fashion to eliminate the possibility of uncontrolled subsidence afterward. Alternatively, tunnels may be back filled with mining spoils or with refuse from the preparation process. A study (see Wilson, 1980; all values shown here have been corrected from $1978 to $1987) has estimated a cost range of $2 to $8/ton for underground mine reclamation.

Health and safety improvements for miners have been required since the 1969 legislation, and improvements have been put into effect. As we can see in Figure 6.5, the number of mining fatalities per 500 workers has fallen off dramatically since the 1960s. Partly, this is the result of better equipment and training. Machinery has been improved and workers are given more training in its use. A great deal of attention is also being paid to preventing roof collapses. Falling rock from the roofs of mines accounted for 40 percent of all mine deaths in the 1970s. Typically, roofs above the seam being mined are secured with roof bolts that are driven deep into the rock. Overall, attention to roof supports in mine tunnels reduced the number

of collapses by two-thirds from 1969 to 1977.

Coal companies have improved mine ventilation as well. In the past, gases (particularly methane) that are found with coal accumulated in underground tunnels and exploded if a spark ignited them. There are ideas for further reducing this risk. One concept for coal mining in the future would eliminate the possibility of explosion by removing all the oxygen from the mine shaft. Since oxygen is necessary for combustion, its absence would make an explosion impossible. However, workers would require individual breathing devices.

Better ventilation also helps to reduce instances of black lung since it lowers the level of suspended coal dust. The use of sprays of water at the coal seam reduces coal dust levels as well. In fact, one study suggested that improvements in dust control could lower pneumoconiosis rates by more than 90 percent (see Chapman 1983).

But better mine safety and health provisions not only have a cost in terms of new equipment, they also have a cost in worker productivity. As Figure 6.5 suggests, the decline in fatalities was accompanied initially by an almost parallel decline in productivity; less coal was extracted for each hour of labor. For example, consider the effects of roof bolting on the production process. In order to prevent a roof collapse, the bolting must take place before the seam is mined, since vibration from the drilling and digging might dislodge loose rock nearby. Consequently, as the coal is extracted, work must stop every few feet so that new bolts can be installed. Collapsing or back-filling tunnels, installing better drainage systems, and other environmental procedures also demand labor that could otherwise be devoted to mining coal and so again reduce productivity.

By the end of the 1970s and in the early 1980s, however, overall productivity turned upward. This resulted, at least in part, from improved mining techniques and machinery. The most significant innovation in underground mining was the longwall technique of mining (Fig. 6.6). This technique uses a large machine that can extract as much as a 5,000-foot-long wall of coal. The cutting machine handles nearly every

step of the extraction process. It has giant cutting blades that slice into the coal seams and, as the coal falls, the equipment gathers it and pours it onto a conveyor system for collection and removal from the mine. Also, the machine typically rides on rails and has its own self-advancing hydraulic roof support system, which eliminates the need for roof bolts (see Dorf 1978).

Though such machinery often is efficient, new machinery and new mine procedures add about $10 per ton (1987 dollars) to the cost of coal. This figure bears some explanation, however. Not all mines use longwall mining and, in fact, it is not always effective. Sometimes other procedures are cheaper, and what is cost effective for one operator may not be cost

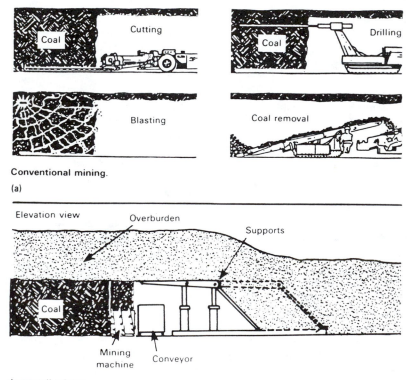

Conventional mining.

(a)

Longwall mining.

(b)

**FIG. 6.6.** Underground coal-mining techniques. (a) Conventional mining. (b) Longwall mining. *Source*: U.S. Department of Energy (1981).

effective for another. Finally, all U.S. mining companies compete in the same market and must charge roughly the same price for the same type of coal. They will tend therefore to find the approach to environmental and safety problems that costs the least for their particular cases.

Many of the negative impacts of surface mining are also controllable and because of the Federal Surface Mining Control and Reclamation Act of 1977, control practices are required.

According to the law, the land that is surface mined must be replanted and returned to its "approximate original contour" and use capability. Yet not all land is easily reclaimable. The climate must allow for reclamation; most importantly, rainfall must exceed 10 inches per year. In certain coal-bearing regions in the West, where rainfall is less than 8 inches per year, the prospects of reclamation are very poor. But other Western coal lands, such as the prairie grass regions of Wyoming, Montana, and North

(a)

(b)

**FIG. 6.7.** Coal Lands Rehabilitation. (a) The area in the center of the photograph is reclaimed surface mined land in the ponderosa pine and mountain shrub zone near Acme, Wyoming. (b) A result of doing nothing. Abandoned spoil banks near Firesteel, South Dakota. Reclamation policy has also called for the restoration of wilderness areas to their natural state and original beauty. *Source*: National Academy of Science (1974). Copyright © by the Ford Foundation.

Dakota and the ponderosa pine area of Colorado, Utah, and Montana, have good prospects if given the proper treatment (see NAS 1974).

The process of land reclamation will depend in part on the kind of land that must be restored. To reclaim farmland, for example, the topsoil needs to be saved during the mining process. In Germany, the subsoil too is stored and returned. This eases the process of reclamation, but adds so much to the cost that it is only cost effective where coal seams are especially thick. Mine pits can be refilled not only with soil but with debris. The coal industry has experimented with using coal ash (a useful technique for mine-mouth power plants), as well as municipal sewage as fill to reclaim lands. Sewage actually may increase the land's fertility and ease revegetation. The results of good wilderness reclamation in Figure 6.7a can be compared with the consequences of a "doing-nothing" policy (Fig. 6.7b).

The overall cost of land restoration from surface mining can vary tremendously. In the West, the cost may be as little as $1,000 per acre, while farmland in Ohio can cost up to $8,000 per acre to restore. The worldwide average costs for land restoration of surface mining have been estimated at $5/ton ($1987, see Wilson 1980).

## Preparation and transportation controls

The impacts from coal preparation are controlled at present in a straightforward manner. Wastes must be separated and disposed of as safely as possible; water used to wash the coal must be filtered or recycled to remove acids and harmful chemicals that otherwise might leach into water supplies. Because of the amount of the waste and the volume of water, these problems, though fairly clear-cut, are considerable. Instead of being dumped indiscriminately into ugly piles, some waste is being used as mine backfill. Experiments have also been conducted to treat waste piles like surface mining areas, that is, the piles are graded and treated so that they will support vegetation.

Yet the problems of the preparation phase

may grow in the future. The utility industry has been interested in developing operations that more thoroughly clean coal for combustion. These plants, referred to as coal refineries, would not only clean the coal of its easily removable impurities, but would remove much of the sulphur. Presently, some coal is washed to remove the sulphur particles that are found on coal, but not the sulphur chemically bound within the coal. Refining techniques, however, would also eliminate a considerable amount of bound sulphur. Such refineries could turn dirty coal into clean coal, reducing the need for other combustion controls. But to remove bound sulphur requires extra preparation, including the treatment of coal with bacteria that feed on sulphur or with chemical solvents. If not handled properly within the plant, either option might create additional environmental threats. At present, however, these technologies have yet to be proven effective and cost estimates are therefore not available.

As for transportation, it seems the least problematic area to control from a technological standpoint, and political and social concerns dominate. Nonetheless, there will be control costs. The need to preserve water purity from slurries as well as the control of the possible impacts from increased rail transport will add an estimated $0.72/ton ($1987).

## Combustion controls

The most important, and costly, problem of coal technology is the control of combustion emissions. There are essentially two related questions. How much air pollution control and abatement is desirable, and how costly will it be? Significant pollution control is possible; indeed, it is required. But the cost of using the technology may rise measurably, depending on the level of control that a community or society demands. As we can see from Table 6.4, abating pollution by 90 percent tripled the cost of a new plant. Not surprisingly then, while control technologies are in place, research efforts are focusing on how to make them cheaper.[8]

There are a number of different ways to reduce pollution from coal burning. Many

**Table 6.4.** Costs of environmental regulation of coal generation

| | |
|---|---|
| *Estimated cost in 1969 for new plant* | $183/KW |
| General inflation | $191/KW |
| Cost of new environmental protection | $795/KW |
| Total cost of new plant, 1981 (Somerset, New York) | $1,169/KW |
| | |
| *Change in air pollution emissions, 1969–81* | |
| Sulphur dioxide | −95% |
| Particulates | −97% |
| Nitrogen oxides | −80% |
| Average reduction | −91% |
| | |
| *Kinds of new equipment now required* | |
| Cooling tower to prevent heat discharge and fish kill | |
| Improved particulate removal | |
| Sulphur oxide scrubbing system | |
| Protected sulphur sludge and ash | |
| Taller stack | |
| Treatment of in-plant waste | |
| Continuing standards revisions: average one per week | |

*Source*: Reprinted from Duane Chapman, *Energy Resources and Energy Corporations.* Copyright © 1983 by Cornell University Press. Used by permission of the publisher.

utilities have relied on low-sulphur coal (Table 6.1), which over the last decade has helped reduce – although not eliminate – the need for new pollution control technology. Because of transportation costs, low-sulphur coal may add to the expense, although when we consider the cost of pollution control (Table 6.4), the savings on capital equipment may more than offset it.

But local, state, and national air quality standards have made investment in pollution control technology mandatory. The simple addition of tall smoke stacks, for example, is required in some cases since they help prevent local concentrations of pollutants (although ironically, high stacks that send pollutants well aloft may have contributed to the acid rain problem). However, tall stacks, which are both straightforward in concept and inexpensive, only dissipate pollution. Other technologies currently in place on coal plants have tried to recapture pollutants from coal smoke or prevent them from being formed in the first place.

Electrostatic precipitators have long been used to remove suspended particulates from coal smoke. The technology works by the simple principle of electrostatic attraction (the same basic idea that accounts for the attraction of small objects to a balloon after it has been rubbed against a sweater). The particles in the smoke become ionized by a strong electric field between two electrodes of an electrostatic precipitator. Once ionized, the particles are drawn to and captured on the electrodes. Periodically, the particles are removed from the electrodes and disposed of. Fabric filters, sometimes called *baghouses*, do a similar job. Essentially, the filters are like vacuum cleaner bags through which the flue gases must pass. The bags then filter out the particulates (see Wilson et al. 1980). Through the use of these two devices, coal burning utilities have captured the larger, more visible particles, and removed the elements that used to cause a smoky appearance in most cities. If we again consider the cost of control in terms of the additional cost of coal itself, on average particulate control adds about $3/ton ($1987).

Other less visible air pollutants have required different technologies. Most important, the present national clean air standards require that sulphur dioxide emissions be reduced 70 to 90 percent.[9] The best available technology for eliminating sulphur dioxide is the *scrubber*, or flue-gas desulphurizer. It was the specified technology for desulphurization in the amended Clean Air Act of 1977.

Most current flue-gas desulphurization (FGD) technology works on roughly the same principle (Fig. 6.8). Coal combustion gases pass through a wet alkaline solution, usually of calcium oxide (lime) or calcium carbonate (for instance, limestone). The solution, which is sprayed through the scrubber as an alkali slurry, reacts with the sulphur and removes it from the gas. While this is effective in lowering sulphur dioxide emissions, it requires large quantities of water (1,000 gallons per minute for a 1000 $MW_e$

**FIG. 6.8.** Flue-gas desulphurization scrubbers. Adapted from FMC Environmental Equipment Company.

treatment that can occur either after or during combustion.

In Europe and Japan, many coal boilers are equipped with a nitrogen oxide reduction system, where ammonia is mixed with the nitrogen oxides in the stack.[10] In the United States, however, efforts to reduce nitrogen oxides center on the combustion process. Nitrogen oxides from through the mixture of air and fuel, but to what extent they form depends on the mix of the two substances and the temperature at which the fuel is burned. New power plants are being designed with low nitrogen oxide furnaces, that is, where the mix and temperature are more carefully controlled to lower nitrogen oxide emissions. The increased capital cost of a low nitrogen oxide furnace is relatively small, but it reduces emissions only by about 40 to 60 percent.[11] Since the passage of the Clean Air Act, nitrogen oxide (as well as sulphur dioxide) control has been required in new plants; as of the late 1980s, there were several bills before the Congress that would require retrofitting old plants as well.

At present there are no operating emission control technologies designed specifically for either carbon monoxide or fine particles in coal-burning plants. It should be noted that most carbon monoxide pollution comes from gasoline-burning cars and trucks, not coal-burning furnaces. The Clean Air Act and its amendments direct that automakers reduce carbon monoxide emissions, which has been accomplished, as with the nitrogen oxides, through more careful control of the combustion process.

Breakthroughs in reducing carbon monoxide and fine particle emissions from coal plants may result from improvements in coal combustion as well. Currently, the most promising development is fluidized-bed combustion, which we will discuss in detail in Appendix C. The fluidized-bed concept permits a more efficient burn at lower temperatures, not only reducing carbon monoxide and fine particle emission but also the nitrogen oxides and sulphur dioxide emissions. Some large demonstration projects have met emissions standards without the need for scrubbers.

power plant) and leaves a sludge that presents a second disposal problem. It is also expensive. Scrubber retrofits on old plants, for example, can cost as much as $300 per kilowatt of capacity (or $300 million for a 1,000 MW$_e$ plant). As Table 6.4 indicates, that probably would be as much or more than the cost of an entire coal plant built in 1969. On an equivalent cost-of-coal basis, the present scrubber technology adds as much as $31/ton ($1987, see Wilson 1980).

FGD is principally designed to remove sulphur dioxide from emissions, but it also removes particulates and nitrogen oxides to some extent. On the whole, though, significant removal of nitrogen oxides requires a different

## Disposal

With the conventional coal technology of the 1980s, coal-burning facilities not only have tons of ash to dispose of, they also have scrubber sludge and particulate dust. Scrubbers leave significant amounts for disposal; if the sludge produced by a single 1,000 MW$_e$ electric generating plant in a year were poured onto the ground it would provide a 1-foot-deep covering of 1 square mile. With landfills at a premium, the electric industry is looking to make productive use of ash and sludge. As we noted, some coal wastes are being used in surface mining reclamation. But utilities are attempting to find other uses. Concepts include mining ash for rare trace metals like cobalt and chromium and refining sludge to remove elemental sulphur and sulphuric acid. Conceivably, disposal mining and recycling could allow utilities to sell the by-products of combustion and pollution control, producing a net financial gain that would reduce environmental control costs overall.

## Total cost of controls

Substantial reduction of the impacts from coal burning is technically feasible. However, control is an expense that business will pay initially, but consumers of industrial goods or of coal-fired electricity will pay in the end. In other words, we all will pay for greater control of coal impacts.

Controversy centers on whether additional emission controls beyond those mandated by legislation from the 1970s are needed. Industries that depend on coal resist paying the cost to reduce impacts further. Environmentalists, on the other hand, feel that the maximum possible effort should be made to control the impacts before they are created, regardless of the cost.

The cost is considerable. As we noted (and as Table 6.4 shows), a coal electric generating plant with emissions controls is far more expensive than one without it. New concepts such as the fluidized-bed furnace may reduce control costs, but as of the late 1980s have not proved conclusively that they will be cost effective on a large scale. With conventional technology, if we add up the costs of the controls for every step in the coal cycle, we find that they increase the cost of coal by anywhere in the range of $25 to $50/ton for underground mined coal and about $18/ton for surface mined coal ($1986, see Wilson 1980).

These costs, while significant to the customer, do not make coal uncompetitive with other fuels, oil in particular. Indeed, for electric generation and some industrial purposes, coal retains a definite advantage. Steam coal had an average price to U.S. utilities of about $46/metric ton in 1985 (see IEA 1986). This reflects a unit price of $1.70/MBTU, which includes the costs of extraction and transportation (and control costs). If we add to this the control costs for emissions and disposal, which are in the range of $13 to $32/ U.S. ton, this gives increments of $0.52 to $1.28/MBTU in the unit costs and results in total unit costs in the range of $2.22 to $2.98/MBTU. The equivalent unit cost of heavy oil (used in electricity generation) was $4.84/MBTU (IEA 1986), which includes extraction, shipping, and refining. Even without adding emission control costs for oil combustion, which are required in a number of urban areas, the comparison is favorable to coal.

As we will see in Chapter 9, coal-burning electric generating plants, even with emissions controls, are much less expensive to build and operate than nuclear power generating plants; consequently, coal maintains a cost advantage over that fuel as well. It is now clear that coal was competitive under the market conditions through the mid-1980s. Should oil prices again rise into the range of $30/bbl, then coal will be even more attractive in price.

As to additional costs to consumers, it is true that health, safety, and environmental controls add to the price of manufactured goods and the cost of electricity. Controls therefore mean that energy costs take a larger percentage of every consumer's income. But at the same time, it can be argued that the increased costs more nearly reflect the true costs of these products and services, because there are costs to society without controls. These costs are *externalities*. In the case of coal, all of the uncontrolled costs to health and the environment were formerly not reflected in the price of coal. But they were real nonetheless.

When a control eliminates an externality, the cost becomes *internalized*, that is, the cost of the control gets reflected in the price of a commodity. For example, in the case of emission and disposal controls, the $13 to $33/ton increment in the price of coal would represent the internalizing of the costs of air pollution, which otherwise would be an externality. There is some usefulness to consumers then in the added cost. If prices reflect spillover costs, consumers know better what they are paying for.

## Transnational effects – acid rain

The internalization of an externality has a clear economic purpose when we consider local air pollution control. The consumer pays more in order to get both the product or service and clean air. But the acid rain problem changes the equation. The costs of cleanup may not be borne by the consumers of the product, but rather by someone else in a different locale or even a different country. With electric power especially, companies are regional or local and a consumer's rates reflect the costs to a company of generating that power. If acid rain is recognized as a problem within national boundaries, the costs of cleanup can be determined and the distribution of the costs settled by legislation or perhaps through national or state taxes. But what if the acids originate in one country and the rain falls in another? Transnational acid rain means that the victims are not even in the same tax base, much less the same electric rate base. As a result, the issue is unlikely to be solved easily.

One important transnational acid rain dispute has pitted Canada against the United States. Increased acidic precipitation in Quebec, Ontario, and the Maritime provinces has provoked protest from Canada. Canadians claim, with at least circumstantial evidence to support them, that the acid rain falling on them originates in the Midwest, and the government of Canada asked the U.S. government to take action to reduce it. However, the Reagan administration refused for several years to enter any kind of agreement with the Canadians on acid rain, on the grounds that full scientific explanations of the phenomenon were not available. Finally, in the summer of 1983, the two countries agreed to study jointly acid rain and recommend strategies for its control.

The studies seemed to bolster the Canadian case. In Ohio, industry was burning coal and tall stacks designed to prevent local pollution were carrying the emissions high into the atmosphere, where winds transported the pollutants into Ontario and Quebec as well as the Northeastern United States. The researchers could not, however, say conclusively how the pollutants were being transformed into the constituents of acid rain. And an official congressional report in 1987 downplayed the importance of acid precipitation. This left room to question the need for any additional control at all.

The problem inevitably became highly politicized. On both sides of the border, interest groups pressured their respective governments not to give in to the other side. In the United States, industrial coal users did not want to make large investments in additional capital equipment (which they claimed would cost $110 billion, a figure widely disputed) and lobbied Congress to block new air pollution control requirements. Most U.S. politicians had more to gain by supporting American industry than by bowing to Canadian demands. The Canadians had no votes and, until 1988, received little political help.

In this instance, however, there was a U.S. constituency siding with Canada. Northeastern states affected by acid buildup in their lakes also sought to force greater emissions control from Midwestern coal users. In June 1988, an agreement in principle was reached between the governors of New York and Ohio to mandate more stringent controls on Midwestern coal-burning plants. However, the externality would be only partly internalized by local consumers under this agreement. It stipulated that part of the cost would be assumed by industry (and consequently, their consumers as well) and part would be paid for by the imposition of national taxes on imported oil.[12] But the proposal by the two governors could not settle the matter conclusively. Congress still faced pressure from industry to block the necessary legislation and the Reagan administration continued to oppose major new controls. As for Canada, though the governors' agreement would meet many Canadian demands on the acid precipitation, and though the administration had promised vaguely to negotiate a transboundary pollution treaty, Canada had limited power to affect U.S. legislation. At the same time, the Canadian government was still under political pressure from its own citizenry to settle the dispute on favorable terms.

It should be noted that agreement has been reached elsewhere on transnational acid rain. In 1979, European countries signed the Transboundary Pollution Convention.[13] Actually, this convention showed the inherent problems in settling such disputes. The agreement was general in tone and lacked mechanisms for enforcement. But it did set down the need for emissions control, particularly of sulphur dioxide, and the parties agreed in principle on basic goals.

## Carbon dioxide's potential global impact

Most of the environmental impacts of fossil fuel technology, including acid rain, can be controlled. Though questions remain as to the cost, we know that the technology is feasible and can work on a large scale. The issue of global carbon dioxide ($CO_2$) concentrations, however, is different.

Carbon dioxide is a universal product in the cycle of life. It is absorbed by plants and generated and released by natural decay processes (see Fig. 6.9). The level of carbon dioxide in the world depends on the balance of these processes. Over time, world levels of carbon dioxide change very slowly.

Carbon dioxide, however, is also a product of the combustion of hydrocarbons in coal, oil, and gas. It cannot be eliminated from combustion and carbon dioxide emissions can be significantly reduced only at great expense and trouble. Essentially, carbon dioxide emissions will take place as long as we use these fuels.[14] But it is argued that the worldwide combustion of coal and other fossil fuels is causing global carbon dioxide concentrations to rise. Careful scientific measurements conducted over decades have confirmed a slow-trend increase in the global concentrations of carbon dioxide (Fig. 6.10). In absolute terms, fossil fuel burning is adding about 20 billion tons/year of carbon dioxide to the world's atmosphere. Although that may seem like a great deal, much more comes from other sources; an estimated 150 billion tons is emitted annually from the oceans and about 65 billion tons from terrestrial plant decay, which itself is being accelerated through deforestation. The increase in carbon dioxide concentration has actually measured a mere 1 part per million per year (see Fig. 6.10). Yet the basic hypothesis is that this addition is enough of a burden on the atmosphere to cause carbon dioxide levels to rise too quickly. With even more massive amounts of fuel to be burned over the decades ahead, carbon dioxide levels are, according to the theory, likely to double or more by the year 2000. If this occurs, then the earth's average temperature may increase several degrees because of what is called the *greenhouse effect*. The greenhouse effect occurs because less heat escapes (as infrared radiation) back into space through an atmosphere containing high levels of carbon dioxide. Like the glass of a greenhouse the atmosphere lets in the solar energy (of all wavelengths), but does not let the heat out. The world's atmosphere, it is feared, will trap an increasing amount of heat and higher temperatures will result globally.

The prospect of a global temperature change is disturbing. Even an increase of a few degrees, on average, could lead to severe regional effects, such as prolonged droughts, crop failures, and increased desertification. Polar ice caps might partially melt, which would in turn result in

**FIG. 6.9.** The carbon cycle. Adapted from Smith (1966).

ocean flooding of major portions of low-lying coastal lands. Cities like New York City might have to construct seawalls to prevent flooding, and coastal countries like Bangladesh could be inundated.

The hypothesis, which blames fossil fuel combustion for threatened changes in global temperature, has been getting a great deal of support and attention in recent years. Leading geoscientists have argued for it,[15] and the

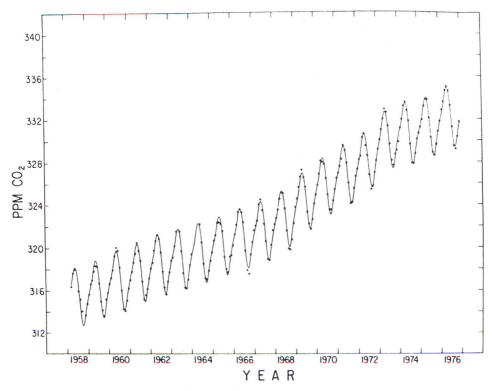

**FIG. 6.10.** Atmospheric concentration of carbon dioxide. Carbon dioxide parts per million (PPM). *Source*: National Academy of Science (1977).

popular press in the United States played up the idea when a severe drought hit the Midwest in the summer of 1988. The drought was portrayed as the start of a carbon-dioxide-induced global warming.

But the hypothesis, in whole and part, has also been challenged. First, analysis of objects tens of thousands of years old suggests that global temperature trends are caused by orbital changes of the earth, not carbon dioxide buildup. Also, some geophysicists point out that as temperatures begin to rise, the cloud cover probably will increase, which could in itself begin to reduce the amount of sunlight reaching the earth's surface.

Whether global warming is in progress or not, some scientists maintain that it is incorrect to blame carbon dioxide alone. Other gaseous releases also would contribute to a greenhouse effect. Methane, chlorofluorocarbons, and nitrogen oxide emissions are estimated to contribute at least half of the problem. Emissions of the latter two would be much easier to reduce than carbon dioxide. Nitrogen oxide emissions, of course, are also related to fossil fuel burning, although as we have seen they can be reduced by technological innovation; chlorofluorocarbons (used in refrigerants, plastics, and aerosols) can be replaced entirely with other chemicals.

Still the largest question remaining about carbon dioxide and the greenhouse effect centers on the nature of biological processes themselves. Just how marine and terrestrial components (Fig. 6.9) respond to increased carbon dioxide input from fossil fuels cannot be determined with certainty. It may be that carbon dioxide will simply be utilized in natural processes; plants and the ocean will begin to absorb more carbon dioxide and no greenhouse effect will occur.

Because the processes and the outcome are not clearly understood, what should the world's response be to the possibility of a greenhouse effect? On the one hand, it can be argued that we should do nothing until we have greater certainty, because action now might be pointless or even incorrect. On the other hand, we could reduce the burning of fossil fuels drastically and cause economic hardship in the process, only to discover that no global warming was in progress or even that an orbital shift was actually leading to a cooler planet. As recently as the 1970s, some climate researchers were warning of a cooling trend for the earth.

Still other scientists claim that we cannot afford to risk such changes, where doing nothing could have a devastating impact. According to this view, we should take preventive action soon because the processes are of unknown magnitude and the potential damage is uncertain. Once a greenhouse effect would be confirmed conclusively, corrective steps taken might come too late to prevent irreversible and catastrophic climatic change. At best, corrective measures at that time might still result in decades of hardship. At worst, the world could face massive famine, drought, and possibly social chaos.

Yet it is not entirely clear what measures should be adopted. Absolute prohibition of the use of fossil fuels would help, but could entail massive economic dislocation until alternatives were found. We might also consider reforesting areas since trees soak up a large quantity of carbon dioxide. But it would require planting an area the size of Australia in order to make a difference – and this presupposes a reversal of the ongoing trend of worldwide deforestation.[16]

It is, of course, possible to slow the use of fossil resources without causing substantial economic hardship. This would mean greater conservation as well as increased use of combustion-free alternatives, such as nuclear fission electric generation, and the development of new sources such as nuclear fusion and solar energy. But as we will see in Chapter 12, the latter two are far from ready for widespread use, and Chapters 7 to 9 will show that a significant increase in nuclear fission power carries risks and costs of a different kind.

Yet the carbon dioxide threat seems considerable enough to warrant action. Some action has been forthcoming. In the United States, in 1988 legislation had been introduced to spur research into alternatives that would be combustion-free. Other industrial countries such as Norway have adopted, or are considering, measures to reduce fossil fuel use. As we saw in Chapter 4, significant reductions can take place within the context of continuing economic growth. Of course, mere reduction in fossil fuel use by the industrial countries may only slow the greenhouse effect. But that has some value, since more time would permit other types of development, such as crops that are more heat- and drought-resistant. But there remains a question of whether sufficient measures on a worldwide basis will be implemented even to slow the process. The effort would need to be widespread, yet the problem would not affect each nation in the same way. In fact, some computer projections indicate that parts of the world would actually *benefit* from a rise in temperatures and a change in global climatic conditions. Some areas, for instance parts of China, would get increased rainfall, making farmlands more productive. Such nations are unlikely to want to reverse processes that will improve the lives and well-being of their own people. Yet if nations act at cross purposes, only nature will be able to forestall a greenhouse effect.

## Ecocentric ethics

The carbon dioxide question, like the other impacts of fossil fuels, concerns harm to the environment. Environmental harm is both a practical concern and an ethical one. For the most part, however, the ethical questions are presented in terms of the protection of human beings. It is an ethical good to protect *people* from the hazard of air or water pollution or the consequences of carbon dioxide buildup. But what of the air or the water or the land? Do they deserve ethical consideration? Is there an ethical imperative to the planet as there is to other human beings?

In recent years, some philosophers have argued in favor of an ethical system that includes inanimate objects; ethics, in this view, should be ecocentric (nature-centered), not anthropocentric (human-centered). The basic concept is that humans are part of an ecological system; they are not above that system. If they act in concert with the system, they are ethically right. If they upset it, they are ethically wrong. Or as one of the principal thinkers of this school of thought put it, "A thing is right when it tends to preserve the integrity, stability, and beauty of the biotic community [that is, ecosystem]. It is wrong when it tends otherwise" (see Leopold 1949). In other words, polluting the atmosphere, increasing carbon dioxide levels, and strip mining are not simply dangerous to humans and in need of some control, they are ethically wrong because they do violence to the ecology of the world.

To accept this view means accepting a very different view of ethics than has generally existed for several thousand years. Humans have traditionally seen themselves as the pinnacle of creation and only other people have been of moral concern. Lower forms of sentient life, much less inanimate objects, have not had inherent ethical standing and we have generally regarded it as a right of humans to use the world (and to a lesser extent, animals) pretty much as we choose, provided that in doing so we do not harm other humans.

Yet there is reason to question a rigid anthropocentric point of view. Is the "superiority" of a species enough to grant it moral preference? It would imply, as one philosopher has pointed out, that if superior beings from another planet came to earth, we would have to regard them as having the right to use humans as disposable property (see Wenz 1988). Of course, human history has many examples where a group of people regarded itself as superior and acted toward other people in ways we now generally believe to be morally abhorrent.

Also, a traditional anthropocentric view appears to misread evolutionary theory. It suggests that humans are the end of evolutionary processes, while biological science sees no end to evolution. In fact, if we reserve for ourselves the right to disrupt the world's ecosystem, we may prevent the development of even higher life forms.

Some would agree that human-centered ethics is incomplete and should include other sentient life – particularly other mammals. But an ecocentric approach argues that all nature has a balance and everything is part of the same community. One cannot make moral distinctions among members – humans, animals, plants, rivers – since they are all interdependent, according to this view. To preserve the rights of any, all must be seen as having rights, even rocks.

Though ecocentric ethics are plausible, they raise many problems. Most important, they suggest that there is a specific way in which we should relate to an ecosystem. It is difficult indeed to say what behavior is proper when interacting with nature. To protect species from extinction seems like one obvious way to preserve the "integrity" and "stability" of nature. Yet nature itself causes extinctions. Indeed stability appears contrary to what happens in the world, since an ecosystem is nothing if not dynamic (see Shrader-Frechette in Repetto 1985).

At the same time, science has tended to endorse the view that protecting the environment is wise. We should try to maintain the diversity of species – saving species from extinction – since diversity maximizes the potential of evolutionary development. (Indeed, some ecologically protective legislation already exists in many countries.) Of course, disrupting Nature clearly has negative consequences or consequences that are unclear but potentially devastating. This view does not necessarily make the planet itself part of an ethical community. Indeed, it can be viewed as strictly *instrumental*. That is, it is good to protect diversity in nature and the ecosystem as a whole because we believe the result of such a policy will be positive (see Wenz 1988). This formulation may have less force than a moral imperative, but it calls forth a similar response.

We also might add that an anthropocentric view can be formulated to lead to the same kind of conclusion as an ecocentric one. For instance, we can condemn the destruction of the environment because it is contrary to most human value systems that condemn greed or rapaciousness. Also, if we grant the right to the use of the earth[17] of all people of the present and of the future, we have little choice but to act in ways that generally protect resources and the basic ecology of the planet. To do otherwise would mean acting immorally, particularly with respect to the generations to come. Yet, the very existence of a new ecocentric formulation suggests that to many people our anthropocentric ethics have not provided adequate safeguards for the environment.

# Notes

1. Prior to World War II, coal was used extensively not only in electric generation and industrial heating, but also in railroad transportation.
2. U.S. coal production in 1986 was actually closer to 900 million tons. We exported 85 million tons; Western European countries bought approximately half that total (*Energy Facts 1986*, Department of Energy).
3. There are other mining techniques in addition to highwall mining. A common alternative is called *open pit mining*. This technique is self-explanatory. Deep pits are dug and, once the coal is extracted, the overburden can be used to fill in the hole.
4. One eyewitness account (quoted in Chapman 1983) of strip mining in the 1950s described how earth and rocks tumbled through fields and into yards and vegetable gardens. The mining operation was so disruptive that, according to this report, three-quarters of the population in the vicinity of the mine moved away.
5. Western farmers have been joined in opposition to slurries by an unlikely alliance of railroads and environmentalists. In this instance, the desire of the railroads to dominate coal transport coincides with the environmentalists' dislike of pipeline development.
6. Sulphur dioxide levels in London increased eightfold over the course of just a few days during this episode (see Wilson et al. 1980). Similar air pollution disasters have occurred in other parts of the world during this century. For example, in October 1948, an air pollution episode killed an estimated 20 people in the town of Donora, Pennsylvania. Another 10,000 – half the population – became ill.
7. Carbon monoxide impairs the oxygen-carrying capacity of blood. This can strain the cardiovascular system, making people with heart disease especially vulnerable to high carbon monoxide levels (see Harrison 1975).
8. For example, at a conference on acid rain held in April 1988, Ian M. Torrens of the Electric Power Research Institute (EPRI) reported on advanced sulphur dioxide control systems that are expected to cost 20 to 50 percent less initially and save up to 40 percent in operational costs over current technologies.
9. This provision for sulphur removal resulted from a legislative compromise in Congress (see Haskell 1982).
10. This technique also requires the use of a chemical catalyst. Called selective catalytic reduction (SCR), the process transforms nitrogen oxides into nitrogen and water (described in the *EPRI Journal* January/February 1988).
11. According to the *EPRI Journal*, low nitrogen oxide boilers add less than $5/KW to the cost of a new coal plant; this works out to less than $5 million for a 1,000 MW plant and would be a tiny fraction of the total cost.
12. The agreement still required congressional action to amend the Clean Air Act. The issue is in doubt because of opposition from some Midwestern legislators and lobbies for Midwest coal interests.
13. The Convention on Long-Range Transboundary Air Pollution was signed in Geneva in November 1979. It was negotiated under the auspices of the United Nations Economic Commission for Europe and was signed by both Western and Eastern European countries (see Ruster 1982).
14. Natural gas produces only about half as much carbon dioxide per BTU of energy than coal, so a switch to natural gas from coal would lower carbon dioxide emissions generally. Conventional natural gas, however, is far less abundant than coal and a switch might not be sustainable for very long.
15. The best known supporter of the greenhouse theory has been Dr. James E. Hansen, director of the Goddard Institute for Space Sciences of NASA. In 1988, he told the press that he was "99 percent" certain that a greenhouse effect was

already occurring. He was, however, quickly disputed by other atmospheric scientists.

16. Some environmental groups have proposed that governments and private organizations buy at a discount the Third World's debts owed to foreign banks. In return, the countries will agree to preserve forest lands, particularly tropical rainforests. These are often referred to as debt-for-forests swaps.

17. This is derived from a theory of the late seventeenth-century English philosopher John Locke. Locke noted in his *Second Treatise on Civil Government* that a person's right to use the earth has a proviso that there be "as much and as good left in common for others." Some have quoted Locke's theories of private property to justify exploitation of the world's resources, but as philosopher K. S. Shrader-Frechette points out this exploitation does not leave "as much and as good" for others. Even if we follow Locke, we can conclude that environmental protection is required from a strictly anthropocentric standpoint (see Shrader-Frechette in Repetto 1985.)

# Bibliography

Barrett, R. N. (ed.). 1982. *International Dimensions of the Environmental Crisis.* Westview Press, Boulder, CO.

Berry, B. J. L. (ed.). 1977. *The Social Burdens of Environmental Pollution: a Comparative Metropolitan Data Source.* Ballinger, Cambridge, MA.

Chapman, D. 1983. *Energy Resources and Energy Corporations.* Cornell University Press, Ithaca, NY.

Dorf, R. C. 1978. *Energy Resources and Policy.* Addison-Wesley, Reading, MA.

Harrison, D. 1975. *Who Pays for Clean Air.* Ballinger, Cambridge, MA.

Haskell, E. H. 1982. *The Politics of Clean Air.* Praeger, New York.

International Energy Agency (IEA). 1986. *Coal Information 1986.* Organization for Economic Cooperation and Development (OECD), Paris, France.

Jones, C. O. 1975. *Clean Air – The Policies and Politics of Pollution Control.* University of Pittsburgh Press, Pittsburgh.

Leopold, A. 1949. *A Sand County Almanac.* Oxford University Press, Oxford.

Murthy, B. N. and A. F. Mann 1985. *Economics of Present and Future Fossil-Based Electric Generations.* Dekker, New York.

National Academy of Science. 1974. *The Rehabilitation Potential of Western Coal Lands.* Ballinger, Cambridge, MA.

Partridge, E. (ed.) 1981. *Responsibilities to Future Generations – Environmental Ethics.* Prometheus Books, Buffalo, NY.

Penner, S. S., and L. Icerman. 1974. *Energy – Volume I – Demands, Resources, and Policy.* Addison-Wesley, Reading, MA.

Repetto, R. (ed.). 1985. *The Global Possible: Resources, Development, and the New Century.* Yale University Press, New Haven.

Ridgeway, J. (ed.). 1982. *Powering Civilization: The Complete Energy Reader.* Pantheon, New York.

Ruster, B. (ed.). 1982. *International Protection of the Environment,* Vol. 28. Oceana Publishers, Dobbs Ferry, NY.

Shrader-Frechette, K. S. 1981. *Environmental Ethics.* Boxwood Press, Pacific Grove, CA.

Stobaugh, R., and D. Yergin. 1983. *Energy Future.* Vintage, New York.

Swartzman, D. (ed.). *Cost-Benefit Analysis and Environmental Regulations – Politics, Ethics and Methods.* The Conservation Foundation, Washington, DC.

Tolley, G. S. (ed.) *Environmental Policy – Elements of Environmental Analysis,* Vol. 1. Ballinger, Cambridge, MA.

Tyner, W. E., and R. J. Kalter. 1978. *Western Coal – Promise or Problem.* Lexington Books, Lexington, MA.

U.S. Department of Energy. 1986. *DOE Facts Book.* Washington, DC.

Wenz, P. S. 1988. *Environmental Justice.* State University of New York Press, Albany, NY.

Wilson, C. L. (project director). 1980. *Coal – Bridge to the Future.* Report of the World Coal (WOCOL) Study. Ballinger, Cambridge, MA.

Wilson, R., S. D. Colome, J. D. Spengler, and D. G. Wilson. 1980. *Health Effects of Fossil Fuel Burning – Assessment of Mitigation.* Ballinger, Cambridge, MA.

# 7

# Nuclear fission technology

## Perspective

### The promise

In 1953 the future of nuclear energy seemed bright, with the launching of the Atoms for Peace program. These promises were to contrast with the dire consequences of the atomic bomb developed at the end of World War II. Commercial nuclear energy would be a clean, abundant, and inexpensive means to generate electricity, according to its proponents. Some early advocates went so far as to claim that nuclear power would "be too cheap to meter."

However, the hopeful vision never materialized. Beginning in the 1960s, nuclear power has been under a cloud of controversy that has intensified in the decades following. What is more, far from being an inexpensive source of energy, nuclear power has often proven too costly to undertake (see Komanoff & Van Loon 1982). By the mid-1980's, construction of nuclear plants had nearly ceased, with no new orders since 1978 in the United States. Many orders for new plants had been cancelled in other countries as well. The accident at Chernobyl in 1986 accelerated the trend away from nuclear energy and its future is in doubt.

### A short history

Following World War II and the successful development of the atomic bomb by the United States, the evolution of peaceful atomic power was slow. The accumulation of more and better weapons was the dominant concern of the Atomic Energy Act of 1946 and was the top priority of the Atomic Energy Commission (AEC), which was organized to carry out this policy. This program was, of course, motivated by pressures of world political/military developments, but initially it ignored the glowing promises for peaceful nuclear energy expressed by leading scientists and the press at the close of the war.

By the 1950s, however, interest in developing nuclear power generation had grown to the point where it was acknowledged by President Dwight D. Eisenhower in the Atoms for Peace proposals at the UN in 1953 and in the Atomic Energy Act passed by Congress in 1954. Several strong advocates of nuclear power included Dr. Glenn Seaborg (a chemist), Dr. Alvin Weinberg (a physicist), and Hyman Rickover (a naval officer and engineer). Specific conceptions of nuclear power reactors had already been proposed at this time.

Nevertheless, the AEC programs of the early 1950s produced no reactor prototypes specifically designed for civilian electric power generation. It took the development of a reactor for military use to lead to the first commercial reactor. Rickover is credited with the successful construction of a small boiling water reactor (BWR), which was used to propel a submarine. He led a design group, in a cooperative AEC –

Westinghouse Electric Corporation effort, which completed the reactor for use by the U.S. Navy in 1954.

In 1957, the first prototype for a commercial reactor, based on the Rickover naval design, was developed. This reactor, a pressurized water reactor (PWR), was installed at Shippingport, Pennsylvania, for the Duquesne Light Company, a privately owned electric utility, and generated only 60 MW (later 150 MW). The first truly commercial design of the Westinghouse PWR was the Yankee plant at Rowe, Massachusetts, in 1958 (Fig. 7.1) and over 20 more such PWRs were installed in the 1960s. A BWR commercial design was introduced by the General Electric Company in 1969, following AEC-supported prototype development at the Argonne National Laboratory; soon after, several more BWRs were in commercial operation.

**FIG. 7.1.** The Yankee Atomic Power Plant at Rowe, Massachusetts. Courtesy of Yankee Atomic Electric Company.

It should be noted, however, that these early commercial reactors were designed for the modest power range of about 100 $MW_e$ (see Chapter 3 and Appendix A). Later reactors were several times more powerful and those designed to come on line in the late 1970s and 1980s were large-scale base load (see Chapter 3) plants in the range of 1,000 $MW_e$. The size of an individual nuclear generating plant is significant not only for the amount of electric load demand it supplies – 1000 $MW_e$ provides enough generating capacity for about a million homes – but also, as we will see, for the implications of accidents.

As of this writing, more than 100 nuclear plants (most of which are somewhat under 1,000 $MW_e$ capacity) are operating in the United States and over 250 more in the rest of the world, with a total world generating capacity of close to 250,000 $MW_e$. Another 157 plants worldwide were being planned or constructed as of 1985, but over 15 cancellations have occurred in the United States and elsewhere due to lower forecasts of electric load growth and rising construction costs. Presently, nuclear generation accounts for about 15 percent of electric energy generated in the United States and about 10 percent worldwide, and these proportions have not changed significantly over the past few years. It should be remembered, of course, that electric generation represents a fraction (about 28 percent in the United States) of the gross energy usage in the economy and that nuclear capacity can only affect the electric sector of gross energy usage (see Chapter 4, Table 4.8).

## The basic physics

Nuclear fission concerns the reaction of nuclei from the heavy end of the periodic table. In the fission process, these nuclei are split into lighter nuclear parts and they release energy in the process. Today's commercial nuclear reactors commonly use the fissionable isotope $^{235}U$, an element found naturally in low concentrations (less than 1 percent) in virgin uranium ore.

Another fuel element in use is plutonium ($^{239}Pu$), which is manmade. It results from

Neutron

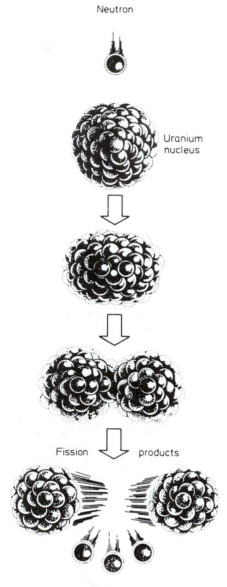

Uranium
nucleus

Fission          products

Neutrons

neutron bombardment in the reactor of $^{238}$U, the major constituent of natural uranium ore. The resulting (nonfission) reactions lead to the *breeding* or *conversion* of $^{238}$U into $^{239}$Pu. Thorium, which exists in nature much like uranium and in similar quantities, may also be a fertile element for these breeding reactions.

## Nuclear fission

Fissionable elements (for example, $^{235}$U) in the fuel rods of a commercial nuclear reactor are split when a neutron collides with one of the nuclei and is absorbed, as depicted in the artist's conception of Figure 7.2. The figure describes the fission process step by step, ending with the splitting of the uranium nucleus into lighter nuclei (called fission products) and neutrons, all accompanied by the release of energy. The energy released through fission is, of course, a manifestation of the famous mass-energy equation of nuclear physics. The sum of the masses of the fission products is less than the aggregate mass of the original $^{235}$U nucleus plus the incident neutron; the discrepancy in nuclear mass is manifested as kinetic energy. This kinetic energy is converted into useful heat in the reactor.

## Chain reactions

The energy released from a single fission reaction would be of no practical importance, because the amount is minuscule, less than $10^{-15}$ kilowatt-hours (KW-hr, see Appendix A, note 8) equivalent electric energy. But the consequences of nuclear fission are significant because of the possibility of a *chain reaction*. In a chain reaction, the neutrons liberated by the fission of one nucleus move on to create other

**FIG. 7.2.** Diagrammatic view of the fission process. Here we see a four-part sequence, in which the neutron is first shown to be moving toward a collision with the uranium nucleus. Next, the neutron has collided and has been absorbed into the uranium nucleus. The nucleus is depicted at this stage as bulging out to the sides to suggest that the nucleus is now unstable and is ready to split apart. In the

third image, we see the start of the splitting – or fissioning – of the nucleus into fragments called *fission products*. These products include the nuclei of lighter elements, such as $^{141}$Ba and $^{92}$Kr, and neutrons. Finally, the fission products fly apart as a result of the energy released by the nuclear fission process. *Source*: Collier & Hewitt (1987). Copyright © 1987 Hemisphere Publishing Corporation.

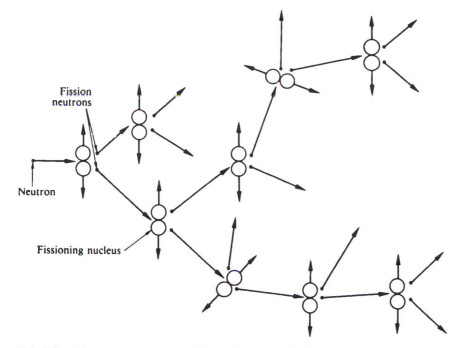

**FIG. 7.3.** Schematic representation of fission chain reactions. *Source*: J. R. Lamarsh, *Introduction to Nuclear Engineering*, © 1983, Addison-Wesley Publishing Co., Inc., Reading, Massachusetts. Reprinted with permission.

fission reactions. Because more than one neutron is usually released in each fission reaction, the fission process may become multiplied (depicted schematically in Figure 7.3). The successive collisions and fission reactions create a self-sustaining (or ever-growing) chain reaction, if the density and total mass of the fissionable fuel are great enough. A volume of fissionable material that can sustain a chain reaction is called a *critical mass*.

The chain reaction is at the heart of both a nuclear bomb as well as a nuclear fission reactor. But there are several distinguishing characteristics that determine whether a body of fissionable material is a bomb or merely a reactor. The most important is that in a reactor the rate of fission reactions can be controlled. The flow of neutrons is limited by *control rods* of materials that absorb neutrons, such as cadmium. These control rods are placed between the fuel rods (see Figure 7.4) and absorb the moving neutrons, thereby preventing the

neutrons from inducing more fission reactions. As we will see, these control rods can be inserted or retracted from the reactor core and can thus control the rate of the chain reaction or even shut it off. A fission bomb, by contrast, assembles a critical mass, which then releases its energy; no further external control of its rate of reaction then is possible.

An important feature of all nuclear chain reactions is the density of the fissionable element, because this determines the likelihood of a moving neutron colliding with a fissionable nucleus. For present-day U.S. nuclear power technology, the fuel element must be *enriched* to a concentration of about 3 percent $^{235}$U for the density of fissionable atoms to be sufficient to sustain the chain reaction.[1]

## Moderating medium

The reactor vessel of present-day reactors contains the reactor core. It is important to

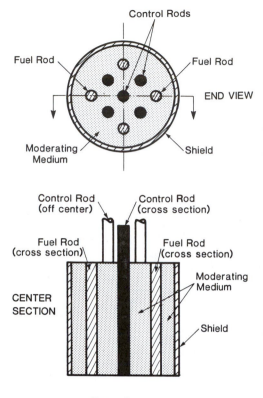

FIG. 7.4.   Idealized nuclear reactor.

understand that the chain reaction in any reactor takes place throughout the entire core from one fuel rod to another, and therefore the entire set of fuel rods determines a critical mass and not any individual rod. Between the rods in most present-day reactors, there is a moderating medium, which for U.S. reactors is simply water. The moderating medium in a typical reactor fills most of the space between the fuel rods (see idealized reactor shown in Fig. 7.4).

The purpose of the moderating medium is to slow the speeds of the fission-liberated neutrons. It is necessary to slow the neutrons in a moderated reactor to enhance their *capture* (collision and absorption) by the fissionable nuclei. Without a moderator a chain reaction could not be sustained because of the low

density of $^{235}U$ even after enrichment. But the early nuclear physicists found that the neutrons at the slow end of the range of thermal speeds had a much better chance of being captured than those at higher speeds and so could better sustain a chain reaction.

In a moderating medium such as water, this is accomplished through multiple (about 35 on average) collisions of the neutrons with the light nuclei, such as hydrogen. In this process, a typical neutron starts out at high velocities upon liberation from a nuclear fission reaction, but after several collisions has been slowed to the speeds of *thermal motion*[2] in the medium. These slowed neutrons are therefore called *thermal neutrons* and the reactors using moderators are often called *thermal reactors*.

A good moderating medium contains atoms with light nuclei, so that each collision of a neutron with an atom causes only a partial loss of the neutron's velocity. In addition, the medium should not contain nuclei that are likely to absorb the neutrons upon collision, causing their loss to the chain reaction. Besides ordinary water ($H_2O$), another good moderator is *heavy water* ($D_2O$, or deuterated water), which absorbs fewer neutrons than ordinary water and is used in Canadian reactors (Canadian deuterated-uranium reactor, called the CANDU). Graphite also has been used as a moderator and was in fact the moderator of the earliest atomic piles; it is still the medium used in U.S. convertor reactors for weapons material production (see Chapter 8). Abroad, notably in the USSR and in Great Britain, a few power-production reactors are graphite moderated. Graphite is actually the best moderator, but it has disadvantages for use in commercial reactors. Its most crucial disadvantage is its flammability; the possible consequences of this disadvantage, as we shall see, were vividly in evidence at the graphite-moderated Chernobyl reactor in the USSR.

## Heat removal

In our description of the fission process, we noted that not only do the fission fragments fly apart, but that there is also a release of energy. About 85 percent of this liberated energy goes

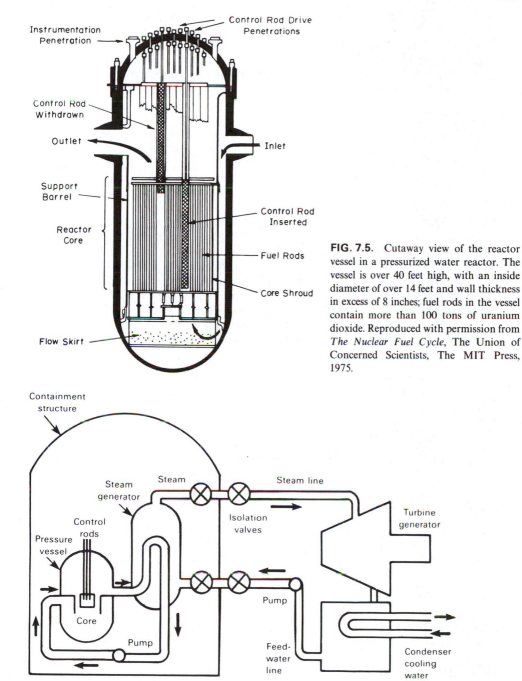

**FIG. 7.5.** Cutaway view of the reactor vessel in a pressurized water reactor. The vessel is over 40 feet high, with an inside diameter of over 14 feet and wall thickness in excess of 8 inches; fuel rods in the vessel contain more than 100 tons of uranium dioxide. Reproduced with permission from *The Nuclear Fuel Cycle*, The Union of Concerned Scientists, The MIT Press, 1975.

**FIG. 7.6.** Boiling water reactor (BWR) – schematic diagram. *Source*: Energy Technologies and the Environment, June 1981, U.S. Department of Energy, Report No. DOE/EP0026. Available from National Technical Information Service, Springfield, VA 22161.

into the nuclear fragments, which rapidly collide with other atoms within the fuel rod before escaping into the surrounding moderator. (Neutrons, on the other hand, mostly escape into the moderator.) When these nuclear fragments collide within the fuel rod, they give up their kinetic energy in the form of heat, thereby heating the rod.

This heat of fission must be removed from the rod to be put to use *and* to cool the rod. In nuclear reactors, this process of heat transfer is accomplished by circulating a coolant past the fuel rods. In *light water reactors* (LWRs) – including both BWRs and PWRs – used in the United States, this coolant is the same water that serves as a moderator. In Canadian reactors, the heavy-water moderator is also the coolant, whereas abroad, for example in Great Britain, some other reactors use circulating gaseous coolants such as carbon dioxide.

A simplified depiction of the coolant flow within a reactor vessel is given in Figure 7.5, showing the input and output ports for coolant

flow. We should visualize the flow as a vertical movement parallel to the fuel rods.

We find in U.S. and Canadian reactors that heat is conducted to the water coolant and heats the water as in a conventional boiler. Indeed, it has been said that a nuclear reactor is merely an exotic way to boil water. There are two alternative ways in which the boiled water from the reactor can be used in a power plant. The coolant can itself circulate as steam through the turbine as shown in the BWR in Figure 7.6. Alternatively, the coolant can circulate through a separate heat exchanger in which the steam used to circulate through the turbine is generated. The advantages of this scheme (Fig. 7.7) are: (1) the turbine is isolated from the radioactive contamination of the reactor and (2) the reactor can be operated under higher pressures and temperatures (see Chapter 3 and Appendix A). A PWR (Fig. 7.7) has a higher thermal efficiency than the BWR because the temperature of the working fluid is initially higher (see Appendix A).

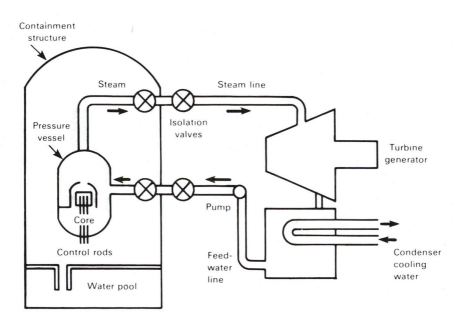

**FIG. 7.7.** Pressurized water reactor (PWR) – schematic diagram. *Source*: Energy Technologies and the Environment, June 1981, U.S. Department of Energy, Report No. DOE/EP0026. Available from National Technical Information Service, Springfield, VA 22161.

## The technology

### The reactor core

The core of an operating reactor is actually more complicated than the configuration in Figure 7.4 suggests. It has not just one set of rods, but a number of fuel assemblies, each one with a cluster of fuel and control rods. The cross section of a typical PWR core (Fig. 7.8) shows the spaces provided for each assembly, of which there are usually hundreds. The insert at the upper right hand corner of the figure shows expanded cross sections of four fuel assemblies, with clusters of control rods dispersed through each assembly.

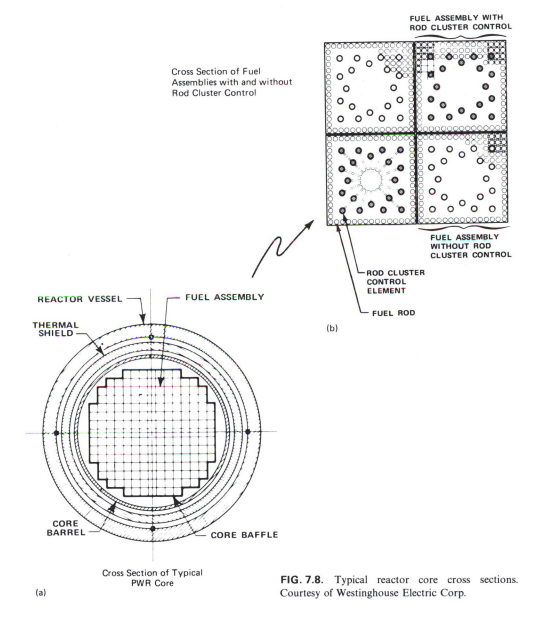

FUEL ASSEMBLY WITH ROD CLUSTER CONTROL

Cross Section of Fuel Assemblies with and without Rod Cluster Control

FUEL ASSEMBLY WITHOUT ROD CLUSTER CONTROL

ROD CLUSTER CONTROL ELEMENT

FUEL ROD

(b)

REACTOR VESSEL — — FUEL ASSEMBLY

THERMAL SHIELD

CORE BARREL — — CORE BAFFLE

Cross Section of Typical PWR Core

(a)

**FIG. 7.8.** Typical reactor core cross sections. Courtesy of Westinghouse Electric Corp.

Figure 7.9 is an expanded cutaway side view of an individual fuel assembly containing a control rod assembly. Each fuel assembly can be loaded or withdrawn individually from the reactor core, as required by refueling operations. The control rods can be inserted or withdrawn during the control operations in a cluster for each fuel assembly. The entire cluster for an assembly is attached to a control-rod assembly structure (Fig. 7.9) and the rods slide in unison, each in a separate tubular guide. As we have noted, water flows in the spaces between and within the assemblies of fuel rods and control rods.

With this layout of the core in mind, we can now better visualize the processes that take place there:

1. Fission *reactions* in a fuel rod result in energetic neutrons and fission fragments (nuclei) being propelled away from the site of the reaction. Most of the high-energy neutrons escape from the fuel rod and penetrate the moderating medium. The fission fragments, however, collide quickly with other nuclei in the same fuel rod and give up their kinetic energy as heat.

2. Fission-liberated *neutrons*, having penetrated the moderator, endure multiple collisions with hydrogen nuclei, each time bouncing elastically and losing a little kinetic energy. The moderating action of the water slows the neutrons so that their capture by a fissionable ($^{235}U$) nucleus in another fuel rod becomes more likely.

3. Some liberated neutrons, in their random paths in the water between the rods, may instead encounter a *control rod* and be absorbed. As a result, these do not create another fission reaction.

4. Finally, the water flowing along the spaces between the rods acts as the *coolant* to carry away the heat generated within the fuel rods by the fission fragments.

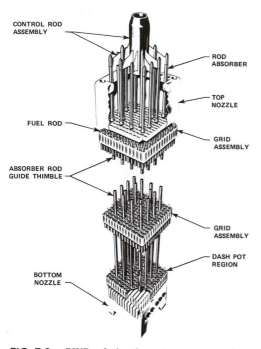

**FIG. 7.9.** PWR fuel element – cutaway view. Courtesy of Westinghouse Electric Co.

## Nuclear reactor kinetics

These are the basic processes within a nuclear power reactor. But to appreciate fully how a reactor works, we must consider one other aspect: the time-dynamic behavior of fission reactions. This behavior, along with the characteristics of neutron production and travel, is referred to as the *kinetics* of the nuclear reactor.

To understand reactor kinetics, we have to appreciate the balance of neutrons over time in a chain reaction. The number of fission reactions in a given time interval depends directly on the number of neutrons moving through each of the fuel rods – neutrons that are thus available for capture by fissile nuclei. These neutrons, of course, have already been liberated by fission reactions, usually from other fuel rods. The possibility of a self-sustained chain reaction exists because an average of 2.5 neutrons are liberated by each fission reaction. The chain reaction may either grow or remain at a constant rate of fissions per time interval, and the issue for nuclear engineers in building a nuclear power plant is to be sure that reactions remain self-sustaining or *critical* without growing out of control in the *supercritical* range. And the key to that question lies in the balance of neutrons available for fission in a given time interval. If the number equals that of the previous interval, the reactor will remain critical; if it is greater, it will go into the supercritical range.

In any time interval, the neutron population accounts for neutrons produced by fissions in the preceding time intervals, neutrons lost from nonfission processes (for example, those captured by $^{238}U$ nuclei), neutrons that are absorbed into control rods, and neutrons lost at the edges of the reactor. Finally, there is another group of neutrons that is crucial to the chain reaction (as well as to an appreciation of nuclear processes) called *delayed* neutrons. These are emitted not by the fission of $^{235}U$ nuclei, but by the decay of unstable nuclei from fission products, such as radioactive bromine and iodine, which are themselves created by the prior fissioning of $^{235}U$ nuclei. These delayed neutrons are distinct from the *prompt* neutrons, which are created immediately in each fission process, and their time delays range from about 1 second to about 1 minute.

In the ordinary LWR, fission results principally from capture of moderated neutrons that have been slowed by the medium between the fuel rods. The average lifetime of a neutron in the moderator is about $10^{-4}$ seconds, which serves as a slight delay in the generation of the next set of fission reactions in the chain reaction. However, since this delay is a small fraction of a second, the number of generations per second could be quite large (thousands, in fact). If this were the case, the chain reaction would have a rapid growth rate and the reactor would be difficult to monitor and control.

But delayed neutrons act as *pacemakers* to keep the reactor from going supercritical. Engineers design a delicate balance of prompt and delayed neutrons into the system by maintaining a flux of prompt neutrons that is just below the density required to take into the critical condition. The delayed neutrons,

however, allow the reactor to *go critical*. Thus, the delayed neutrons for the most part determine the reaction rate of growth when the control rods are withdrawn or the rate of decay when the rods are inserted. The long time delays (an average of 14 seconds for $^{235}U$ fission) determine the reactor kinetics.

The consequence of these long average delay times is a much slower growth rate of fission reactions than those that would result from prompt neutrons alone. An LWR requires several minutes to go from low to high levels of the chain reaction (and power). With such dynamics, the LWR operator can monitor the build-up of the reactions, override automatic controls (if necessary), and manually cut back on the criticality (the supercritical growth). It is fair to say as Enrico Fermi, a pioneer of modern nuclear physics, once observed, that without delayed neutrons, there would be no nuclear power (see Segre 1965).

There is the possibility, however, that under some conditions the delicate balance of delayed versus prompt neutrons in the reactor may be upset. This is the reason often given for the violent accident at the Chernobyl reactor. Some have argued that the reactor went into a *prompt criticality*, whereby supercritical growth was determined solely by the prompt neutrons. The reactor was reported to have soared from a thermal power output a fraction of its rated value to 100 times its rated value in 4 *seconds*.

It should be noted that it is possible to build a working power reactor designed to operate from unmodulated neutrons. As we will see in the next chapter, the *breeder reactor* is one example. Its fuel rods are packed more densely than those in a LWR, thus increasing the chance of neutron capture shortly after fission, while the neutrons are still at high speeds. However, the fast reactor, too, is paced by a minority of delayed neutrons and again the long delay times of these neutrons is controlled by slowing the changes of the chain reaction. In the event of a *prompt* criticality, however, the fast reactor's doubling time is much shorter than that of the moderated reactor, because the average lifetime of prompt *fast* neutrons is 1,000 times shorter than delayed moderated neutrons.

## Fissionable fuel resources

Fissionable resources consist of naturally occurring deposits of uranium and thorium. Resources for uranium and thorium may be estimated in much the same way as fossil fuel resources. Reasonably assured resources (RAR)[3] of uranium for the world are about 2.9 million tons. About 0.9 million is in the United States. Current global usage is about 200 metric tons per year (m.t./yr) for each 1,000 MW$_e$ of nuclear power plant capacity and world production is about 55,000 m.t./yr, since installed world nuclear-generating capacity is about 275,000 MW$_e$. These estimates result in an $R/P$ ratio (see Chapter 2) of a little over 50 years for uranium. Using the other measure of natural resources (see Chapter 2), remaining recoverable resources throughout the world are about double the RAR estimates. Similar estimates exist for thorium, which remains untapped as a fuel resource.

A comparison of these estimates of fissionable resources with fossil fuel resources is made on Table 7.1 on the basis of equivalent thermal energy measured in quads (see Chapter 4). The

**Table 7.1.** Comparison of nuclear and fossil fuel resources in the United States (quadrillion Btu)

|  | Reserves[a] | Additional recoverable resources[b] | Remaining resource base[c] |
|---|---|---|---|
| Petroleum | 410–530[d] | 1,100–2,200 | 16,000 |
| Natural gas | 440–570[d] | 1,100–2,200 | 7,300 |
| Oil shale | 900–3,400[e] | 7,600[e] | 150,000[f] |
| Uranium |  |  |  |
|   Thermal reactors[g] | 310[i] | 630[j] | 8,000[j] |
|   Breeder reactors[h] | 22,000[i] | 44,000[j] | 560,000[j] |
| Coal | 5,000 | (unspecified) | 78,000 |

[a] *Reserves* are economically recoverable resources in identified deposits, with the extent of the resource measured or inferred on the basis of geological evidence.

[b] *Additional recoverable resources* are resources judged economically recoverable on the basis of broad geological evidence, but for which insufficient, detailed knowledge is available to classify the resources as reserves.

[c] *The remaining resource base* is the total amount of the resource estimated to be left in the ground, usually in deposits having some minimum grade or better.

[d] As of 1982, U.S. petroleum reserves were about 154 quads and U.S. natural gas reserves about 160 quads.

[e] In deposits having 30 gallons of oil or more per ton of shale.

[f] In deposits having 10 gallons of oil or more per ton of shale.

[g] With thermal reactors, a metric ton of uranium oxide (yellowcake) yields one trillion Btus.

[h] With breeder reactors, a metric ton of uranium oxide yields 70 trillion Btus.

[i] For uranium oxide up to $10/lb.

[j] For uranium oxide up to $20/lb.

*Source*: S. D. Freeman, *A Time to Choose*, Ballinger, Cambridge, MA. Copyright © 1974 by the Ford Foundation.

table shows that U.S. uranium resources are comparable in thermal energy magnitude to oil and natural gas. This estimate is valid on a worldwide basis as well (see Chapter 5).

A breeder reactor can increase the yield from uranium ore. The breeder enhances the creation of plutonium $^{239}U$ – which is fissionable and can be used as a fuel – from $^{238}U$ through neutron bombardment that is more intense than in a moderated reactor. The breeding process can create almost 100 times the amount of fissionable fuel from a given amount of natural uranium. Breeding, as a result, can theoretically provide energy resources comparable or larger in energy value than those of coal. A similar breeding reaction can transmute natural thorium ($^{232}Th$) into $^{233}U$ (which also is fissionable) and thereby provide energy resources comparable to coal.

## Nuclear plants – operation

### Normal operation

Let us imagine the operation of a PWR under normal conditions. Technicians start the reactor by retracting the control rods. (Refer to Figures 7.4 and 7.5.) As the rods slide up toward the top of the reactor vessel, the flux of neutrons grows; soon, the chain reaction commences.

The more the control rods are retracted, the more are fission neutrons able to collide with other fissile nuclei in the chain reaction. The reactor as a whole passes from the subcritical stage (below the number of neutrons needed for a sustained chain reaction), through the critical stage, and temporarily into the supercritical stage to sustain growth. As the number of nuclear reactions increases, the thermal power rate grows as well. This growth continues until the full power level is attained, at which time controls are adjusted to keep the reaction rate steady in time – called the *steady state.*

Maintaining a steady reaction rate is a delicate control problem and typifies the precise demands of nuclear technology. Not only must the balance of reactions be kept at a constant rate, but also temperatures must be controlled by the rate of circulation of the coolant. Maintaining a sufficient cooling rate is a critical requirement and makes an absolute demand for an emergency cooling system.

Under normal operations, the reactor is operated in the critical range and generates thermal energy. This thermal energy, in turn, is carried away in the PWR (Fig. 7.7) by the circulation of the coolant water. The water travels in a loop – the *primary loop* – from the reactor vessel to a steam generator and back to the reactor. The steam generator is a heat exchanger through which the reactor heat is transferred into the working fluid (steam) in a secondary loop, which is physically isolated from the coolant of the primary loop. The steam, in turn, flows to the blades of the turbine and the turbine acts as the prime mover for the electric generator. In the BWR (Fig. 7.6), the steam from the reactor directly turns the turbine blades.

The process, as designed, is therefore self-contained (see Fig. 7.10, containment and shielding) and discharges virtually no effluent to the atmosphere other than steam exhaust. There are, furthermore, no unsightly piles of fuel or wastes outside the plant, nor any large lineups of railroad cars or barges around the plant. As such, the image of the fission plant is considerably different from that of the coal-fired plant and these features are among the advantages enumerated by proponents of the technology. (The popular image of a nuclear power plant is typically of its cooling towers, which we learned in Chapter 3 are not peculiar to nuclear plants.)

During normal operations, virtually all radioactive materials are contained within the reactor system, and small amounts only of short-lived radioactive elements are allowed to vent into the atmosphere. Shielding of the reactor and associated components is required by law to protect the health and safety for workers inside the plant. This shielding, along with the limitations on radioactive discharges, is also designed to meet public health standards for the population outside the plant. The United States Nuclear Regulatory Commission (USNRC) is charged under the Atomic Energy Act with enforcing all standards of

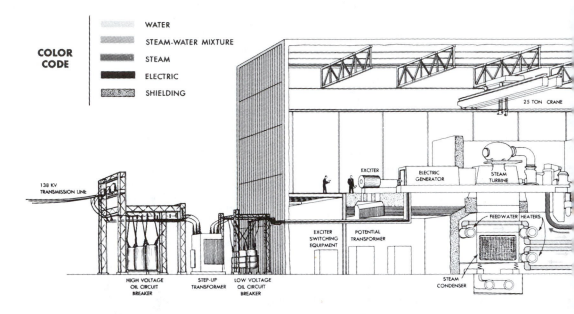

**COLOR CODE**

WATER
STEAM-WATER MIXTURE
STEAM
ELECTRIC
SHIELDING

25 TON CRANE

138 KV
TRANSMISSION LINE

EXCITER

ELECTRIC GENERATOR

STEAM TURBINE

FEEDWATER HEATERS

EXCITER SWITCHING EQUIPMENT

POTENTIAL TRANSFORMER

STEAM CONDENSER

HIGH VOLTAGE OIL CIRCUIT BREAKER

STEP-UP TRANSFORMER

LOW VOLTAGE OIL CIRCUIT BREAKER

**FIG. 7.10.** Nuclear power plant – cutaway view. Courtesy of Consumers Power Co.

health and safety for nuclear power plants and related activities (see Ford 1982 and Keeny 1977). The USNRC has set a standard for the maximum radiation exposure level of the general public from a nuclear plant at less than 25 millirems per year (mrem/yr) whole-body dose.[4] Thus, some have argued that a person living *near the fence* of the plant would receive additional radiation dosages less than the ambient natural dosage at sea level of cosmic ray radiation of about 50 mrem/year.

Radiation standards for normal operations concern low-level radiation. This term generally denotes dosage at or below ambient levels. At these levels, the impacts on occupational or public health do not have a clear-cut cause and effect relationship. Nevertheless, these standards have provoked controversy regarding thresholds

of hazard. The concept of thresholds has already been encountered in our discussion of air quality standards (see Chapter 6). The question in both cases is, to what extent should environmental controls be required to be below levels of contamination because of clear-cut cause and effect relationships? We will discuss the implications of this issue in Chapter 10. But we should note here that no uniform regulation policy exists today in the United States with regard to thresholds.

### Impacts of normal operation

Leaving aside the question of low-level radiation standards, let us consider the comparative benefits and costs of nuclear-powered generation.[5]

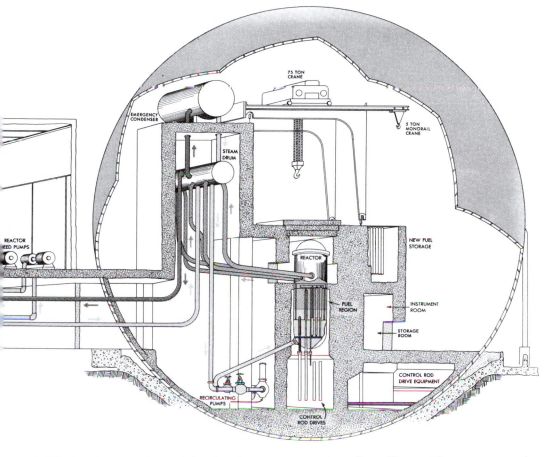

EXTRACTION. Uranium mining is evidently not as accident prone an occupation as coal mining, even though much of it is underground. There is evidence, however, that radiation hazards have affected the overall mortality rates of miners.[6] In addition, uranium mining may have public health consequences; uranium ore *tailings* (residues) have often been left in open piles that may present a health hazard.

TRANSPORTATION (PREPROCESSED URANIUM). No measured impacts have been reported as a result of the transport of the ore from mine to processing plants. This step might be considered comparable to the shipment of coal to its site of preparation for combustion. In the case of nuclear fuel, however, there is another step: the shipment of processed fuel rods to the nuclear-

generating plant. Here again no measured environmental impacts have been reported. Still, concerns have been expressed by some municipal officials and others over possible traffic accidents that could cause spills of radioactive material.

PROCESSING. Nuclear-fuel processing consists of uranium enrichment and fuel-rod fabrication. Enrichment and fabrication can be compared logically to the *preparation* step in the coal cycle, although nuclear-fuel processing is a vastly more sophisticated technological operation. Occupational hazards for nuclear-fuel industry workers appear to be lower than for coal miners, although controversies will likely continue over standards, especially as they relate to the question of acceptable levels of risk (see Chapter 10).

THE REACTOR.   As a step in the nuclear-power process, this corresponds to the combustion step for coal. Here, we must be careful to specify that we are considering *normal* operating conditions, in contrast to abnormal or accident conditions. Under normal operating conditions, the environmental or public health impacts of nuclear plants are less than those of an equivalent coal-fired plant. Hazards for operating personnel, however, may be (or may have been) greater at nuclear plants than at coal-fired plants. Historic data on nuclear plant personnel dating back to the early days of the industry indicate radiation dosages averaging slightly less than one rem/year (1,000 mrem/year), but the range of dosages runs up to several hundred rems/year. During these periods a statistically significant[7] number of excess cancers was recorded for nuclear plant workers. Since 1969, however, the average annual dosage to nuclear workers has been reduced by a factor of two and at the present time federal regulations permit no more than 5 rem/year (5,000 mrem/year) for any individual nuclear worker – still well above the ambient level of radiation exposure.

DISPOSAL.   In this chapter, we will consider only the issue of *short-term storage* (on-site) of nuclear waste (particularly spent fuel rods), rather than *long-term storage* (away from the site), or reprocessing or *permanent disposal* – unresolved issues that we will discuss in Chapter 8. At the reactor site, the 25 mrem/year dosage standard is for the most part maintained, and we can conclude therefore that on-site storage of spent nuclear fuel appears to have a lower impact on the local environment than the disposal of coal wastes.

Another environmental concern is the thermal discharge from both nuclear plants and coal-fired plants (both are thermal-type plants). Discharged heat has caused some ecological concerns because it can raise temperatures in natural bodies of water significantly, killing plant and animal life. The problem is controllable by the use of cooling towers (see Chapter 3), which help limit discharge temperatures to within regulated limits. There is little difference in the thermal effects caused by nuclear and coal-fired plants, although the LWRs do discharge more heat and thus require more cooling of the discharge to meet standards.

## Unresolved issues – safety, disposal, and proliferation

Proponents of nuclear technology emphasize normal operating characteristics of nuclear power plants. By using that state, they are able to demonstrate that nuclear plants have lower local environmental impact and a better local visual impression (Fig. 7.1) than fossil-fired plants. Proponents have also claimed that only a massive shift away from the combustion of fossil fuels, primarily to nuclear power, can provide a technical response to the long-term dilemma of the carbon dioxide–greenhouse effect (see Chapter 6). Weighed against these arguments, however, are the controversies surrounding nuclear plant safety, long-term waste disposal, and weapons proliferation.

### Nuclear plant safety

When we talk about nuclear plant safety, we mean simply the prevention of accidental release of radioactive materials outside the plant, which would expose the public at large to the hazards of radiation sickness and cancer. Although plant safety is a complicated issue, we want to focus on the most important type of nuclear accident, called the *loss of coolant accident* (LOCA). As we have learned in our discussion of reactor operation, the coolant must circulate through the reactor to remove the heat generated by the fission reactions. If coolant circulation is interrupted for any reason, the accumulation of heat will increase the temperature, which, if allowed to continue, can lead to the *meltdown* of the reactor core. If not totally confined within the reactor containment building (Fig. 7.1 and Fig. 7.10), a core meltdown can release a devastating inventory of radioactive materials into the atmosphere. The radioactive inventory measured in curies (Ci, see Chapter 8) of a large commercial reactor is thousands of times more than the amount released by a Hiroshima-sized fission bomb, and the potential for catastrophe is great indeed.

A LOCA can have any one of a number of initiating causes, many of which are mundane matters of plumbing – quite surprising given the "high-tech" image of the operation. For example, malfunctioning valves were reported an average of 7 times per year for a group of 17 plants in 1972, early in the era of large nuclear plants. A malfunctioning valve figured significantly in the accident in 1979 at Three Mile Island. Since then, quality control on valves has reduced such incidents, but a probability of such malfunction still exists and can never be reduced to zero.

There are other mechanical failures that could initiate an accident sequence: high-pressure pipes could rupture or pumps could fail. These have all occurred at nuclear plants. Electrical equipment could also fail; failures of circuit breakers, generators, batteries, transformers, and electric instruments or indicators have been recorded at plants as well, and could initiate or contribute to plant accidents.

The important aspect to recognize at the outset in assessing such technological malfunctions is that their occurrence can never be reduced to zero. The basic question, therefore, is not how to prevent individual malfunctions absolutely, but rather how to deal with them within the entire system in an acceptable way. The approach then is how to engineer the system so that as an entity it functions in an acceptably safe manner. Such approaches have been used in aircraft safety and in mission planning for space flights.[8]

Safety of operation became a major concern of the U.S. nuclear program during the early 1970s. It was soon recognized that equipment failures could occur and that even the safety equipment designed to deal with such failures could fail. Consequently, studies of plant safety were initiated, although there was controversy over what constituted an adequate safety study (see Ford 1982). The best-known reactor safety study was conducted by Professor Norman Rasmussen of the Massachusetts Institute of Technology for the reactor safety part of the

Atomic Energy Commission (AEC, which was reorganized into the NRC). The study, a probabilistic analysis of the complex workings of the plant and its safety systems, was reported in 1975 as the AEC document number WASH-1400. This report figured heavily in the safety debate over the following decade. It was known variously as *WASH 1400* or the *Rasmussen report*. (Our discussion of reactor safety will refer frequently to this report.)

## The engineered safety system

For nuclear plants, the approach to safety generally has taken the form of the engineered safety system (ESS). The ESS executes five basic actions, as indicated schematically in Figure 7.1. Briefly, these actions are:

1. *Reactor trip* (*RT*) – This is the rapid automatic shutdown of thermal power generation in the core by the quick insertion of the control rods. This shutdown is often called a *scram* in the industry.

2. *Emergency core cooling* (*ECC*) – This system is designed to provide an independent means of supplying cooling water to the core by flooding the reactor vessel. It requires an entirely independent pressurized supply of water that can be circulated through the reactor vessel, even if other systems (primary coolant, pumps, electricity) have failed.

3. *Postaccident radiation removal* (*PARR*) – This system of pumps and filters collects and removes radioactive material from the containment building. The atmosphere and surfaces are sprayed and the contaminated wash is collected and passed through filters.

4. *Postaccident heat removal* (*PAHR*) – This provides the means for continued removal of heat from a reactor core after the main power reaction has been stopped by RT. PAHR is necessitated by the *decay heat* of spontaneous radioactive reactions, which generates thermal energy at about 7 percent of full thermal power immediately after a shutdown. The heat level returns to safe levels only weeks later.

5. *Containment integrity* (*CI*) – CI refers to

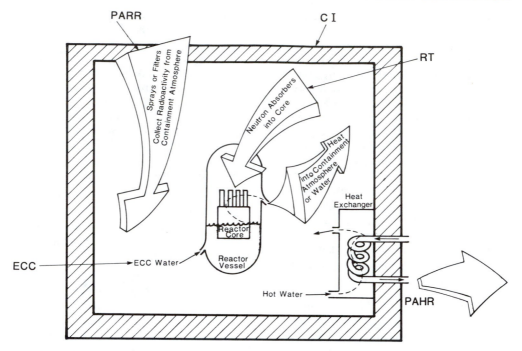

**FIG. 7.11.** Schematic diagram of engineered safety system (ESS) for a water-cooled reactor. *Source*: U.S. Atomic Energy Commission (1975). WASH-1400 Washington, DC.

the containment building (see Figures 7.11 and 7.10), whose structural integrity provides the *last line of defense* against the release of radioactive materials to the atmosphere at large, if all other ESS functions fail.

In considering whether this entire ESS functions in an acceptably safe manner, analysts have reduced the system to a probabilistic assessment of the failure of its parts. The possible failures of the various parts must be analyzed in a sequence of distinct steps, since the consequences of any single failure are conditional upon what prior failures have occurred. Such a sequential, probabilistic analysis is called a *fault tree* analysis and is represented on a treelike chart (Fig. 7.12) called an *event tree*. This was the methodology of the Rasmussen safety study.

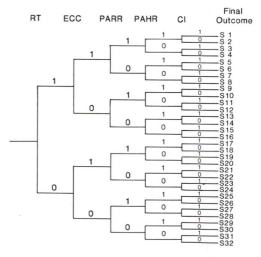

**FIG. 7.12.** Illustrative event tree for LOCA functions. *Source*: U.S. Nuclear Regulatory Commission (1980).

The event tree is a means to trace the sequences of possible failures, and it provides thereby the logical framework for assessing the combined probabilities of sequential failures. Each branching junction in the tree represents the binary outcome (for example, 1 = success or 0 = failure) for the functioning of a step in the ESS. Thus, for example, the first branch represents the success or failure of the reactor trip (RT) to properly function. Given the outcome of this step, there are then two possibilities for the ECC step for each of the two RT outcomes, or a total of four outcomes possible at the outset of the PARR step. For the whole system as depicted here there are 32 possible final outcomes (see Fig. 7.12, right hand side).

Outcome S1, for example, means that all five functions operated successfully; the chain reaction was shut down (RT), the core received emergency cooling (ECC), the postaccident radiation was safely removed (PARR), the postaccident decay heat was removed (PAHR), and the containment building allowed no radiation leakage to the atmosphere (CI). Outcome S13, on the other hand, represents partial failure of the ESS. Following successful RT shut down, the emergency core cooling (ECC), radiation removal (PARR), and heat removal (PAHR) all fail, leaving the containment (CI) as the last defense against catastrophic discharge of radioactive materials. Finally, of course, outcome S32 represents failure of all systems, including the reactor trip and the containment building, and so means almost certain massive radioactive discharge into the atmosphere.

Each step in the event tree must have a probability of failure assigned to it in order to assess the overall safety of the ESS. It is reasonable to expect that probabilities would be based on the statistics of long-term operating experience with the devices or subsystems of each step. Thus, for example, the probability of failure of the emergency core cooling (ECC) system would be based upon actual operating experience over many years of such systems in nuclear plants or, at the least, upon extensive repetitive testing of prototypes. In fact no such operating experience existed and no testing had been performed by the time the reactor safety study (WASH 1400) was issued.[9]

In considering the probabilities of failure for each step in such an event tree, all possibilities of failure must be anticipated. For a technological system as complicated and sophisticated as a nuclear plant, it is indeed a formidable task to anticipate *every* possible route to failure. Critics of the fault/event tree approach contend, as a matter of fact, that the task is impossible. Operating experience, or extensive testing, it has been said, is even more important for an overall assessment than for assessments of the parts, since confidence in the overall absolute probability predictions are contingent on the confidence that no possibilities have been omitted from the tree analysis.

## The chances of an accident

Let us next examine how the probabilities were assembled in the Rasmussen safety study for the prediction of a major accident and its consequences.

1.  The probability of an *initiating event* – The estimate is supposed to include all possible events that could lead to a LOCA. An estimate should include causes internal to the plant (for example, valve failures) as well as external causes (such as earthquakes). The Rasmussen probability estimate taken theoretically over all conceivable causes was 1 in 100 reactor years[10] (expressed as: $10^{-2}$ per reactor year).

2.  The probability of the failure of the ESS, *exclusive of CI* – Here all major steps (RT, ECC, PARR, and PAHR, that is, outcome S31 in Fig. 7.11) have failed and major radiation release is prevented only by the integrity of the containment building. The Rasmussen estimate was 1 in 100 reactor years or $10^{-2}$ per reactor year.

3.  The probability of *containment integrity* (CI) failure – In this case, the integrity of the building has been broken due to forces developed in the accident sequence itself (for example, extremely high steam pressures) as

represented by outcomes S8 or S24 in Figure 7.12 or due to the other causes (for example, earthquake stresses). The estimate was 1 in 100 reactor years.

4. The probability of *worst weather* – This refers to that combination of wind direction, precipitation, and so forth that would promptly deposit the released radioactive material on the largest population in the vicinity of the plant. This was estimated as a chance of 1 in 10 ($10^{-1}$) at any time of occurrence of the accident. This probability follows the failure of the entire EES and is therefore outside the event trees of Figure 7.12.

5. The probability of an accident occurring at a site of the highest *population density* – The Rasmussen study averaged all nuclear plant sites in the United States to estimate the chances of such an accident occurring at a site with the highest population density. The estimate was 1 in 100 ($10^{-2}$) sites. (Subsequent to the Rasmussen study, accident probabilities and consequences have been estimated for specific plants.)

These probabilities of failure may logically be combined into composite failures simply by multiplying the individual probabilities, provided the individual *probabilities are independent* of one another. The Rasmussen report proceeded on this assumption and came up with an overall probability of an accident that the study group was convinced[11] would appear acceptably low to the public. They concluded:

1. The chances of a reference accident (or meltdown)[12] were:

$$
\begin{array}{ccc}
\text{initiating} & \text{ESS} & \\
\text{meltdown} = \text{event} & \times \text{failure} & \\
\text{probability} & \text{probability} & \text{probability} \\
10^{-4} = 10^{-2} & \times & 10^{-2}
\end{array}
$$

That is, the chances were 1 in 10,000 reaction years or 1 in 100 years for 100 reactors in operation.

2. The chances of a worst possible accident were defined as:

$$
\begin{array}{cccc}
\text{worst} & & \text{CI} & \\
\text{accident} = \text{meltdown} & \times \text{failure} & \times \\
\text{probability} & \text{probability} & \text{probability} & \\
10^{-9} = 10^{-4} & \times & 10^{-2} & \times \\
\text{worst} & \text{population} & & \\
\text{weather} & \times \text{high} & & \\
\text{probability} & \text{probability} & & \\
10^{-1} & \times & 10^{-2} &
\end{array}
$$

That is, 1 in 1,000,000,000 reactor years or 1 in 10,000,000 years for 100 reactors.

The executive summary of the Rasmussen report then claimed that the "chances of being killed by a nuclear accident are as remote as those of being hit by a meteor." Such claims, proponents of the technology contended, would lead a "reasonable" person to conclude that the risks of a nuclear plant accident are so low that they could be ignored. However logical these claims may have seemed, they were not universally accepted by the public.

## Consequences of an accident

Included in the Rasmussen safety study results were estimates of the casualties of the various possible accidents, ranging from the reference accident to the worst possible accident. These casualties would arise, of course, solely from the effects of exposure to radioactivity (since there are no long-distance shock, blast, or heat effects from a runaway reactor of the moderated type as there would be from a fission bomb). Radiation exposure at high levels causes death and illness due to prompt somatic effects[13] and can induce cancer over the longer term. Somatic effects include *prompt fatalities* from massive doses of radiation.

Radiation doses above the level of 30 rem have a definite cause and effect relationship to human health. Doses above 100 rems cause radiation sickness and 200 rem is the threshold for promptly fatal doses. Doses in the range of 1,000 to 3,000 rems cause death within 2 weeks and exposures above 3,000 rems results in death within hours. Besides the real danger from radiation exposure, psychiatrist Robert J.

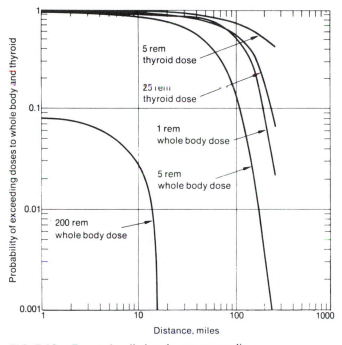

**FIG. 7.13.** Expected radiation dosage versus distance from a nuclear accident. Copyright © 1980 by IEEE.

Lifton (1976) has noted that the mere idea of such exposure can have a profound impact on people. Lifton described "primal fears about the integrity of the human body, as threatened by the invisible poison of radiation" (also see Weart 1988). Some engineers and scientists believe that psychological impact of the use of nuclear energy is irrational, and they have argued that it should be dismissed in the debate over nuclear energy. But even if it were irrational, should that alone determine its relevance? If people fear something for whatever unscientific reason, should they be forced by scientists or others in society to endure that fear? Should the rationality of the expert be the final word in societal debates? These questions are more philosophical than technological, and as we will see in Chapter 10, they have no pat answers.

The consequences of a major nuclear plant accident to the surrounding area are illustrated in Figure 7.13. The graph, which was projected by the Rasmussen study in 1975, shows the spread of expected radiation dosages according to the radial distance from the plant. The thyroid dose curves are especially important because of the susceptibility of the thyroid gland to cancer-causing isotopes. Therefore, they are commonly used as indicators for emergency measures for the population. Although these curves were estimated before the actual experience at Chernobyl, they provided a fairly good picture of the dosages from the damaged reactor. The actual data from that accident (Fig. 7.14) show thyroid doses of 1 rem and more extending to a range of about 1,000 miles. Other more detailed Chernobyl data suggest that an unprotected person living about 50 miles northwest of the accident site would have received a cumulative 30 rem thyroid dose over the 6 days following the accident.

The potential for the rapid spread of radiation from a nuclear accident site has spurred much debate over the issue of public

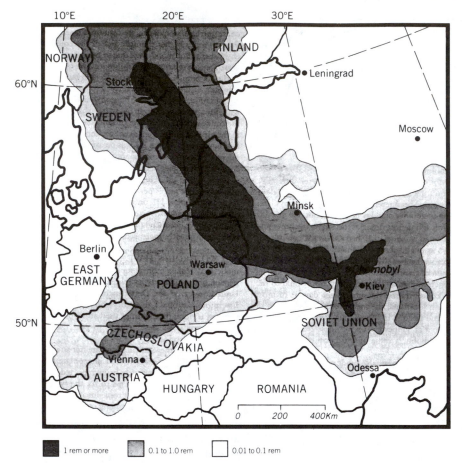

**FIG. 7.14.** Estimated thyroid inhalation doses, 4 days after the accident at Chernobyl. Reprinted by permission of the *Bulletin of the Atomic Scientists*, a magazine of science and world affairs. Copyright © 1986 by the Educational Foundation for Nuclear Science, 6042 South Kimbark Avenue, Chicago, Illinois 60637.

evacuation. Following the accident at Three Mile Island, the NRC adopted a rule requiring public protection, including possible evacuation, within a 10-mile radius. But the range has been subject to dispute. Critics of the industry want the radius extended to 20 miles or more, while industry proponents have argued that 10 miles is unnecessarily great.

Even with an evacuation, the Rasmussen study group estimated heavy casualties for the worst possible accident. They projected 2,300 fatalities, which critics called too low. A more recent calculation (*The Washington Post*, November 1, 1982), on behalf of the NRC, supports the critics. It suggests that fatalities might reach the tens of thousands. According to Soviet reports on the Chernobyl accident, approximately 30 "prompt" fatalities due to radiation occurred, but experts have estimated that delayed cancer fatalities will number at least a few thousand and possibly more than 100,000 over the next decades. Early in the nuclear program it was recognized that a major nuclear accident could cause huge legal claims for indemnity for loss of life, illness, and property damage against a plant operator. It was also clear that few, if any, privately owned electric utilities would opt for a nuclear plant with such a possibility, because the potential size of the claims could lead to bankruptcy. During the period of the initial promotion of nuclear power in the 1950s, there was thus a strong motive to remove this barrier, and this resulted in the Price-Anderson Act, passed by the U.S. Congress in 1957.

The Price-Anderson Act has been controversial since it was enacted and the controversy has been renewed every time Congress considers its extension. The act limited the liability of a nuclear plant owner to a fraction of the possible claims for a major accident and therefore appeared to leave the public at risk of great uncompensated losses. The measure initially limited liability to $60 million for the operator, with the U.S. government responsible for additional amounts up to a maximum of $500 million; possible claims even then were anticipated to be as much as 30 times that figure. A subsequent extension of the act in 1975 increased the maximum liability of the operator to $800 million, but again this figure was far short of the possible claims.


## Public dispute

### The experts disagree

Beginning in the 1970s, prior to the WASH 1400 study, many members of the technical community became concerned with nuclear reactor safety. There was a growing fear that standards of safety for nuclear reactors were being sacrificed to promote the fast expansion of nuclear power (see Ford 1982).

Upon the publication of the draft of WASH 1400, several critical studies were made of the results. The American Physical Society (APS) sponsored the most prominent: the APS Study Group on Light-Water Reactor Safety. This committee, made up of professional physicists, criticized the method, omissions, and results of WASH 1400. Salient among these were:

1. Failure probabilities were estimated with virtually no *operating experience* or large-scale experiments to back them up. This was especially important in view of the complexity of the large base-load plants that were built.

2. The WASH 1400 report did not account sufficiently for *human error* on the part of operating personnel at a plant. The APS study stated that this factor may set an *irreducible minimum* to accident probabilities.

3. Event/fault tree analysis was only reliable for estimating *relative* probabilities. It had not been proven reliable for predicting failure probabilities on an absolute basis, especially when low absolute probabilities are involved.

4. The consequences (fatalities and illness) of an accident were underestimated; for example, the APS study group predicted 23,000 to 36,000 fatalities for the worst possible accident.

5. The Rasmussen study ignored the possibility of sabotage, and the APS report doubted the adequacy of reactor security provisions.

6. There was insufficient *outside review* of reactor safety studies at the AEC, and the APS group recommended that the results of the study be published in the open scientific literature.

Other groups inside and outside the government[14] pointed out shortcomings of the WASH 1400 draft as well. Notably:

7. The Rasmussen study was said to have overlooked the high potential for *common-mode* failures.[15]

8. There was a challengeable assumption that populations near an accident site would be *evacuated promptly*; at the time, no public plans existed.

9. There was no accounting for the aging of nuclear plant components and the possibility of such problems as embrittlement and corrosion.

The nuclear industry along with the AEC/NRC and several distinguished scientists attacked the critics. But the debate, which pitted eminent scientists, such as Professor Robert Pohl against Hans Bethe, appeared to confuse the public and many elected officials. After all, they asked, how could there be disagreement on a matter of scientific fact? But there was no agreement and no *fact* available to resolve issues of estimates and probabilities. As we will discuss in Chapter 10, experts have their limitations – a point made abundantly clear to the public by the actual failures of nuclear technology at Three Mile Island and Chernobyl.

## Three Mile Island (TMI)

A major event in the history of nuclear power took place on March 28, 1979, when a high-pressure relief valve opened in reactor number 2 at the Three Mile Island (TMI) nuclear plant near Harrisburg, Pennsylvania. Although the valve opening was the *initiating event* of the TMI accident sequence, it does not indicate the fundamental problem. As an official investigation revealed, the accident sequence started because of human error. A major feedwater valve had been left closed erroneously at the end of routine maintenance the day before. Critics had warned years before of the inevitable potential for human error.

As the TMI accident sequence progressed, hour by hour early that March morning, some steps of the ESS functioned and other steps did not. For example, the RT (insertion of control rods) functioned as designed, and the ECC (emergency core cooling) functions were initiated as designed. But the erroneously closed valve prevented the flow of cooling water for the PAHR function. More important, plant operators were unaware of the closed valve for many critical minutes and received a false indication on the proper setting of another valve. These two faults caused the operators to take the counterproductive corrective measure of shutting off cooling water pumps, resulting in a drop of the coolant level in the reactor core. Before the operators recognized the errors and corrected them, there was a near meltdown of the reactor core.

Even though there were no prompt casualties, the consequences of the TMI accident were catastrophic. The reactor core was severely damaged, and the entire area within the containment building had become prohibitively radioactive, hampering the means to repair it. The cost of the cleanup has already been so great as to nearly bankrupt the operating utility – General Public Utilities (GPU). It has been estimated that the cleanup and repair of the station will cost in the range of $1 billion and remains incomplete a decade later. While this has been going on, GPU has had to purchase electric power from neighboring utilities at an added cost over generating its own power.

A review of the safety systems at TMI[16] showed that human failures were at least as important as mechanical failure. Actually, the human errors committed by the operators during the accident have been attributed to inadequate training, which could be called an *institutional* failing. However, the critics of the technology contend that human error can be held only to an irreducible minimum level and that this minimum will determine the lower limit of accident probability at a nuclear plant.

It should be reemphasized that this accident was of the type that can occur even if the reactor is *scrammed* successfully (that is, the control rods are inserted and the reactor is made subcritical). Even with the chain reaction shut down, heat continues to be generated and a loss of coolant will cause core temperatures to rise. This type of accident, a simple LOCA, is distinctly different from the accident that occurred subsequently at Chernobyl.

## Chernobyl

The Chernobyl accident has been called a case of a runaway reactor. The implication of this characterization is not only that the reactor's chain reaction was not shut down, but that a supercritical condition took place in its core, causing rapid growth of the chain reaction.

Not only was the Chernobyl accident

different from the TMI event in terms of the critical state of the reactor, but also as to the initiating cause. Whereas the TMI accident was initiated in the course of routine operation, the Chernobyl event occurred during a special test of the plant. The causes of the accident have been attributed largely to the nonroutine and ill-advised reactor conditions created for the test. Nonetheless, as we will see, the operator errors and misjudgments that occurred could happen elsewhere.

A great deal has been made of the design of the reactor at Chernobyl, which is typical of a number of operating nuclear plants in the Soviet Union designated as the RMBK design (see Fig. 7.15). Indeed, this design has a problem-plagued history. According to reports, at times engineers have had difficulties in controlling dangerous concentrations of reactivity in RBMKs.[17]

The poor ability to control arises in part from the design combination of graphite as a moderating medium and water as a coolant. This combintion leads to a positive feedback effect on reactivity when a burst of heat causes steam *voids* in the water coolant. This comes about since the coolant water in the RBMK acts more as an absorber of neutrons than a moderator. The nuclear fuel mixture in this design has a lower enrichment (about 2 percent $^{235}U$) than a Western-designed LWR (about 3 percent $^{235}U$) and depends on the superior moderating properties of graphite to achieve the chain reaction. Thus, when a void occurs in the water, more neutrons are allowed to penetrate the graphite moderator and, ultimately, reach another fuel rod for fission. The overall affect is to have bursts of chain reaction and heat production, leading to further increases in the chain reaction. A reactor in which this can occur is said to have a *positive void coefficient*.

The RBMK was also known to be susceptible to concentrations of high reactivity occurring at a point where one region of the core is in the supercritical state while the rest remain in the subcritical condition. Some control rods in fact were included in the design to stabilize these localized excursions of reactivity, especially

along the long length of fuel rods in the 7-meter-high core.

Graphite itself has several undesirable features for use in high-temperature applications. First, it burns at temperatures above $700°C$ when exposed to air. Normally, the operating temperature of the graphite is about $600°C$ and the entire core is filled with inert gases, but the potential for combustion remains. Furthermore, it is possible that the combination of steam and carbon will generate hydrogen and carbon monoxide (components of combustible *producer gas* – see Chapter 12). The potential for hydrogen generation is enhanced by an oxidation-reaction of steam with the zirconium tubes in the fuel-rod channels (see Figure 7.15 insert) if the temperatures rise above $1,000°C$.

Ironically, the Chernobyl disaster resulted from a special test to improve safety responses to emergencies. In the test that initiated the disaster, the plant operators were obviously very intent upon achieving the test objectives, so much so that they violated safety rules of operation, several of which were strict prohibitions. Their purpose, generally stated, was to test the ability of the turbogenerator to supply electric power to the station for a brief interval following the cutoff of steam from the reactor. The supply of electricity was important in such an event to trigger the emergency core-cooling system and other engineered safety operations for a loss of coolant accident. There lies the irony of the outcome.

In conducting the test the operators took the RBMK reactor into a very low-power operating condition, where the positive void effect is even more pronounced than near full-power operation. In addition, when they encountered a momentary drop in reactivity, they retracted more control rods than recommended to reach the prescribed conditions of the test. Finally, the operators disabled two automatic reactor trip (RT) functions of the safety system and shut off the emergency core cooling system (ECCS), so that the reactor was not only operating in a highly unstable condition, but also had most of its automatic safety features turned off.

Later analysis by a group of reactor experts, convened by the International Atomic Energy

# Inside the Chernobyl reactor

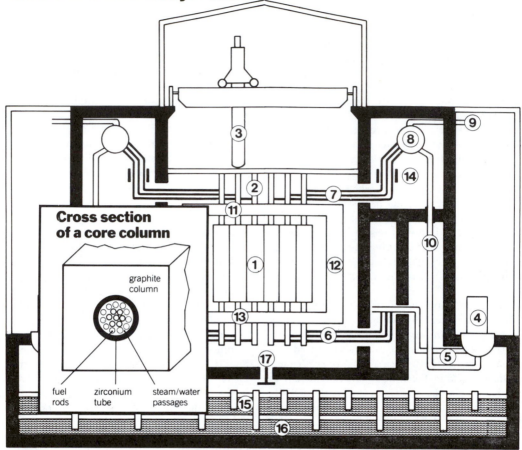

**Cross section of a core column**

graphite column

fuel rods | zirconium tube | steam/water passages

**Legend:**

1. Reactor core, composed of graphite columns (see inset for cross section)
2. Fuel and control rod channels
3. Refueling machine
4. Main circulation pumps
5. Distribution pipes for coolant
6. Coolant inlet pipes for water
7. Coolant outlet pipes for steam and water
8. Drums for steam separation
9. Steam pipes (to turbines)
10. Water return lines (from steam drums to circulation pipes)
11. Upper biological shield
12. Lateral biological shield
13. Lower biological shield
14. Cladding failure detecting system
15. Upper bubbler pool
16. Lower bubbler pool
17. Pressure relief valve

☐ Biological shields ■ Reinforced concrete

The Chernobyl reactor differs from most nuclear reactors in the United States in two respects: it uses graphite to moderate the nuclear reaction in the core (1), and the core itself uses many pressure tubes instead of a single pressure vessel. The inset in the lower right corner of the diagram shows a cross section of one of the 1,661 graphite tubes holding uranium fuel rods. A highly complex system of "control rods" starts, controls, and stops the nuclear reaction (2). A machine (3) can refuel the reactor while it is operating by inserting new fuel rods into the graphite columns. Spent fuel is stored in pools (not shown), as is done at U.S. reactors.

Under normal operating conditions, the controlled chain reaction in the core heats water sent from the main circulation pumps (4) via many distribution pipes (5) to narrow coolant inlet pipes (6), and from there into the steam/water passages within the graphite tubes (see inset). The heat generated by the nuclear reaction is kept under control by the coolant water; it also turns some of the water to steam.

This steam-and-water mixture travels through coolant outlet pipes (7) to tanks which separate the steam from the water (8). Steam from the tanks passes through pipes (9) to power turbines which generate electricity; water separated in the tank passes through pipes (10) to return to the main circulation pumps, where the water cycle begins again.

A series of biological shields (11, 12, 13)– so called because they protect workers from radioactivity–surrounds the core. The space inside the shields is filled with inert gases. The cladding failure detecting system (14) is another protection device which detects radioactivity released into the coolant in the event of leaks in the metal tubes (cladding) around the fuel rods. In addition, if one of the large coolant pipes (located within concrete-walled compartments) should rupture, the escaping steam will be condensed in the bubbler pools (15, 16). Steam released into the reactor compartment can pass through pressure relief valves (17) and also be condensed in the bubbler pools.

**FIG. 7.15.** The Chernobyl reactor – a schematic view. Note in the cross section of the figure that the water coolant flows along the fuel rods within a tube contained inside a graphite moderator block. Reprinted by permission of the *Bulletin of the Atomic Scientists*, a magazine of science and world affairs. Copyright © 1986 by the Educational Foundation for Nuclear Science, 6042 South Kimbark Avenue, Chicago, Illinois 60637.

Agency (IAEA) in Vienna, concluded that a prompt critical *excursion* (where the chain reaction grows and is controlled by prompt neutrons) occurred, causing a steam explosion that ripped open the reactor core. The roof of the containment building collapsed and massive amounts of radioactive debris spewed into the air.[18]

The amount of radioactive material discharged into the atmosphere was, in fact, unprecedented and much larger than most nuclear safety experts had predicted, approaching 100 megacuries (MCi). This included the volatile radioactive isotopes of iodine, tellurium, and cesium, which were carried great distances from the accident. Also, larger amounts than had been expected of heavier isotopes, such as strontium and plutonium, were detected in the fallout. The quantity and composition of the fallout from this accident was of course attributed to its violent nature. The accident at TMI, in contrast, resulted in the discharge of an estimated 15 Ci, because it was a meltdown type accident and was contained. Even if containment had failed at TMI, however, the discharge of radio nucleides would most probably have been a fraction of that at Chernobyl.

The radiation dosage to the surrounding population is depicted in Figure 7.14. About 135,000 people were evacuated from the immediate area (30-km radius) and about 25,000 were estimated to have received doses of at least 35 rems. Vegetables and dairy products were tested throughout Europe, and large quantities were confiscated. Contamination by cesium ($^{137}Ce$), with a half-life of 30 years, may make the immediate area around Chernobyl uninhabitable for decades.

Soon after the accident, experts in the United States and Europe began arguing about the implications of the accident for the future of nuclear energy. Proponents of nuclear power contended[19] that such an accident could not happen in U.S. reactors and that the accident was unique to the RBMK design. Even though these contentions were true for commercial power reactors in the United States, it was soon pointed out that some graphite-moderated reactors are used for military production of plutonium (see Chapter 8). Moreover, nuclear critics did not accept the arguments concerning dissimilarities between U.S. and Soviet reactors, noting that certain aspects of reactor operation are generic to the technology.

For example, all presently operating power reactors worldwide have the same:

1. Chain reaction and kinetics, which make control an exacting task and there is always the possibility of high-power excursions, including *prompt criticalities*, which are remote but cannot be entirely eliminated.
2. Absolute requirement for *heat removal* from a reactor in operation or immediately after shutdown, therefore exacerbating the threat of a LOCA.
3. Possibility of *human error*, carrying with it an irreducible probability of occurrence with a technology that is unforgiving of error.

Some of these characteristics of the technology have been addressed by the industry. Future reactors, for example, can probably be designed to be inherently safe against the LOCA meltdown. Such designs immerse the entire reactor in a coolant at all times, so that emergency cooling is always present and can flow naturally through convection. In one such proposal, called the PIUS (process inherent ultimately safe) reactor, the coolant pond is filled with borated water, which would act as a strong absorber of neutrons in the event of a rupture of the normal coolant/moderator of pure water. Even if such new designs work as predicted and are adopted for future use, the issue of reactor safety will undoubtedly remain, because TMI and Chernobyl reactor types will still be in operation. Worldwide investment in reactors has been enormous (see Chapter 9) and it is unlikely that presently operating reactors will be scrapped for decades.

The U.S. industry has also taken steps to reduce the probability of human error by setting up an operator training program, the Institute of Nuclear Power Operations (INPO). At INPO, an organization established by the industry following TMI, operators are drilled in their responses to various hypothetical emergency situations using simulated reactors. Although this training will doubtlessly improve safety responses at the operating plants, human error can never be eliminated entirely. Indeed, as we have noted, human error determines the lower limit to accident probabilities in this and other technological risks.

### Emergency evacuation

The TMI incident dramatized the need for emergency management of the surrounding population. At the time, some experts and the media raised questions about evacuation and about the need to have thyroid protection pills[20] available if radiation were released. With the massive radiation release at Chernobyl in 1986, of course evacuation became a clear necessity.

Since 1979 the NRC has required emergency plans for all operating or planned nuclear plants. The formulation of such plans (not required at the time of the WASH 1400 study) resulted in heated public controversy at several nuclear plant sites. In the case of the Shoreham plant in New York State, for example, political leaders and citizens groups opposed the Long Island Lighting Company's evacuation plan on the grounds that safe evacuation was impossible because of the geography of Long Island (see pg. 162, Shoreham: Why a Nuclear Plant was Cancelled). A similar situation developed at the Seabrook plant, located on a narrow strip of land on the coast of New Hampshire near the border to Massachusetts.

### Transnational fallout: a global issue

The accident at Chernobyl also focused attention on a global issue of nuclear safety: transnational radioactive fallout. Countries surrounding the Soviet Union recorded high levels of radioactivity. Indeed, the world first became aware of the disaster when ambient levels of radioactivity rose significantly in countries neighboring the Soviet Union. An international uproar followed. Much of the rest of the world criticized the Soviet government for not issuing warnings about radioactive emissions and for limiting information on conditions at Chernobyl.

The Soviets responded to these charges defensively at first. Officials said that the West was using the accident to attack the Soviet system and to undermine political detente. However, soon after, the Soviet government acknowledged the gravity of the issue and became more forthcoming in providing information on the disaster. Leaders also tacitly recognized that earlier news restrictions were due to official embarrassment. In the end, top officials of the nation's nuclear establishment were dismissed for gross breaches of approved procedures and those directly responsible for the disaster were sent to prison.

Once the initial furor over the Soviet response had died down, there was widespread recognition by all sides that nuclear accidents posed an international problem of significant dimension. In that regard, scientific and political leaders declared that radiological accidents needed to be dealt with on a multilateral basis by the entire world community. As Soviet leader Mikhail Gorbachev told the director of the IAEA, nuclear safety had become an "obligation of all states." He suggested that the Soviet Union would accept some – unspecified – supervision of its nuclear facilities by the IAEA.

For its part, at its meeting soon after the disaster, the IAEA tried to begin the process of internationalizing nuclear safety regulation. Though there was considerable agreement about the problem, not much progress was made in adopting new supervisory procedures. As we will discuss in Chapter 8, international supervision of nuclear power has its limitations since it raises challenges to national sovereignty – the right of countries to determine their own courses of action.

But in view of the magnitude of the Chernobyl accident and its transnational spillover effect, experts from many countries hope that an agreement on safety standards and regulation can be implemented. It has also been suggested that an international agency, such as the World Meteorological Agency, monitor radiation levels worldwide to provide quick warning to surrounding nations of a radiological accident. Other ideas may emerge from this debate, but in any case the need for cooperative effort among nations has been recognized, and there is reason for some optimism that countries will cooperate on the question of cross-boundary fallout.

## Criteria for safety

At the heart of the issues of nuclear plant safety, is a question common to many modern technologies: How safe is safe enough? (See Fischhoff, Lichtenstein, Slovic, Derby, and Keeney 1981.)

Nuclear safety is but one of several technological risks of modern society. (As we saw in Chapter 6, another concerns hazards due to pollution from fossil-fired plants.) In Chapter 10 we will discuss criteria for safety and health standards on a more general basis. We will look at risk from the point of view of costs and benefits. Such an approach, though fraught with difficulties, provides the starting point for evaluating the societal benefits of improved safety/health standards, or conversely, the societal costs of lower standards. The economic question, though, overlays a moral dilemma. The standard of safety cannot simply be cost effective, but must have some qualitative standard that must be met. That standard, however, is inevitably difficult to put into a clear, unambiguous form. For example, NRC policy (see *The New York Times*, January 10, 1983) on nuclear plant accident prevention states: "individual members of the public should be provided a level of protection from the consequences of nuclear plant operation such that individuals bear no significant additional risk to life and health." Such a policy is grounded in the concept of *acceptable risk*, which has its origins in the earlier safety studies (for example WASH 1400, which formed the basis of Rasmussen's 1975 report). But of course, there are questions about the acceptability of the risk, its significance, and what measures are required to provide adequate protection. The

controversy surrounding the NRC's policy will likely continue.

Indeed, the controversy may only grow. Not only have critics challenged the nuclear industry on whether it has met the basic goal, they have disputed the formulation of the goal itself. It seems objective, but risk is by definition probabilistic and its acceptability can never be objective. As long as there is room for debate, it is sure to continue to be an issue where the stakes – both in terms of money and risk to health – are so high.

## Technical problems remaining for safety

After the official inquiry by the Kemeny Commission into the TMI accident, the NRC and the U.S. nuclear plant industry began a reexamination of nuclear safety. As part of this evaluation, they reconsidered the NRC's safety goals. They also surveyed the hardware of nuclear plants – equipment and design.

Actually, unresolved hardware issues were not new. The NRC already had a list of unresolved safety issues prior to TMI. But the accident impelled officials to take another look at some basic components. Among others, they investigated PWR steam generator tube integrity, PWR water-hammer vibrations, BWR pipe cracks, and reactor vessel material toughness. Also the NRC studied embrittlement of the reactor vessel, which could cause failure under the stresses of normal operation in older nuclear plants, and the study led to a new consideration of containment integrity (CI).[21]

Soon after TMI, the NRC declared that some of the lingering hardware problems had been solved through the issuance of new regulations and specifications. For example, the NRC increased the standards of quality for welds on the reactor vessel in response to the embrittlement problem.

When such design changes are adopted by the NRC or when it reaffirms existing regulations and specifications, it characteristically does so with the announcement that despite the need for improvement, the prevailing condition poses no undue risk to the public. In other words, the NRC is claiming for itself the right to say how safe is safe enough. This established approach is nonetheless subject to fundamental philosophical questions. But it also raises a practical question: How can the NRC make such reassuring safety assessments when it has clearly failed to recognize problems that the new regulations are designed to overcome? Indeed, the critics of nuclear technology have not always accepted that new regulations actually offer solutions to specific problems much less settle overall safety issues.

## Nuclear waste disposal and weapons proliferation

Aside from the problems of plant safety, there are also unresolved problems having to do with the by-products of the reactors, nuclear wastes.

These wastes result from the fissioning of the uranium fuel in the reactor. Some of these fission products are long-lived and highly radioactive, and they pose a potential danger to the environment and to society for generations. They present a technical dilemma too, both in handling after removal from the reactor and with their ultimate disposal. Disposal requires isolation for many thousands of years, presenting not only technological problems but unprecedented institutional problems as well.

Another critical consideration with nuclear reactor wastes is the potential for diversion to use as weapons. The specific waste product of concern is $^{239}$Pu ( a weapons material), which can be separated from other spent fuel elements by chemical means. These considerations are taken up in more detail in the next chapter.

---

### Shoreham: why a nuclear plant was cancelled

Evacuation plans for a nuclear power plant at times can be exceedingly difficult to formulate. In the case of the Shoreham power plant on Long Island, New York, the creation of an acceptable evacuation plan proved impossible. As a result, the multi-billion dollar plant has not been allowed to open.

The Shoreham facility was planned in the 1960s by the investor-owned Long Island Lighting Company (LILCO). Construction of the 810 MW plant began in 1970. But the building process was a managerial and financial disaster. By the time Shoreham was completed, it had cost $5.5 billion and had taken 20 years to build.

Public opposition to Shoreham was led initially by environmentalists. That changed, however, with the accident at Three Mile Island in 1979. Large-scale anti-Shoreham demonstrations began, and by the next year, the Shoreham Opposition Coalition was formed and given official intervenor status for nuclear regulatory hearings.

But TMI had a different consequence that ultimately led to Shoreham's demise. After the accident, the NRC required evacuation plans for areas within a 10-mile radius of nuclear plants. However, a utility could not simply impose a plan on a community. Instead, any plan had to be formulated with the cooperation and participation of the local and state governments that had the legal authority to protect health and safety. In May 1981, with Shoreham 85 percent complete, the Suffolk County Legislature, the governmental body with jurisdiction for the Shoreham site, was asked to authorize an evacuation plan.

This was not done. Instead, opposition grew, with the evacuation plan suddenly the center of the debate. Because Long Island is long and narrow, a 10-to-20 mile danger zone effectively trapped all of the population living to the east of the plant. Furthermore, the road system of Suffolk County was not suited for a mass evacuation. Government officials soon concluded that evacuation simply was not feasible, and in 1983, both New York and Suffolk County authorities announced that they would not cooperate in the formulation of an evacuation plan for Shoreham.

LILCO, meanwhile, was growing increasingly anxious to get its near-finished plant into operation. Faced with government opposition, LILCO decided to train its own employees to conduct evacuations, and it asked the U.S. government to eliminate the requirement that state and local authorities participate. In 1988, President Reagan obliged (although his executive order raised constitutional questions). LILCO then conducted two drills (involving only its employees, not the public) that the company and federal authorities pronounced successes. However, not everyone agreed with this assessment. One federal emergency official objected so strongly that he resigned.

But even as it was defying state and local officials, LILCO executives began to consider a different course. Burdened by the ongoing costs of building Shoreham, the company was at the edge of bankruptcy and had lost public support. The company appeared willing to abandon the project if the terms were acceptable.

But the terms raised controversy as well. LILCO wanted to retrieve most of its investment through rate increases. While opponents differed in their demands, some not only disputed LILCO's right to any money, they demanded the dissolution of LILCO itself. Control of electric power production and distribution was to be given to a public power authority. In fact, a group constituting itself as the Long Island Power Authority made a bid for all of LILCO's stock.

But LILCO was not dissolved. Instead, negotiations toward a settlement began in early 1984 under the auspices of New York Governor Mario Cuomo; the issue took 5 more years to resolve. In the end, the settlement provided for abandonment of the plant, along with rate increases to restore LILCO's financial position. Groups protested that rate payers were being made to pay for LILCO's misjudgements and poor management, but there was little they could do to change the outcome.

Yet the abandonment of Shoreham was in itself an extraordinary result. A decade earlier it probably would not have occurred. There was no precedent for the cancellation of a multibillion dollar project based largely on people's perceived fears of an accident – accidents that in the 1970s were deemed virtually impossible. But after TMI and Chernobyl, the fear of nuclear accidents no longer seemed irrational.

## Notes

1. Even higher levels of enrichment – close to 10 percent – had to be achieved before the creation of the atomic bomb in 1945. Enrichment is still a hurdle for countries that want to develop atomic weapons.

2. The speeds of thermal motion in a medium are the expected velocities for a particle in the medium at a given temperature as it moves about randomly from one collision to the next. When a neutron moves through a moderating medium, the collisions are mostly with the nuclei of the moderator, such as carbon nuclei in a graphite moderator. The fission neutron starts at about half the speed of light as it enters the moderator in a particular direction. It soon endures several collisions with moderator nuclei, each collision deflecting it in a different direction and slowing it down. After several collisions, the neutron speed is in the *thermal range*, which is a factor of $10^5$ or less of its initial (fission) speed. At such thermal speeds, the neutrons, like the other particles in the medium, move about randomly from collision to collision with speeds that vary about the *average thermal speed* for particles at the given temperature (see Collier & Hewlett 1987 and La Marsh 1983).

3. RAR for these minerals is roughly equivalent to reserves for fossil resources (see Chapter 2),

implying the amounts estimated to be recoverable at or below a given price (here $50/1b in 1979, see Chapter 5).

4. Radiation dosage is measured usually in *rems* (roentgen equivalent man) to quantify the impact of ionizing radiation on humans. It can be specified for the entire human body (a whole-body dose) or for specific organs that are sensitive, such as the thyroid gland. Ionizing radiation, such as alpha, beta, or gamma radiation/particles from nuclear reactions, carries a high microscopic density of energy that is damaging to living tissue. The rem is determined by the energy absorbed microscopically per unit of mass of the absorbing tissue, as adjusted by a quality factor for soft versus hard tissues (bone).

5. Note that the impacts to be discussed here for nuclear technology will parallel those considered for fossil fuel technology in Chapter 6. The two technologies have been compared in these parallel steps in he following study: National Academy of Sciences, 1977, National Research Council, *Implications of Environmental Regulations for Energy Production and Consumption*, Vol. 6, Analytical Studies for the U.S. Environmental Protection Agency, Washington D.C.

6. Accident injury and death rates for uranium miners are estimated to be as low as one-tenth that of underground coal mining. Figures on illnes and death from uranium mining are controversial, however, due to the nature of radiation-related casualties, which must be accounted for statistically as *excess cancers*. The National Academy of Science (1977) estimates of such occupational diseases are about one-fifth the rate for pneumonoconiosis (black lung), the comparable occupational disease for coal mining.

7. 0.02 *excess* cancers per 1,000 MW$_e$ reactor year have been recorded.

8. There was no collaboration of the AEC/NRC-WASH 1400 reactor safety study with the safety programs of the FAA or NASA, which faced similar design concerns. See Ford (1982).

9. Prototype testing on a small (55 MW) PWR was planned by the AEC/NRC. This was called the LOFT (loss of fluid test) and was expected to be held at a site in Idaho. But the test was not carried out by the time of the nuclear accident at Three Mile Island in 1979, after which the NRC decided to reevaluate the entire nuclear safety program. Subsequently, NRC officials failed to agree on the need for, and the role of, prototype testing, and LOFT was not rescheduled.

10. This means 1 in 100 years for one reactor. If we consider 100 reactors, it is 1 per year.

11. In Ford (1982), we are given a vivid picture, documented by files from the AEC/NRC, of a bureaucracy prejudging the conclusions of the scientific study that it was charged to carry out.

12. A *reference accident* was originally defined as a major accident, but not with catastrophic release of radioactive materials. More recently, this class of accident has simply been termed a meltdown.

13. These effects damage the body and/or degrade bodily functions.

14. Government groups included the EPA. Prominent outside critics included Professor Robert Pohl (Cornell University), Professor Frank Von Hippel (Princeton University), and a group called the Union of Concerned Scientists (UCS) led by Professor Henry W. Kendall (Massachusetts Institute of Technology).

15. In common mode failures, one event (or failure) in the event tree depends on another, thus violating the assumption of independent (multiplicative) probabilities. The ESS designs presumably took meticulous care to ensure independence of function for every step, and yet an accidental fire at the Brown's Ferry nuclear plant on March 22, 1975, uncovered common mode dependencies that drastically altered the fault tree probabilities of that plant.

16. President Carter's Commission on the Accident at Three Mile Island was headed by John G. Kemeny (former president of Dartmouth College) and termed the Kemeny Commission. It investigated the causes of the accident and made recommendations for the prevention of recurrences.

17. The RMBK reactor core is physically large (30 m × 40 m) and has a large inventory of fissionable material in its 1,661 fuel rods. It is possible for one region of the core to go supercritical, creating a hot spot. Overheating or partial meltdown in such a region can be more difficult to control than a uniform condition over an entire core of smaller dimensions.

18. These conclusions about the sequence and cause of the accident were, and must remain, somewhat qualified. Scientists were unable to inspect the remains of the Chernobyl plant closely because of the extremely high radiation levels. Soon after the accident, Soviet authorities decided to entomb the plant, making inspection of the reactor impossible.

19. In the weeks following the Chernobyl accident, nuclear industry groups, such as the Atomic

Industrial Forum, the Edison Electric Institute, and the American Nuclear Society, issued public statements claiming that U.S. nuclear plant safety systems and radiation containment were superior to those in the USSR. These statements all implied, some more strongly than others, that such an accident could not happen here. More caution in drawing such conclusions, however, was evident in expert opinions expressed by scientists at the Brookhaven National Laboratory and the International Atomic Energy Agency. Opinions on both sides were to be found among commissioners of the USNRC. (See *The New York Times*, May 19, 1986.)

20. Iodine tablets prevent the absorption into the thyroid gland of radioactive iodine from the contaminated water and food supplies.

21. *The New York Times*, September 2, 1981.

# Bibliography

## Books

Collier, J. G., and G. F. Hewitt. 1987. *Introduction to Nuclear Power*. Hemisphere Publishing, New York.

Fischoff, B., S. Lichtenstein, P. Slovic, S. Derby, and R. Keeney. 1981. *Acceptable Risk*. Cambridge University Press, Cambridge.

Ford, D. 1982. *The Cult of the Atom – The Secret Papers of the Atomic Energy Commission*. Simon & Schuster, New York.

Ford, D. (Executive Director). 1975. *The Nuclear Fuel Cycle*. Union of Concerned Scientists, MIT Press, Cambridge, MA.

Grenon, M. 1981. *The Nuclear Apple and the Solar Orange – Alternatives in World Energy*. Pergamon, New York.

Kaku, M., and J. Trainer. 1982. *Nuclear Power: Both Sides – The Best Arguments For and Against the Most Controversial Technology* Norton, New York.

Keeny, S. M. (chairman). 1977. *Nuclear Power Issues and Choices*. Ballinger, Cambridge, MA.

Lamarsh. J. R. 1983. *Introduction to Nuclear Engineering, Second Edition*. Addison-Wesley, Reading, MA

Lipschutz, R. D. 1980. *Radioactive Waste – Politics, Technology and Risk*. Ballinger, Cambridge, MA.

Penner, S. S. (ed.). 1976. *Energy, Nuclear Energy and Energy Policies, Vol. III*. Addison-Wesley, Reading, MA.

Perrow, C. 1984. *Normal Accidents – Living with High-Risk Technologies*. Basis Books, New York.

Stobaugh, R., and D. Yergin. 1983. *Energy Future, Third Edition*. Vintage Books, New York. (See Chapter 5, Nuclear Power: The Promise Melts Away.)

Segre, E. R. 1965. *Collected Papers of Enrico Fermi*, Vol. 2. University of Chicago Press, Chicago.

Weart, S. 1988. *Nuclear Fear – A History of Images*. Harvard University Press, Cambridge, MA.

## Reports

Kemeny, J. G. (chairman). 1979. *The Accident at Three Mile Island – Report of the President's Commission on TMI*. U.S. Government Printing Office, Washington, DC., October.

*Reviews of Modern Physics*. 1975. Vol. 47, Supplement No. 1, "Report to the APS by the Study Group on Light-Water Reactor Safety," Summer. (contains critique of the AEC report WASH 1400).

## Magazines

*Bullletin of the Atomic Scientists*.
   J. Beyea and F. Von Hippel. "Containment of a Reactor Melt-down." Aug./Sept. 1982, pp. 52–9.
   R. J. Lifton. "Nuclear Energy and the Wisdom of the Body." September, 1976, pp. 16–20.

*IEEE Spectrum* (published monthly by the Institute of Electrical and Electronic Engineers).
   "Analyzing Malfunctions of Nuclear Power Plants." June 1983, pp. 53–58.
   "Finding the Flaws in Nuclear Power Plants." September 1982, pp. 55–60.
   "The Puzzle of Chernobyl." July 1986, pp. 34–41.
   N. Rasmussen. "Rasmussen on Reactor Safety." August 1975, pp. 46–55 (based on WASH 1400).
   Special Issue. "Three Mile Island and the Future of Nuclear Power." November 1979.

*Nucleas*.
   C. Komanoff and E. E. Van Loon, 1982. "Too Cheap to Meter or Too Costly to Build: How Nuclear Power Has Priced Itself Out of the Market". A Report to the Union of Concerned Scientists, Sponsors, Vol. 4, No. 1, pp. 3–7.

*Science* (weekly publication of the American Association for the Advancement of Science, AAAS).
   F. R. Mynatt. "Nuclear Reactor Safety Research Since Three Mile Island." April 9, 1982, Vol. 216, pp. 131–135.
   News and Comments. "Assessing the Effects of a Nuclear Accident." April 5, 1985, Vol. 228, pp. 31–3.

News and Comments. "Ultra Safe Reactors, Anyone?" January 21, 1983, Vol 219, pp. 265–67.

J. J. Taylor. "Improved and Safer Nuclear Power." April 21, 1989, Vol. 244, pp. 318–25.

*Technology & Culture.*

J. G. Gunnell. "The Technocratic Image and the Theory of Technocracy." July 1982, Vol. 23, No. 3, pp. 392–416.

*Technology and Society* (magazine published quarterly by the IEEE Society on Social Implications of Technology). Spcial Issue. Debate on Nuclear-Electric Power Generation. March 1984, Vol. 3, No. 1.

# 8

# The nuclear fuel cycle

## Introduction

Nuclear fuel undergoes a cycle of extraction, preparation, use, and disposal. Throughout the course of that cycle there are hazards that threaten health and property and that, in some instances, present society with enormous social and ethical questions. Fuel cycle problems have two key aspects. First, the hazards of nuclear fuel fabricated today are not only potentially great, they will persist for generations to come. Second, the same processes of the nuclear fuel cycle (and many of the elements generated in that cycle) can be used in the making of nuclear weapons.

A diagram of the steps in the nuclear fission fuel cycle for commercial power plants is shown on Figure 8.1. In the cycle of uranium use, we can distinguish the following steps:

1. Mining of uranium ore in the form of uranium oxide ($U_3O_8$).
2. Conversion to gaseous form (Uranium hexa-fluoride or $UF_6$).
3. Enrichment of the $^{235}U$ concentration.
4. Fabrication of the processed uranium into reactor fuel rods.
5. Reactor use for power-producing fission.
6. Storage (and possibly reprocessing) of the spent reactor fuel, including uranium, plutonium, and other radioactive waste products.[1]

7. Disposal of radioactive and fissile products of spent fuel.

Some of these steps were discussed in Chapter 7. The environmental impact of the nuclear plant, and the front-end steps (1–5) of the cycle were the prime concern in that chapter. Here we will examine the back end of the cycle, the question of what happens to spent fuel that comes out of the reactor. We will consider, too, the dangers the waste poses, how the problems might be solved, and how proposed breeder reactor technology might increase the back- (and front-) end problems enormously.

## Management and disposal of nuclear wastes

Nuclear reactor (LWR) fuel is considered *spent* (or used up) when about 75 percent of the $^{235}U$ has been fissioned. During its cycle of use as a fuel, however, uranium (both $^{235}U$ and $^{238}U$) produces many other elements through fission and other transmutation processes. One of these elements, plutonium, can itself fission and release useful energy, but the remainder of the reactor products are nonfissionable and contribute nothing more to power generation.

Yet many of these products are extremely toxic and their handling, storage, and ultimate disposal are difficult. Not only are they highly

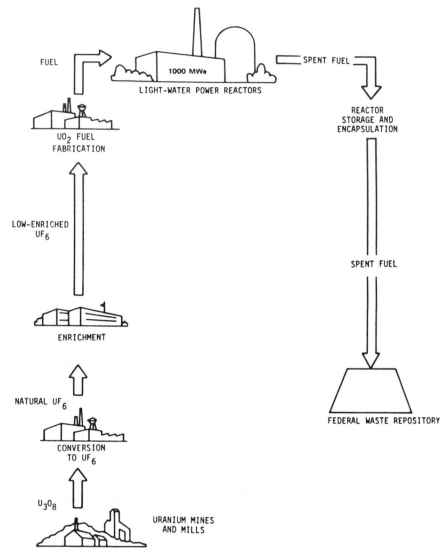

**FIG. 8.1.** The nuclear fuel cycle. *Source:* U.S. Nuclear Regulatory Commission, 1976, NUREG – 0016, October Washington, DC.

radioactive, but they continue to generate heat. Both the reactivity and the heat decline through the processes of radioactive decay. However, the time required for decay varies greatly among the various elements found in spent fuel. Tables 8.1 and 8.2 show the various fission products and actenides[2] that would be typical for a ton of spent LWR fuel. The radioactive half-lives (the decay times to one-half the radioactive levels) of the elements shown range from 285 days to 17 million years, with the important fissionable plutonium isotope $^{239}P$ reaching its half-life only after 24,400 years, a time greater than all of recorded human history. Table 8.1

**Table 8.1.** Typical yields of actinides[a]

| Element | Half-life (yr) | Decay interval[b] | | | | | | | | | |
|---|---|---|---|---|---|---|---|---|---|---|---|
| | | (q) 0.3 yr | (Ci) 0.3 yr | (q) 10 yr | (Ci) 10 yr | (q) 500 yr | (Ci) 500 yr | (q) 10,000 yr | (Ci) 10,000 yr | (q) 100,000 yr | (Ci) 100,000 yr |
| $^{237}$Np | $2.1 \times 10^6$ | 760 | 0.59 | 760 | 0.59 | 786 | 0.61 | 810 | 0.63 | 790 | 0.61 |
| $^{238}$Pu | 86 | 5.8 | 105 | 5.5 | 100 | 0.1 | 1.8 | — | — | — | — |
| $^{239}$Pu | 24,400 | 27.5 | 1.7 | 27.5 | 1.7 | 32 | 2.0 | 59 | 3.6 | 4.5 | 0.3 |
| $^{240}$Pu | 6,580 | 8.5 | 2.0 | 19.2 | 4.5 | 38.4 | 8.8 | 13.9 | 3.2 | — | — |
| $^{241}$Pu | 13.2 | 4 | 464 | 2.4 | 273 | — | — | — | — | — | — |
| $^{242}$Pu | 379,000 | 2 | 0.009 | 2 | 0.009 | 2 | 0.009 | 2 | 0.009 | 1.7 | 0.007 |
| $^{241}$Am | 462 | 54 | 189 | 55.6 | 198 | 29.5 | 103 | — | — | — | — |
| $^{243}$Am | 7,370 | 82 | 17.0 | 82 | 17.0 | 77 | 16.0 | 31 | 6.5 | — | — |
| $^{244}$Cm | 17.6 | 30 | 2,570 | 19.7 | 1,700 | — | — | — | — | — | — |
| Total grams/metric ton fuel | | 974 | | 974 | | 965 | | 916 | | 796 | |
| Actinides, including daughter products (approx.) | | | 200,000 | | 10,000 | | 900 | | 100 | | 50 |

[a]Includes intemediate and long half-lives in the waste stream from processing 1 metric ton of light water reactor fuel irradiated to $33 \times 10^9$ watt-days (thermal) per metric ton of fuel.

[b]g = grams per metric ton of fuel. Ci = Curies

*Source:* Reproduced with permission from *The Nuclear Fuel Cycle*, The Union of Concerned Scientists, The MIT Press, Cambridge, MA, 1975.

**Table 8.2.** Typical yields of fission products[a]

| Element | Half-life | Curies remaining | | | |
|---|---|---|---|---|---|
| | | 10 yr | 100 yr | 500 yr | 1000 yr |
| $^{144}$Ce/$^{144}$Pr | 285 days | 300 | — | — | — |
| $^{106}$Ru/$^{106}$Rh | 367 days | 1,100 | — | — | — |
| $^{155}$Eu | 1.8 yr | 160 | — | — - | — |
| $^{134}$Cs | 2.1 yr | 8,300 | — | — | — |
| $^{125}$Sb/$^{125}$Te | 2.7 yr | 980 | — | — | — |
| $^{90}$Sr/$^{90}$Y | 28.1 yr | $1.2 \times 10^5$ | $1.32 \times 10^4$ | 0.6 | — |
| $^{137}$Cs/$^{137}$Ba | 30 yr | $1.6 \times 10^5$ | $2.1 \times 10^4$ | 2 | — |
| $^{151}$Sm | 90 yr | 1,100 | 520 | 30 | 0.4 |
| $^{99}$Tc | $2.1 \times 10^5$ yr | 15 | 15 | 15 | 15 |
| $^{93}$Zr | $9 \times 10^5$ yr | 3.7 | 3.7 | 3.7 | 3.7 |
| $^{135}$Cs | $2 \times 10^6$ yr | 1.7 | 1.7 | 1.7 | 1.7 |
| $^{107}$Pd | $7 \times 10^6$ yr | 0.013 | 0.013 | 0.013 | 0.013 |
| $^{129}$I | $17 \times 10^6$ yr | 0.025 | 0.025 | 0.025 | 0.025 |
| Total curies (approx.) | | 300,000 | 35,000 | 53 | 22 |

[a]Includes intermediate and long half-lives from processing 1 metric ton of light-water reactor fuel irradiated to $33 \times 10^9$ watt-days (thermal) per metric ton of fuel.

*Source*: Reproduced with permission from *The Nuclear Fuel Cycle*, The Union of Concerned Scientists, The MIT Press, 1975.

includes the amount of radioactivity measured in curies (Ci)[3] for each element over time intervals as long as 100,000 years. (Thermal power decay over time per ton of spent fuel is shown graphically in Fig. 8.2a.)

If during this tremendous time span, there is an uncontrolled release into the biosphere of these waste materials, the public will be endangered. The hazard may be measured by the relative toxicity of the wastes compared to

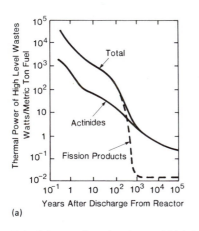

(a)

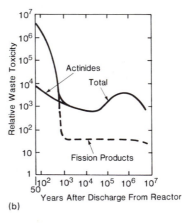

(b)

**FIG. 8.2.** Radioactive decay of high-level nuclear waste from light-water reactors. (a) Thermal power per ton of reactor wastes versus years after discharge.

(b) Relative toxicity (compared to uranium ore) versus years after discharge. Adapted from Ford (1975).

the toxicity of uranium ore, which is radioactive in its natural state. We see in Figure 8.2b that it takes about 1,000 years for the reactor waste to decay in toxicity from over a million times to a few thousand times that of natural uranium ore. After 1,000 years, the decay never drops much below a relative factor of 1,000. Indeed, it actually rises again after 1,000,000 years or so because of the spontaneous creation of *daughter elements* from the radioactive decay process.[4]

At the present time, countries operating nuclear plants temporarily store, rather than permanently dispose of, wastes. This situation exists because there are substantial doubts about the feasibility of the schemes proposed to date for permanent, or long-term, disposal. In the United States, for example, all spent fuel is stored at reactor sites in ponds that resemble swimming pools (the ponds absorb the radiation and dissipate the decay heat). There could be alternative interim methods of dealing with waste; for example, it could be stored in central depots or spent fuel could be reprocessed to take out fissionable elements for use again in fuel rods. But as wastes accumulate from the operating reactors, the basic dilemma of long-term disposal grows.

Long-term disposal must meet acceptable criteria of safety, given the dangers of accidental radioactive releases into the environment. Due to the extraordinarily long lifetimes of fission products, the disposal of these wastes presents unprecedented technological and societal problems. Technologically, the method of disposal must ensure a high degree of isolation for many thousands of years, thus requiring containment materials and disposal facilities that are known to be stable for periods of this magnitude. Furthermore, unless the disposal is terminal (i.e., requires no further human intervention), society may be faced with the problem of guarding or monitoring the wastes. But it would have to do so for a span of time beyond that of any civilization in history.

U.S. government policymakers – Congress and the Department of Energy – have turned to the concept of *geologic isolation* as the sole objective for radioactive waste disposal. They have adopted a policy to achieve reliable, long-term safety against accidental radioactive release of the most dangerous nuclear wastes called *high-level wastes* (HLW).[5] If they can do this, then their obligation to public safety will be met. By geological isolation, they mean the placement of wastes in rock or sedimentary formations in the earth's crust that have remained intact – that is, have not been subject to seismic or tectonic movements – for millions of years.

Such formations exist both on land and beneath the oceans, although how easy it is to identify and use them remains unclear. In the United States, for example, critics have argued that sites under consideration as disposal repositories may not be as stable as the government has claimed. Nevertheless, the plan calls for the placement of wastes deep within formations through superdeep drilled holes or in mined vaults, or by rock melting. This last technique would use the decay heat of the wastes themselves to melt the rock until the wastes were deep within the earth.

Still, attempts to develop an acceptable disposal technique have proven difficult. The continued generation of decay heat may cause unstable molten conditions in some of the disposal media. There have been concerns that these conditions might lead to rock fractures that in turn could permit migration of radioactive materials into groundwater. Questions remain, too, about the containment of wastes. As of the mid-1980s, U.S. planners were considering placement of HLW in canisters or in solid vitrified form. But there has been debate among experts over the ability of these containment techniques to withstand chemical activity, pressures, and temperatures for so long a time period (see Lipschutz 1980).

The problems of long-term disposal are greatly magnified when we consider not just the *accidental* release, but also the *intentional* release of radioactive material. By intentional release we mean the retrieval of wastes for the fabrication of weapons. It might seem unlikely that wastes placed deep in the ground could be easily recovered, but there is no reason to believe that over the thousands of years involved, retrieval could not be simplified. Wastes could be used to create explosive nuclear devices, or they could be fabricated more easily into radiological weapons, where the release and diffusion of toxic particles – such as through an aerosol sprayer – could lead to widespread death and social upheaval.

The most dangerous substance in HLW is plutonium ($^{239}$Pu). Unfortunately, separation of plutonium from spent fuel rods is a relatively straightforward chemical procedure, and plutonium is produced constantly in nuclear reactors. About 8 kilograms of $^{239}$Pu is bred in a typical LWR during every month of operation. A few micrograms of plutonium ingested or inhaled by a human are fatal; 15 kilograms or less may be sufficient to fabricate a crude fission bomb. Because plutonium is fissionable, some have suggested reprocessing spent fuel rods to extract the plutonium for reuse in reactors. However, reprocessing would also present an additional short-term danger: large quantities of separated plutonium that could be readily assembled into a weapon.

An alternative fuel cycle, based on thorium, has been proposed as a way out of the plutonium problem. As we noted in Chapter 7, plutonium is bred from nonfissile $^{238}$U in the process of $^{235}$U fission reactions. Thorium, which also exists naturally as an ore and is as abundant as uranium, can be used in a fuel cycle that does not leave a potential weapons material in its waste products. The thorium cycle would breed fissionable $^{233}$U for a reactor fuel, but $^{233}$U fuel can be protected against diversion for weapons in the short as well as the long term by a process called *denaturing*. Denaturing is a form of dilution that would require an entire process of enrichment to overcome. Thus, spent thorium fuel could be left in a state equivalent to natural uranium ore and would not be readily useful in the production of weapons. There is no movement in the world toward the thorium cycle, however, and even if there were, the problem of disposal of plutonium (and other HLW) would not be solved. As of the late 1980s, in the United States alone more than 10,000 metric tons of spent fuel sat in commercial nuclear plant storage pools in need of permanent disposition.

In order to defeat intentional retrieval at any time in the remote future, long-term disposal must have what Gene I. Rochlin has referred to as a *high irreversibility*,[6] that is, the disposal method must have slim prospects for retrieval. According to his concept, disposal techniques may range from more or less irreversible up to complete irreversibility. At least two disposal techniques would be irreversible or essentially so: the disposal of wastes in outer space and the transmutation of waste elements into more benign elements. The first is not a serious option; no responsible official would risk the dispersal of massive amounts of radioactivity into the atmosphere due to the failure of a rocket booster, especially after the *Challenger* accident.

A process of transmutation is possible, in principle, using particle accelerators or nuclear fusion reactors, in which nuclei are bombarded by high-energy particles. The most fruitful of such transmutation operations would be on the long-lived actinides in the spent fuel, which would first require a separation process (*partitioning*). However, at present and for the foreseeable future, this option is not likely to be practical technologically, since it is apt to consume more electric energy than the original reactor fuel produced.

That leaves the option of remote geological isolation. This plan, while most feasible, can never be completely irreversible. Various degrees of irreversibility can be achieved, however, according to Rochlin (1977). He rates deep-rock-drilled emplacement (and subsequent melting from decay heat) on land and deep-sea-bed emplacement as "high in irreversibility."

The greater the irreversibility of retrieval that society chooses for disposal, the greater will be both the technological requirements and the cost. Deep-sea-bed entombment, for example, would be far more difficult and more expensive than land-based burial. Yet the evaluation of the total costs and benefits of any waste-disposal technique presents a problem beyond the capacity of any current methodology. Certainly standard cost-benefit calculations, such as those we will discuss in Chapter 10, are essentially useless to the waste-disposal issue because the ultimate costs – including those of accidental or intentional release – are unknown and would be incalculable anyway given the timescale involved (see Grossman & Cassedy 1985). But at the same time, the possible costs of an accident cannot be ignored. Indeed, they are so monumental in terms of potential human suffering, that they must be given highest priority from an ethical standpoint.

Since the impacts may come millennia in the future, we are dealing with an ethical issue of vast intergenerational dimensions – one where the very societies that may be affected are unimaginable to us. This only compounds our dilemma, in deciding *to what lengths present-day society should go to limit the risks to its descendants thousands of years into the future.*

We should note here that U.S. government policy states that technological projects should not endanger future generations any more than they endanger the present. A few philosophers have argued that we actually have a greater obligation to protect the future since they will not have had a say in choosing the risks we "export" to them. Yet the argument has another side, that is, since we cannot know the thinking and the character of future generations, we are being presumptuous in attempting to choose for them at all. It is possible, for example, to imagine that the future would want readily retrievable wastes to use as fuel for power production, and thus irreversibility would be undesirable from the standpoint of the future. What is more to the point, we really cannot know what would be best for them. As a consequence, according to this view, the best we can do is make the most ethically acceptable choices for ourselves and our children, and let the future make its own decisions.[7]

If our society decides to go to great lengths to protect even the generations of the not-so-distant future, we will have to pay more now. The cost will be reflected in the overall generating cost of nuclear power, because disposal will become part of the cost of the power. This is not always so explicit in the debate over waste, which tends to focus on safety. But safety has a cost that in this case will affect the cost of the energy we use.

## Nuclear energy – peaceful and nonpeaceful uses

As we stated at the outset, although intended for peaceful purposes, nuclear power technologies are tied inextricably by physics with nuclear weapons technologies. As a result, the world is confronted with the awesome threat of nuclear weapons proliferation.

The chain reaction described in Chapter 7 is the basic process for both nuclear power and nuclear weapons. This holds true at two levels:

1.  The chain reaction is the means by which energy is released for both peaceful and violent purposes.
2.  The chain reaction is a prime means of breeding (see Chapter 7) plutonium for weapons or nuclear fuel.

Historically, of course, nuclear fission was first applied in the development of the atomic bomb during World War II. After the first experimental proof of the chain reaction,[8] U.S. physicists worked on developing the means of suddenly assembling within a bomb a super-critical mass to create a rapidly growing chain reaction and a violent release of energy. They used two different elements – enriched uranium ($^{235}U$) or plutonium ($^{239}Pu$) – which had previously been bred in fission reactors.

They demonstrated in 1945 that either element can be used in fabricating weapons and later that the same chain reaction in its controlled form could produce nuclear power. In other words, current processes of nuclear power and its fuel cycle are physically indistinguishable from those of nuclear fission weapons. Inevitably, then, in the process of spreading nuclear power around the world, we increase the prospect for nuclear weapons proliferation as well.

Both weapons and power depend on having fissionable isotopes (either $^{235}U$ or $^{239}Pu$) in sufficient concentration to sustain a chain reaction. The achievement of separating the fissionable isotope in sufficient concentration was the *technological hurdle* to be overcome during the production of the atomic bomb.

In Chapter 7, we briefly discussed the enrichment of natural uranium ore. Natural concentrations of $^{235}U$ in uranium are too low to sustain a chain reaction in an LWR and most constituent $^{238}U$ is nonfissionable. The $^{235}U$ concentration can be raised by enrichment through a process called diffusion or through other high-technology means. The diffusion process, which was first developed on a mass scale at Oak Ridge, Tennessee, during World War II, represents a major industrial effort of great sophistication that few other countries have duplicated. The enrichment process remains a technological hurdle to any country that is trying to achieve a chain reaction with uranium.

Actually, there is an alternative to enrichment. The CANDU or heavy water reactor permits a chain reaction with natural uranium ore. The deuterated water achieves its moderation of neutron energies more efficiently by capturing fewer neutrons, thereby leaving more to participate in the fission chain reaction.[9] But the elimination of the enrichment process does not solve the problem of weapons proliferation, because heavy water reactors as well as LWRs breed plutonium. Since the separation of plutonium from spent reactor fuel is a relatively simple process,[10] a country would have no further technological hurdles to obtain *weapons-grade* materials (that is, materials with high concentrations of fissionable nuclei).

In World War II, Germany attempted to build a fission weapon by producing heavy water to breed plutonium. Had the Axis powers overcome this hurdle first, they would have had the means to produce sufficient plutonium for weapons and conceivably could have won the war.

Today, either of these two technologies, enrichment or heavy water, is a key to nuclear energy for either peaceful or nonpeaceful purposes. And several developed nations – France, Great Britain, the United States, and the USSR – have the capability to manufacture both nuclear power reactors and nuclear weapons. A few developing nations, notably China and India, have exploded nuclear weapons, yet find it necessary to purchase commercial power reactors from developed countries. Still other countries, such as Libya, Israel, Iraq, Iran, Pakistan, South Korea, South Africa, and Taiwan, are thought to be nuclear weapons aspirants. Finally, several developed countries – Canada, West Germany, Italy, Switzerland, the Netherlands, and Sweden – have highly developed nuclear capabilities, some even for export, but have so far chosen not to manufacture nuclear weapons.

The acquisition of nuclear technology by a nation appears to have a variety of motivations (see Greenwood, Feiveson, & Taylor 1977). In some instances, it seems to be merely a question of national prestige; in others, nations are spurred by concerns over national security needs or economic well-being. This appears to be true both for weapons and power reactors. For example, a country without fossil fuel resources might feel its independence threatened by too much reliance on oil and coal exporters. Whatever the motivation, the proliferation of nuclear facilities, and by implication the potential for weapons creation, has alarmed the international community.

In 1968, following several years of negotiations and debate, 89 countries signed the Nuclear Non-Proliferation Treaty (NPT). The NPT includes the major nuclear powers and now includes over 125 signatories. Each nuclear weapons nation that signs the treaty agrees not to supply weapons or the means to fabricate them to nonnuclear nations and all nonweapons states agree not to receive or acquire such devices or technology. In order to monitor compliance with the treaty, all signatory states have agreed to a system of international inspection of nuclear facilities.

This inspection system has been set up by the International Atomic Energy Agency (IAEA) with the following objectives: "Timely detection of diversion of significant quantities of nuclear materials from peaceful nuclear activities to the manufacture of nuclear weapons... and deterrence of such diversion by early detection."

The key phrase in this statement is *timely detection*. Timely detection refers to a country's imminent ability to assemble a critical mass of fissionable material. The imminent capability of a country or clandestine organization to do so depends on whether it has overcome the major *technological hurdles* to creating a weapon. If, for example, it has uranium in enriched concentrations or reprocessed plutonium, then the only remaining question is whether it has sufficient weapons grade material to fabricate one or more weapons. How much is the significant amount that is required?

1. For plutonium ($^{239}$Pu)
   a. Ideally, it would take only 4 kilograms of pure $^{239}$Pu inside a neutron-deflecting shell of beryllium to make a critical mass. The diameter would be the size of a baseball.
   b. More practically, the presence of fission poisons ($^{240}$Pu) requires a larger mass, but the presence of oxygen creates pressures that somewhat reduce the required mass. Thus, 15.7 kilograms of plutonium dioxide ($PuO_2$) is a critical mass.
   c. The IAEA considers 8 kilograms of $^{239}$Pu a "significant amount."
2. For highly enriched uranium ($^{235}$U)
   a. An ideal (100 percent enriched) uranium would require 15 kilograms for a critical mass and would be the size of a softball.
   b. At 20 percent enrichment of $^{235}$U, the critical mass of the uranium mixture is 250 kilograms.
   c. The IAEA considers 25 kilograms of 25 percent $^{235}$U uranium a "significant quantity."
3. For uranium $^{233}$U, the fissionable isotope of thorium, the IAEA considers 8 kilograms a "significant quantity."

The scope of the IAEA's safeguarding responsibilities may be appreciated by referring to Tables 8.3 and 8.4. Note that the IAEA is responsible for over 700 nuclear installations, including enrichment facilities and reprocessing

**Table 8.3.** Nuclear installations under IAEA safeguards or containing safeguarded material

| Nuclear installation | End of 1980 | | |
|---|---|---|---|
| | NPT | Non-NPT | Total |
| *Facilities* | | | |
| Power reactors | 103 | 24 | 127 |
| Research reactors and critical assemblies | 147 | 28 | 175 |
| Conversion plants | 3 | 1 | 4 |
| Fuel fabrication plants | 31 | 7 | 38 |
| Reprocessing plants | 4 | 3 | 7 |
| Enrichment plants | 4 | 0 | 4 |
| Separate storage facilities | 15 | 6 | 21 |
| Other facilities | 40 | 0 | 40 |
| Subtotal | 347 | 69 | 416 |
| *Other locations* | 340 | 18 | 358 |
| Total | 687 | 87 | 774 |

*Source*: 1980 Annual Report of the IAEA. Reprinted by permission of the *Bulletin of the Atomic Scientists*, a magazine of science and world affairs. Copyright © 1986, 1982 by the Educational Foundation for Nuclear Science, 6042 South Kimbark Avenue, Chicago, Illinois 60637, USA.

operations. As we can see in Table 8.5, several tons of separated plutonium and highly enriched uranium, both of which are weapons-grade materials, are included in the international safeguard system. The major concern exists with the bulk-handling facilities (Table 8.4), where most of the material resides.

The basic approach of the IAEA is to verify quantities of nuclear materials at each facility it monitors. A system of material accounting is operated by each participating country. The IAEA maintains a running evaluation of

**Table 8.4.** Representative bulk-handling facilities under IAEA safeguards in 1979[a]

| Country | Abbreviated name | Location | Type |
|---|---|---|---|
| Argentina | Pilot Fuel Fabrication Plant (HEU) | Constituyentes | Fuel fabrication |
| Belgium | Belgonucleaire-BN-MOX | Dessel | Fuel fabrication |
| Brazil | Resende Fuel Fabrication Plant | Resende | Fuel fabrication |
| Canada | ENL Port Hope | Port Hope | Conversion |
| | Westinghouse Fuel Fabrication Plant | Varennes | Fuel fabrication |
| | Metallurgy | Chalk River | Fuel fabrication |
| China, Republic of | INER Fuel Fabrication Plant | Lung Tan | Fuel fabrication |
| | INER Uranium Conversion Pilot Plant | Lung Tan | Conversion |
| Denmark | Metallurgy Department | Risø | Fuel fabrication |
| Germany, Federal | NUKEM | Wolfgang, Hanau | Fuel fabrication |
| Republic of | GWK-WAK | Leopoldshafen, Karlsruhe | Reprocessing |
| | Uranit | Jülich | Enrichment |
| India | Nuclear Fuel Complex | Hyderabad | Fuel fabrication |
| | PREFRE | Tarapur | Reprocessing |
| Italy | Fabnuc-Bosco Marengo | Alessandria | Fuel fabrication |
| | EUREX | Saluggia | Reprocessing |
| Japan | PNC Reprocessing Plant | Tokai-Mura | Reprocessing |
| | NFI (Kumatori-1) | Kumatori, Osaka | Fuel fabrication |
| | SMM (Tokai-1) | Tokai-Mura | Conversion |
| Netherlands | Ultra-Centrifuge | Almelo | Enrichment |
| Spain | Metallurgical Plant | Madrid | Fuel fabrication |
| | Juan Vigon Research Center | Madrid | Reprocessing |
| Sweden | ASEA-ATOM | Västerås | Fuel fabrication |
| United Kingdom | PFR Reprocessing Plant | Dounreay | Reprocessing |

[a]Conversion plants, fuel fabrication plants, enrichment plants, and chemical reprocessing plants and pilot plants with an annual throughput or inventory exceeding 1 effective kilogram.

discrepancies in these accountings, called material unaccounted for (MUF), which would indicate a possible diversion for weapons purposes. Ideally, the system works to produce timely warnings that a significant amount of potential weapons material is unaccounted for. (A "timely warning" to the international community assumes that it will take a minimum of 7 to 10 days to fabricate the materials into an explosive device.)

Critics have challenged the assumption that the NPT safeguards provide effective assurance against proliferation of nuclear weapons. On technical grounds, they claim the IAEA safeguards are ineffective because of omissions (see *Bulletin of Atomic Scientists*, October 1981 and December 1988). For example, it does not inspect uranium ore processing facilities. Some have also taken the IAEA to task for its statistical methodology. It has been charged that incomplete measures do not allow, logically, the conclusion that the absence of statistical evidence of MUF means that *no* diversion has taken place.

Critics have also pointed out institutional and political problems with the NPT. These relate to the overriding issue of national sovereignty. Thus, a host country can delay on-site inspections to defeat timely warning. More important, countries determine which

**Table 8.5.** Quantities of nuclear material under IAEA safeguards in nonnuclear weapon states (in tons)

|  | 1975 | 1976 | 1977 | 1978 | 1979 | 1980 |
|---|---|---|---|---|---|---|
| *Plutonium* | | | | | | |
| Separated | 2 | 3 | 6 | 7 | 8 | 5 |
| In irradiated fuel | 15 | 23 | 30 | 44 | 60 | 78 |
| Total | 17 | 26 | 36 | 51 | 68 | 83 |
| *Uranium* | | | | | | |
| Enriched to 20 percent or more | 4 | 5 | 11 | 11 | 11 | 11 |
| Enriched to less than 20 percent | 3,091 | 3,613 | 7,849 | 10,495 | 11,714 | 13,872 |
| Source materials (natural or depleted uranium and thorium) | 4,440 | 5,336 | 12,234 | 13,150 | 15,399 | 19,097 |

*Source*: 1980 Annual Report of the IAEA. Reprinted by permission of the *Bulletin of the Atomic Scientists*, a magazine of science and world affairs.
Copyright © 1986, 1982 by the Educational Foundation for Nuclear Science, 6042 South Kimbark Avenue, Chicago, Illinois 60637, USA.

facilities the IAEA can inspect, thereby allowing the possibility of separate clandestine weapons materials projects (research reactors producing plutonium weapons using noninventoried uranium ore). The IAEA is also limited in its ability to enforce a violation of the treaty, to prevent a signatory nation from denouncing the treaty, or to compel any nation to sign.

Because of the problems of enforcement, some experts have urged a technical fix – a switch to the thorium fuel cycle. As we have noted, thorium is an abundant natural ore that produces no wastes with separable fissionable isotopes equivalent to plutonium and that permits easy denaturing to discourage use in weapons. The thorium cycle originally had enthusiastic support. But rapid adoption has not followed. The enormous investment worldwide in the uranium fuel cycle has been the principal barrier, but also the thorium fuel cycle is not foolproof. It is vulnerable to weapons diversion at the stage following breeding and preceding denaturing. What's more, experts believe that many other countries will inevitably develop new enrichment capabilities[11] over the next couple of decades, thus defeating the denaturing protection in the thorium cycle.

A technical fix probably can never solve this problem because it is less technological than political. The NPT has been viewed primarily as a political achievement and the present deficiencies of the treaty are likewise political. Their solutions, if they are to be found, probably will be political.

It should be noted that one major incentive for nonnuclear nations to agree to the NPT was the assurance of the transfer of nuclear-power technology to them. These nations did not want to remain nuclear have-nots as a result of not signing the treaty. This telling fact illustrates well the political nature of the issue (see Yager 1974 and Greenwood et al. 1977).

## Breeder reactor technology

The proliferation problem could be greatly complicated if many nations opt for a still-experimental fission technology, the *breeder reactor*. The typical LWR that we saw in Chapter 7 is what we call a *burner* reactor. Fissionable nuclei are split, liberating energy for a power plant, and the nuclei are thereby *spent*, used up in much the same way as a combustible fuel is burnt up. The breeder, on the other hand, creates new fissionable nuclei as a by-product in the process of fission-energy production and does so faster than these nuclei are used up to generate power.

The creation of fissionable $^{239}$Pu from $^{238}$U in nuclear reactors is known more strictly as the process of *conversion* and some reactors are designed as *converter reactors*, meaning that they are intended principally for the production of plutonium from uranium. These converter reactors have been designed for the production of weapons material and several exist in the United States.[12]

We can therefore distinguish three reactor types of fission reactors, and although they all generate heat and breed plutonium, each has a particular design emphasis:

1.  The (LWR) *burner* reactor – designed for most efficient power production.
2.  The *breeder* reactor – designed for net growth of fissionable material along with power output.
3.  The (military) *converter* reactor – designed to maximize plutonium output without regard to power output.

In any of these reactors, breeding occurs through the bombardment of $^{238}$U nuclei by fission-liberated neutrons. The fissioning of either $^{235}$U or $^{239}$Pu in any nuclear reactor creates high-energy neutrons, some of which will collide with a $^{238}$U nucleus and be absorbed. The $^{238}$U nuclei will then become transmuted into $^{239}$Pu after 2.3 days by the process of radioactive decay.

In the burner reactor, conversion takes place incidentally to power production. In a converter reactor, on the other hand, the objective is to create a fissionable isotope (plutonium) without the need for enrichment. As we have noted, the enrichment process is complex, whereas the separation of plutonium is relatively simple.

In both convertor and breeder reactors, more new $^{239}$Pu nuclei are created than $^{235}$U nuclei are used up. This is possible because in uranium there are many more $^{238}$U nuclei than $^{235}$U and more than one neutron is liberated, on the average, per fission. Alternatively, $^{239}$Pu, previously converted from uranium, can be

loaded along with the uranium as an additional fissionable material, in which case we have plutonium breeding more plutonium. In either case, the breeder mixture of uranium and plutonium is removed from the reactor when the mixture of fissionable material (both $^{235}$U and $^{239}$Pu) has grown to the desired fraction.

The cross section of a breeder core is depicted in Figure 8.3a. Each small hexagonal area represents an assembly of rods of one sort or another, as indicated. The *blanket assemblies* are rods of uranium that are subjected to bombardment by fission neutrons. The rods of the blanket assembly can be loaded with natural uranium to start the breeding process. Notice that these blanket rods are surrounded by successive rings of *fuel assemblies*, which contain the fissionable material necessary to subject each blanket assembly to intense fluxes of fission neutrons. There are also *control assemblies*, which absorb neutrons and can be inserted or withdrawn as in an LWR. Surrounding the entire set of fuel, blanket, and control rods is a ring of *radiation shielding rods* to absorb the high-energy neutrons that have escaped capture within the core.

The core of a breeder reactor can be cooled by the flow of a coolant along the sides of the rods in much the same manner as the water is used in an LWR. For a breeder, however, the coolant is not likely to be water. Breeder designs in the United States and France use molten sodium as the coolant and the reactors are called liquid metal fast breeder reactors (LMFBR). For a LMFBR core with a cross section of the type shown in Figure 8.3, the sodium coolant would flow along the length of the rod assemblies (see Figure 8.3b).

Sodium was chosen as the coolant for these reactors because of its excellent heat-transfer properties and because it can be operated at high temperatures (600 to 700°C)[13] in the molten state without the high pressures required for steam. The lower pressures thereby lessen the chances and severity of high-pressure pipe ruptures. Sodium, on the other hand, has several disadvantages. It is intensely radioactive after neutron bombardment, and it is chemically active. It can, for example, burst into flames on

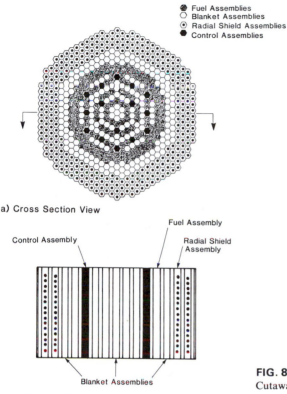

Fuel Assemblies
Blanket Assemblies
Radial Shield Assemblies
Control Assemblies

a) Cross Section View

Control Assembly

Fuel Assembly

Radial Shield
Assembly

Blanket Assemblies

b) Cutaway View

**FIG. 8.3.** Fast breeder core. (a) Cross section. (b) Cutaway view. Adapted from U.S. Energy Research and Development Administration (1975).

contact with air and it can explode when mixed with water.

The entire set, or even a subgroup, of fuel rods in a breeder constitutes a critical mass with all control rods withdrawn. The fuel mixture for a breeder is more highly enriched than an LWR, although the enrichment is not as high as that for a nuclear weapon. Higher enrichment and a dense packing of the fuel and blanket rods are necessary to allow criticality without a moderating medium and to provide the high neutron flux density required for breeding. This high density of fissionable material, however, taken together with a background of prompt fission neutrons (see Chapter 7) can have devastating consequences.

Early in the development of fast breeder reactors, it was recognized that fast breeder cores held far greater potential for destructive accidents than moderated reactors. An early

U.S. experimental breeder reactor (EBR-I) suffered a partial meltdown of its core in 1955 when it was inadvertently allowed to slip into a condition whereby a rise in core temperature caused an increase in the rate of reaction (i.e., an increase in neutron flux). This effect, called a *positive temperature coefficient*, occurred when thermal expansion pushed the fuel rods closer together. Another experimental FBR, the Enrico Fermi I, also suffered a partial meltdown in 1966 when the flow of sodium coolant was accidentally blocked by a loose metal plate. A radiation emergency was declared at the plant near Detroit, but almost all of the radioactive contamination was kept within the containment shell (see Collier & Hewlett 1987).

During this period of FBR development, it was soon recognized that breeder reactors of larger sizes than the EBR-I or the Enrico Fermi I could conceivably suffer an even more violent criticality called a *core collapse* or a *core disruption accident* (CDA, see Collier & Hewlett 1987). A CDA can occur, it is theorized, when a blockage of coolant flow causes melting of fuel rods in one portion of the core, thus allowing a collapse of the fissionable mass into a lower portion of the core. Using Figure 8.3b, we can imagine that several of the fuel rods melt in adjacent portions about half-way up in the reactor. The melted rods become a molten mass in that region and begin to flow into the lower region. The result of joining two subcritical fissionable masses without the inhibition of control rods creates a *prompt* critical mass. Most likely, this would not explode with the force of a nuclear weapon (which is designed for maximum supercritical growth). But it would be a much more violent event than a nonexplosive meltdown, which is the worst accident an LWR can suffer.

It should be noted, too, that the sodium coolant in the LMFBR creates a possible positive void coefficient such as that found in graphite-moderated LWRs like the one at Chernobyl. Thus, an unexpected surge in heat generation in some part of the core causes a local boiling of the sodium coolant, which in time leads to even higher reactivity – a *positive feedback* effect on heat generation.

In addition to the operational hazards of fast breeder reactors, there is an enormously complicated issue of safeguards. After all, the whole point of the breeder is to create more plutonium, but this would vastly increase the potential for diversion to weapons. Inventories of already separated plutonium would soar, thus taxing systems of monitoring and control.

Breeder reactors also have a higher capital cost than LWRs, but some advocates have seen economic advantages in breeder technology nonetheless. Indeed, early advocates envisioned a *plutonium economy* (see Häfele 1981). The plutonium economy was to be a comprehensive system of breeder reactors that supplied surplus plutonium as fuel for burner reactors. The additional capital cost of breeder plants was to be offset by income from the sale of these fuel surpluses; each breeder would be able to create approximately 25 percent more fuel than it

consumed in generating power. The plutonium from the fuel rods of breeders would be extracted at reprocessing plants, which would also separate this fissionable element from burner reactor spent fuel (see Figure 8.1). With these two sources of plutonium, a great deal of power would be generated with very little input of virgin uranium.

The advocates of the plutonium economy saw this system of nuclear plants as an answer to many problems, some of which appeared acute in the 1960s and 1970s (see Staufer et al. 1977). One such problem was the high projected growth of electricity demand, which implied an urgent need for huge increases in generating capacities for the rest of the twentieth century. The need for breeders seemed even greater when the world oil crisis made it apparent to all that oil-fired generation had become far too expensive. Coal-fired generation, the breeder advocates argued, was not the alternative to oil because of the many drawbacks of the coal cycle, including the occupational hazards of coal mining and the carbon dioxide dilemma (see Chapter 6).

Advocacy of the plutonium economy, it should be recognized, was not merely a proposal to add nuclear power generation capacity. Rather, it envisioned a much larger expansion of nuclear technology in the world economy overall. First, proponents argued that low-priced uranium would be soon exhausted if large new burner reactors were built. Second, they claimed that the price of uranium-based fuels could be made low and kept low by going into *large* systems of breeders and burners. The adoption of breeder technology would lead to a huge increase in nuclear power development.

But after the proposals for a plutonium economy were advanced, other experts de-murred. They recognized the inherent danger of vast inventories of plutonium and widespread reprocessing operations throughout the world. The move toward breeders soon lost momentum. Besides the risks, it was apparent that many of the assumptions of the early advocates were not coming to pass. Demand for electricity slackened in the late 1970s (see Chapter 4). The cost of constructing nuclear plants escalated above the general inflation rate greatly and breeder plants were already known to be more costly than ordinary LWRs. In the United States, for example, President Carter sought to eliminate both nuclear fuel reprocessing and breeder reactor development.[14]

On the other hand, France did proceed with breeder technology development, with its Phénix series of prototype reactors. The first Phénix plant, rated at 270 MW$_e$, has been in operation since 1973 and the larger Super-phenix, at 1,242 MW$_e$, since December 1986. These units are demonstration prototypes and their costs are several times those of conventional burner reactors of comparable sizes. At the time these breeders were planned and built, the expectation was that many more LMFBRs would follow. But instead, technical problems combined with high cost led France to suspend large-scale construction of breeders for the foreseeable future.

Should there be a revival in development of breeder technology, it is likely to happen in France. The French have pursued a national policy to limit drastically that country's dependence on imported oil. They have no oil and little coal of their own; as they say of their nuclear program: "we have no coal, no oil, no choice." There have been few critical opponents or public protests against the French nuclear program, as there have been in many of the other Western democracies. A rare public dispute took place in France after the Chernobyl accident, concerning the reporting of radioactive fallout.

But the dangers and uncertainties of breeder technology remain. In the event of a serious accident, it is doubtful that any breeder program could be sustained.

## Notes

1. Policies for disposing of radioactive wastes are not completely settled in the United States and several other countries. In the United States, the Department of Energy, operating under the Nuclear Waste Policy Act of 1982, planned to open the first depository in 1998, but the date has been pushed back beyond 2000. Indeed the evaluation and selection of sites for these depositories will probably not be completed until the late 1990s or later, and all proposals have thus far proven to be very controversial.

2. The actinides belong to a group of isotopes following actinium in the periodic table and include the transuranium elements (which themselves include the various plutonium isotopes).

3. The rate of disintegration of radioactive material – 1 Ci = 3.7 × 10$^{10}$ disintegrations/second. This is a measure of the amount of radioactivity regardless of mass. For example, 1 gram of radium has 1 Ci and 1 gram of spent fuel has about 0.5 Ci.

4. *Daughter elements* are the new elements spawned from the spontaneous splitting of nuclei during the radioactive process. As fissionable nuclei in the waste split, they create new elements with unstable nuclei. In the case of reactor wastes, the daughter elements have higher radioactive emissions than the elements they emerged from.

5. High-level wastes (HLW) are officially defined as "the waste streams that result from the reprocessing of spent reactor fuel" (Lipschutz 1980). However, in practice, HLW primarily consists of spent fuel rods themselves. In any case, it is distinct from materials that have been contaminated but were not initially radioactive – for example, reactor pipes.

6. Rochlin (1977) makes another important point. He distinguishes irretrievability by humans as *social* irreversibility, in contrast with the *physical* irreversibility of stable geological formations.

7. For example, economist S. A. Marglin once noted that the present generation's welfare has precedence as a necessary consequence of democracy. It is, he claimed, "axiomatic that a democratic government reflects only the preferences of the individuals who are presently members of the body politic" (in S. A. Marglin, "The Social Rate of Discount and the Optimal Rate of Investment," *Quarterly Journal of Economics*, 77:95–112, 1963). Also philosopher Martin P. Golding wrote that "the more distant the generation we focus on, the less likely it is we have an obligation to promote its good" (in Partridge 1981). Golding's point is that we cannot know what will constitute "the good life" for people of the distant future and therefore we cannot promote it. However, it seems doubtful that any age would find radioactive contamination anything other than a social evil.

8. The carbon-moderated *pile*, a primitive version of a reactor, was demonstrated by the famous physicist Enrico Fermi at the University of Chicago in 1941 (see Penner 1976). This

achievement fit the classic pattern of evolution of a new technology; it was the scientific *proof of principle* (see Chapter 11).

9. The production of heavy water is itself a difficult technological hurdle.

10. The plutonium separation process is a far simpler undertaking than the enrichment process. It only involves chemical reactions to separate the plutonium, a distinct atom that will combine into distinct compounds, whereas enrichment is required to separate isotopes of the same atom, which would enter into the same chemical reactions.

11. A laser-driven enrichment technique received great interest from nuclear research scientists in the early 1970s. The concept was to achieve separation of the two isotopes ($^{235}U$ and $^{238}U$) through selective ionization or dissociation by intense laser beams. For example, the laser beam could illuminate a vapor of uranium hexafluoride ($UF_6$), the same uranium compound already used in the gaseous-diffusion technique. The scheme works because the laser wavelength is sufficiently pure to distinguish between the slightly different atomic spectra of the isotopes (see Sproull 1963). Other light sources, even of the same high intensity, have too wide a spread of wavelength, and therefore do not distinguish between the isotopes. If the technique is found feasible, it could reduce the size of the enrichment process operation, and this has provoked fears of clandestine enrichment plants springing up throughout the world. The technique, however, has required extremely sophisticated processes that have confined it to research laboratories. But with the state of the art of laser technology advancing, there are no guarantees that the laser enrichment technique might not soon become easily accessible.

12. The U.S. government operates reactors to create plutonium for nuclear weapons in Hanford, Washington, and at Savananah River, South Carolina. One of the reactors at Hanford, called the N-reactor, is graphite moderated and water cooled like the reactor that exploded at Chernobyl. Several of these weapons production plants have been shut down in recent years because of safety concerns.

13. Sodium is liquid above 98°C and boils at 903°C.

14. An experimental LMFBR was started in 1972 at Clinch River, Tennessee. But six years later, with concerns rising over the issue of nuclear weapons proliferation, President Jimmy Carter announced his intention to cease plutonium recycling and to press for the termination of the Clinch River LMFBR project. Congress continued to fund it, however, and the Reagan administration in the early 1980s spoke of maintaining a commitment to the breeder project. But Congressional funding slowed and, in 1983, the project was suspended indefinitely.

## Bibliography

Carter, L. J. 1987. *Nuclear Imperatives and the Public Trust – Dealing with Radioactive Waste.* Resources for the Future, Washington, D.C.

Collier, J. G., and G. F. Hewitt. 1987. *Introduction to Nuclear Power.* Hemisphere Publishing, New York.

Deese, W. 1983. *Nuclear Nonproliferation: The Spent Fuel Problem.* Pergamon Press, New York.

Feiveson, H. A., and F. von Hippel. 1982. "Cutting Off the Production of Fissile Material for Weapons." *Public Interest Report,* newsletter of the Federation of American Scientists, pp. 10–11. June.

Greenwood, T., H. A. Feiveson, and T. B. Taylor. 1977. *Nuclear Proliferation – Motivations Capabilities and Strategies for Control.* McGraw-Hill, New York.

Grossman, P. Z., and E. S. Cassedy. 1985. "Cost-Benefit Analysis of Nuclear Waste Disposal: Accounting for Safeguards" *Science, Technology and Human Values,* Vol. 10, No. 4, pp. 47–54, Fall.

Häfele, H. 1981. *Energy in a Finite World – Paths to a Sustainable Future.* Ballinger, Cambridge, MA.

Kasperson, K. E. (ed.). 1983. *Equity Issues in Radioactive Waste Management.* Oelgeschloger, Gunn & Hain, Cambridge, MA.

Lipschutz, R. D. 1980. *Radioactive Waste – Politics, Technology, and Risk.* Ballinger, Cambridge, MA.

Lovins, A. B., L. H. Lovins, and L. Ross. 1980. "Nuclear Power and Nuclear Bombs." *Foreign Affairs,* Vol. 58, Summer, pp. 1137–1177.

Murdock, S. H., F. L. Leistritz, and R. R. Hamin (eds.) 1983. *Nuclear Waste: Socioeconomic Dimensions of Long-Term Storage.* Westview Press, Boulder, CO.

Partridge, E. (ed.). 1981. *Responsibilities to Future Generations.* Prometheus Books, Buffalo, New York.

Penner, S. S. (ed.). 1976. *Energy,* Vol. 3. Nuclear Energy and Energy Policies. Addison-Wesley, Reading, MA.

Rochlin, G. I. 1977. "Nuclear Waste Disposal: Two

Social Criteria." *Science*, pp. 23–31, January 7. ("The Politics of Nuclear Proliferation." 1980. *The Center Magazine*. University of California at Santa Barbara, May/June.

Sproull, R. L. 1963. *Modern Physics: The Quantum Physics of Atoms, Solids and Nuclei*. Wiley, New York.

Staufer, T., H. Wyckoff, and R. S. Palmer, 1977. "To Breed or Not to Breed," *Mechanical Engineering*. Vol. 99, No. 2, pp. 32–41. February.

Willrich, M., and T. B. Taylor. 1974. *Nuclear Theft: Risks and Safeguards*. Ballinger, Cambridge, MA.

Yager, J. A. 1974. *Energy and U. S. Foreign Policy*. Ballinger, Cambridge, MA.

See also *Arms Control and Disarmament Agreements, Texts and Histories of Negotiations*. U.S. Arms Control and Disarmament Agency. Washington, DC, 1980 edition.

In addition, several articles appeared in *The Bulletin of the Atomic Scientists*:

"IAEA Safeguards and Non-proliferation," S. Moglewer, pp. 24–29. "Testimony from a Former Safeguards Inspector," R. Richter, pp. 29–31; "The IAEA on Safeguards," S. Eklund, pp. 36–6; "Osirak and International Security," A. Fainberg, pp. 33–6, in the October 1981 issue.

"Plutonium for All: Leaks in Global Safeguards," R. Bolt, pp. 14–19, in the December 1988 issue.

See also special issue of *The Progressive* entitled "The H-Bomb Secret," November 1979.

# 9

# Power plant economics

## Introduction

In public discussion of electric power generation, the issues raised are most often about health, safety, and environmental impact. Electric utilities may, and often must by law, consider these matters in making choices about the means they will use to meet current and future energy demand. But finally, their decisions are usually made on economic, not social, grounds, and their choice is usually determined by the answer to this question: Which type of electric power plant will be the least costly to build and operate? The answer would seem to be straightforward because the elements are quantitative and well-defined rather than moral and abstract. Yet power plant economics are not so clear-cut. In order for utilities to decide on the lowest cost alternative, they must make economic forecasts that extend through the entire lifespan of a power plant – 30 or more years into the future. Such forecasting to date has provided uneven results at best, and virtually every conceivable cost factor of electric generation – including the cost of protecting the environment and reducing risk – has been subject to debate.

A consideration of power plant economics should properly compare the costs of all reasonable possibilities, not just central power plants but also alternatives to central plants.

Environmentalist groups and some independent energy analysts[1] have argued that such alternatives as cogeneration and improved energy efficiency may have cost advantages, and so their costs should be weighed against those of central power plant generation. As we saw in Chapter 4, efficiency gains alone can produce large savings. But even with large-scale efficiency and cogeneration programs, we probably would only reduce our need for central power plants, not eliminate it. Central plants still provide the bulk of U.S. electric power and are likely to do so for the foreseeable future. As a result, most of the debate has focused on the costs of alternatives among central plants, and that will be our focus here as well.

Further, the argument we will consider primarily examines the linchpin of our electric system, large base-load plants of several hundred megawatts or more (see Chapter 3). There is little doubt that among base-load plants, large hydrofacilities on balance have the cost advantage. Of course, though the United States could expand its hydrocapacity to some extent, there is not enough hydropower to meet most of our electric demand.

That leaves two possible types of central generating plant: fossil-fired and nuclear. In the United States, most fossil-fired plants use coal, the most abundant and most widely used (for electric power) fossil fuel resource. Oil is seldom considered anymore for base-load generation. The price hikes of the 1970s made oil a far more expensive alternative than either coal or nuclear plants. Natural gas may become an alternative in the near future. It is at present used extensively as a fuel in gas turbines for peak-load generation. Proponents of gas technology hold out hope that new high-efficiency combined-cycle turbines (see Chapter 3) can also be used for base-load generation, particularly in regions where natural gas supplies are reliable and abundant.[2] Still, the central debate of power plant economics has not involved oil or gas. Rather it has focused on the comparison between coal-fired generation and nuclear power.

The argument is of no small consequence. It is important to understand that the cost of building and operating large power plants is enormous. A typical 300-$MW_e$[3] coal plant, for example, costs a couple of hundred million dollars to build and millions more each year to run. Meanwhile, a 900-$MW_e$ nuclear plant might cost several billion dollars to build and millions more thereafter. A mistake that leads to huge cost overruns can bankrupt a utility and raise the cost of electric power significantly, affecting the economic development of a state or a region of the country for years to come. Consequently, utilities as well as the public have a vested interest in trying to forecast the ultimate cost of a plant before building it.

The huge initial construction cost of even a successfully completed nuclear plant does not, it should be emphasized, settle the issue of power plant economics. Indeed, in the 1960s and 1970s, nuclear power plants, though somewhat more costly to build than coal plants during that period, apparently held the cost advantage. The point is not which type of plant costs more to build, but which will be more economical over the plant's life. Initial construction cost is a major factor in that equation, but also at issue are two questions: How much will it cost to produce electricity? How much electricity will the plant actually produce over the next three decades?

## Factors of power plant economics

The life-cycle cost of an electric power plant is calculated by combining the initial cost of construction – the capital cost – with ongoing and future charges: operation and maintenance costs and fuel. Also, in the case of nuclear power there will be a cost to *decommission* the plant when it is closed down; it needs special handling because its components are highly radioactive. There is another factor to consider as well, one that is technical but, as we will see, has important economic consequences: the operational efficiency with which the plant operates, which is measured typically by its capacity or availability to generate its designed power output, quantified as a *capacity factor* or as an *availability factor*. All of these life-cycle elements must be considered in determining the cost of generating electric power, as expressed in terms of cents per kilowatt-hour (cents/KW-hr). However, each of these factors is subject to great uncertainties for forecasting even a few years into the future.

The capital costs – construction, equipment – would seem on the surface somewhat less problematic to predict because the time horizon of the forecast is shorter than that for, say, fuel costs over the plant's lifetime. For the most part, coal plants are designed and built within about 5 years, whereas nuclear plants have taken anywhere from 6 to 15 years to go from the drawing board to the operational stage. But even within a 5-year span, uncertainties abound and factors can change so much that capital cost estimates can prove wildly inaccurate only a few months after they are made. New environmental and safety regulations, for example, can necessitate design changes and add millions of dollars to the capital cost and years to construction time. Labor and material costs can also change suddenly. And above all, interest rates – the cost of money – can change suddenly and drastically.

Interest costs are extremely important to this discussion. The capital costs are customarily not one-time outlays of huge amounts of cash, but rather are paid over time. Typically, utilities must borrow most of the money for new plant and equipment and pay back the debt over decades. This is true for investor-owned utilities (IOUs) as well as public power authorities. For virtually all plants, then, amortization of the debt becomes an ongoing expense, expressed on a cents/KW-hr basis as a *capital fixed charge*. Of course, that fixed charge is affected by the cost of borrowing money.

To raise the necessary financial capital, a utility will get loans from banks and/or sell debt to investors.[4] Both kinds of debt have to be paid back with interest by the utility over varying terms from a few months to 30 or 40 years. So a utility that builds a nuclear power plant costing $2 billion probably will have to pay interest charges well into the next century as well as paying off the debt principal at some specified point in time.[5] Of course, if interest rates are 15 percent, the cost of a plant will be much greater than if rates are 8 percent. In fact, a 15 percent rate means a 71 percent increase in total costs of construction – interest and principal over 30 years – compared with an 8 percent rate.[6]

A utility knows the rate it must pay for debt only at the moment it borrows. (Most utility debt has a fixed rate of interest, so the company knows the rate for the entire term of the loan.) If it could borrow the full cost of construction at one time, the utility would have a good idea of the capital fixed costs over the life of the plant. But borrowing all the money at once is neither feasible nor sound. For one thing, the company would have to know exactly the final cost of construction. Also, it would be pointless for a company to pay interest today on $1 billion when it would need, say, $100 million this year and the rest 5 years later during an extended construction period. So instead, it will borrow $100 million now and the rest as needed. That means that interest costs become increasingly subject to forecast and uncertainty. In the period between 1978 and 1983, a utility might have paid as much as 18 percent or as little as 9 percent on a debt. A utility embarking on construction in 1978 would have tried to estimate rates throughout this period and

probably would have been wrong; during the entire century, rates had never before reached 18 percent in the United States. Obviously, the timing of a project can make an important difference in cost, but utilities have to make decisions based on current rates and projections both of interest rates and of needed electrical capacity. At the same time, a utility wants some confidence that it will be able to pay back what it borrows. If it is wrong, it may lose its credit rating for borrowing, need to raise its electric rates, and/or go bankrupt.

Also, the cost of borrowing – the interest rate on debt – will differ from utility to utility. Power companies, given the fact that they build and operate many power plants, frequently carry huge amounts of debt. Commonwealth Edison of Illinois, for example, had almost $7 billion in long-term debt as of 1987. As a rule, banks and investors are not troubled by large debts on a power company's balance sheet because there is lots of cash coming in from customers. But there is always a limit to how much debt a company can carry for a given cash income. Some companies are close enough to that limit that any new debt they incur carries a risk that it will never be repaid. If these utilities want to borrow more, they have to offer an inducement to investors in the form of higher interest rates – a *risk premium*. So at the same time that one utility pays 10 percent on debt, another must pay 12 percent, meaning that the second utility will have a higher fixed capital charge even if it builds a new plant as efficiently and cheaply as the first utility.

Utilities building new nuclear plants generally need to pay risk premiums. In fact, they might pay a triple premium. First, there is a cost overrun premium. In the 1970s, a number of utilities that began to construct nuclear plants made cost estimates that were very far from the actual cost. In some well-publicized cases, such as the Shoreham plant on Long Island and the Seabrook plant in New Hampshire, the final cost was many times the initial estimate even adjusting for inflation. The principal owner of Seabrook was technically bankrupt by the late 1980s and investors holding debt in that company risked losing their money. Even more discouraging to investors was the failure of the Washington Public Power Supply System (WPPSS), a consortium of municipal and rural public power entities. In the early 1970s, WPPSS began constructing five nuclear plants, selling debt on each unit separately. In 1982, faced with huge cost overruns and a reduced growth in electric demand, WPPSS abandoned two of the plants. A year later the utilities in the group defaulted on $2.2 billion in bonds sold to the public to build those two units. This incident more than any other made investors cautious about nuclear power.

Utilities building nuclear plants also have to pay an interest rate premium on debt to cover the financial consequences of an accident. This premium also applies to the debt of utilities that operate existing nuclear facilities even if those plants have operated efficiently for years. The meltdown at Three Mile Island demonstrated how costly a nuclear accident can be. The accident cost its owner General Public Utilities about $1 billion in cleanup costs.

Finally, there is what might be called a special problems premium for nuclear builders. Some plants are partly or fully built, but sit unused because of unresolved safety questions or delays in regulatory action. The problem here is not only that such plants produce no revenue, but they also continue to require payments for debt service and maintenance. The more crucial problem for the utilities and investors is that state public service commissions, which determine local electric rates for customers, usually allow utilities to pass along construction costs to ratepayers only after a plant has opened. This is the reason cost overruns can lead to insolvency; a utility cannot get any of its money back until the plant comes on line, according to the public utility regulations of most states. As long as the plant is not put on line, financial problems loom for the plant's owner. Taken together, the building of a nuclear plant can add an interest premium of between .5 percent and 1.5 percent to a company's cost of funds.

An additional source of uncertainty during recent years for both coal and nuclear construction has been the cost of public and environmental safety. Legislative and regulatory authorities have at times changed the regulatory rules in response to new scientific data or to events like TMI and Chernobyl (see Chapter 7). Coal smoke emission control devices for coal plants and new safety features for nuclear plants have added millions of dollars to construction costs. In the case of nuclear power, rule changes have even required that plants already under construction be redesigned.

The spectacular overruns of construction costs at a few nuclear plants have been well publicized. Other new plants, however, have not experienced such financial disasters and nuclear power need not bring a utility to the brink of insolvency. But can it be cost competitive with coal-generated power? In theory, it can be, but we must make two assumptions. First, because nuclear power enjoys some operational cost advantages over coal generation, nuclear capital costs can be higher by a modest percentage without causing nuclear power to lose its overall cost advantage. Expressed in terms of dollars per kilowatt installed capacity, nuclear plant construction costs must not be more than 50 percent greater than those for a coal plant. Indeed, most analysts would say the nuclear/coal capital ratio must be 1.4:1 or less. Second, if nuclear power has higher capital costs, we must assume moderate interest rates. The higher the rates, the lower the ratio must be.

Two operational costs are most important in considering the economic advantage of any type of electric generating plant: fuel, and operations and maintenance. Fuel costs are perhaps the least predictable variable of all. Who would have guessed, for example, that oil prices would increase 1,000 percent in less than a decade as they did in the 1970s? Since the United States is self-sufficient in coal, it is unlikely to face the same sort of escalations as it experienced with imported oil, which were in part politically motivated. Still, the price of coal 30 years hence is uncertain. The same is true for uranium oxide (see Chapter 7). Yet forecasters generally agree that coal plants will face significantly greater fuel costs – possibly over 2:1 – per KW-hr than nuclear plants (see Komanoff 1980 and Perl 1982).

Operations and maintenance costs (O&M)

have also been subject to debate although they are expected to be roughly equal at coal and nuclear facilities. Again, a great deal of uncertainty remains. In the early 1980s, for example, O&M costs at nuclear facilities rose far more rapidly than had been forecast only a few years earlier.

Finally, we must consider the operating efficiency of the power plant measured by the *capacity factor* (CF). By capacity factor, we mean the quantity of electricity actually produced annually compared to the amount that would have been produced if the generators were run at full power continuously. Ideally, a base-load plant would operate continuously to supply that irreducible level of electric demand that never ceases (see Chapter 3). However, all operating machinery must be shut down periodically for maintenance. Also, unscheduled interruptions will occur randomly due to mechanical failure and accidents. The better the design and quality of operation, the higher the fraction of hours of the year the plant will operate, but of course it can never approach 100 percent. Still, a generator that runs at a high capacity factor will produce more revenue than a generator with a low CF.

We calculate the capacity factor of a plant by dividing the number of actual kilowatt hours produced in a year by the number of kilowatts the plant is capable of producing times the number of hours in a year:

$$CF = \frac{\text{actual KW-hr/yr}}{\text{KW} \times 8,760 \, \text{hr/yr}}$$

Using the capacity factor, we can calculate the electric rate (¢/KW-hr) necessary for the actual energy (KW-hr) generated, which is annually required to pay off the capital fixed charges. The lower the capacity factor, the higher the electric rate required. Or put another way, if a plant operates at low capacity, it will not produce enough revenue to pay for itself.

Actually, it may be better to use a slightly different measure than the CF in determining the efficiency of a plant. In recent years, some utilities have reported misleading CF results. Because they overestimated demand, they had

excess capacity in both coal and nuclear plants. Though both were available for operation, in some instances they continued to use their nuclear plants for continuous operation base load, but used the coal plants only intermittently. Since CF is a measure of actual use, the nuclear plants were credited with higher CFs than the coal plants. But, in fact, many of the coal plants had a better *availability factor* (AF). The AF is calculated simply by dividing the hours available – that is, those hours that are ready for service if needed, even at partial output – by the number of hours in a year:

$$AF = \frac{\text{actual hours available annually}}{8,760 \, \text{hours per year}}.$$

Because the AF accounts for the time that a plant could have operated, it may be a truer measure than the CF for the potential performance of a power plant type. Certainly, it is an important indicator of a type's dependability of service and should be a factor in any discussion of power plant economics.

## The coal/nuclear debate

In the early 1970s, nuclear power seemed to enjoy a clear cost advantage over coal. By the end of the decade, however, that advantage was no longer so evident. In fact, the rapid increase in the cost of constructing nuclear plants appeared to swing the advantage to coal – if not for plants coming on line at that time, certainly for those of the future. This perception increased as interest rates soared in the early 1980s to unprecedented levels and new reports surfaced of horrendous cost overruns at several nuclear construction sites.

Yet at this time, several studies argued that nuclear energy remained and would continue in the future to be a more cost effective means of generating base-load electric power than coal. Even after Three Mile Island, a study (Table 9.1) by Lewis Perl of the National Economic Research Associates purported to show that new nuclear plants would enjoy a significant cost advantage over coal.[7] This study

**Table 9.1.** The Perl study – average projected costs (1982 constant dollars)

|  | Nuclear | Coal |
|---|---|---|
| Unit size | 1,000 MWe | 600 MWe |
| Capital cost | $1,406/KW | $975/KW |
| Interest rate | 10.4% | 10.4% |
| Capacity factor | 59% | 57.5% |
| Capital fixed charges | 2.10 ¢/KW-hr | 1.36 ¢/KW-hr |
| Decommissioning[a] | — | — |
| Fuel | .91 ¢/KW-hr | 1.91 ¢/KW-hr |
| Operating and maintenance | .32 ¢/KW-hr | .40 ¢/KW-hr |
| Total | 3.33 ¢/KW-hr | 3.67 ¢/KW-hr |
| | Nuclear/coal cost ratio: .91 | |

[a]Perl makes substantial allowances for the cost of decommissioning, a charge embedded in capital fixed charges.

*Source*: Derived from "The Economics of Nuclear Power" by L. J. Perl, National Economic Research Associates, Inc., June 3, 1982.

was touted by the Atomic Industrial Forum (AIF), a nuclear industry group, in what appeared to be an effort to convince utilities to renew their commitment to nuclear energy.

The Perl report and prior analyses issued in the late 1970s by industry and government study groups based their pronuclear conclusions on important assumptions. First, they argued that the growth of nuclear power construction costs would end because with experience utilities would get the cost of building plants under control. At the same time, these studies claimed that the cost of building coal plants would rise dramatically because of new government regulations concerning emissions control. The pronuclear argument assumed that the nuclear/coal capital cost ratio would stabilize at the break-even point, 1.4:1, or less – a point favorable to nuclear power. The studies also assumed that nuclear facilities would regain control of O&M costs, which had also risen much more rapidly than expected. Finally, the projections anticipated higher CFs at nuclear plants than actually existed. Plants that opened in the late 1970s had proven disappointing in

this respect, maintaining a capacity factor of less than 55 percent on average. The studies assumed CFs of 60 percent or better.

Industry critics, especially Charles Komanoff, attacked these projections. In fact, Komanoff argued in 1980 that these assumptions were incorrect and he made his own estimates of average costs at coal and nuclear facilities that showed (even after regulatory compliance) the advantage to coal (Table 9.2). it soon became apparent that Komanoff and the other critics were correct. The nuclear proponents were right about improved capacity factors, and O&M costs showed some improvement (although nuclear O&M costs still lagged behind those of coal). But overall they were wrong because capital costs did not come under control. Even excluding celebrated disasters such as WPPSS and Shoreham, nuclear construction costs grew at rates far above expectations even when adjusted for inflation. The cost of coal smoke emission technology proved much more predictable and initial cost estimates were in line

**Table 9.2.** Projected costs of 1988 plants (U.S. average, 1979 constant dollars)

|  | Nuclear | Coal |
|---|---|---|
| Unit size | 1,150 MW | 300 MW |
| Capital cost | $1,460/KW | $838/KW |
| Decommissioning | $138/KW | |
| Real fixed charge rate | 10.3% | 9.8% |
| Capacity factor | 60% | 70% |
| Capital cost fixed charges | 2.86 ¢/KW-hr | 1.34 ¢/KW-hr |
| Decommissioning fixed charges | 0.21 ¢/KW-hr | |
| Fuel | 1.09 ¢/KW-hr | 1.96 ¢/KW-hr |
| Operating and maintenance | .62 ¢/KW-hr | .62 ¢/KW-hr |
| Total | 4.78 ¢/KW-hr | 3.92 ¢/KW-hr |
| | Nuclear/coal cost ratio: 1.22 | |

*Note*: Komanoff excluded the prospective cost impacts of the Three Mile Island accident from this table.

*Source*: C. Komanoff, *Power Plant Cost Escalation*, Table 12.1, Van Nostrand Reinhold, New York, 1981.

with reality. It was clear that nuclear power capital costs were nowhere near the break-even ratio and averaged more than 2:1.

Actually, the exact ratio was difficult to calculate because of the way utilities account for capital costs. Utilities typically state costs in mixed dollar values. As a result, it can be misleading to say that because a nuclear plant costs $3,000 per kilowatt and a coal plant $1,000, the former is three times more expensive than the latter. The reason is the *time value of money*. A plant costing $1 billion and paid for this year has the *identical* economic cost of a plant that costs $1.1 billion and paid for next year assuming an inflation rate of 10 percent. Because they do not take this time value into account, the accounting system used by the utilities[8] can lead to an overestimation of the real economic cost of one plant and an underestimation of the cost of another plant.

Economists get around this accounting problem by choosing a baseline dollar value – constant (real) dollars of any one year – and deflating costs to that baseline year. It is difficult, however, to find industry data of construction costs in constant dollar terms. In the mid-1980s, Komanoff, using his own proprietary data base, estimated in constant 1984 dollars a nuclear/coal ratio of about 2.5:1, which is far above the break-even point. Even in the unlikely event that Komanoff's figures were incorrect by 20 percent on both sides of the equation, nuclear power would still be at a cost disadvantage.

Around this time, the nuclear industry finally admitted that among new plants coming on line, coal was more cost effective. But in 1984, the Atomic Industrial Forum (AIF) prepared a new study that argued that nuclear plants built in the 1990s would regain their advantage. The new study restated many of the assumptions of its predecessors – most importantly a capital cost ratio approaching 1.4:1 – that had failed to materialize in the 1980s. Not only did this study fail to convince the critics, the U.S. electric industry too seemed skeptical. There were no new orders for nuclear plants in the United States (continuing a trend since 1979), while coal plant construction continued.

But this state of affairs did not end the coal/nuclear debate, because nuclear power was enjoying a cost advantage – not in this country, but in other nations, particularly France. The French nuclear industry, however, was quite different from its U.S. counterpart. Most notably, French nuclear power was under the control of one single authority – Electricité de France (EDF) – whereas U.S. reactors were built and operated by many different utility companies and power authorities. And where U.S. plants were all custom designed, French plants were constructed from standard models. In other words, the French chose a couple of designs only and built them – not once or twice, but over and over again. The question that U.S. electric industry officials began to ask was: Could the French system be replicated here, if not exactly then in some modified version, and would it give nuclear power the cost advantage again?

## Standardization of U.S. nuclear plants

In 1974, France made nuclear power plant standardization its national policy and over the next 11 years the EDF ordered, built, and for the most part began operating 54 new pressurized water reactors. Of these, 34 were 900-megawatt units of a standard design and 20 were 1,300-megawatt units that were also nearly identical to one another. (The differences were almost exclusively site specific, reflecting local geologic, climatic, and other natural conditions.)

In the French system, EDF operates all of the plants, hires and trains the staff, and acts as its own architect-engineering firm for construction. It is responsible for all quality control and implementation of safety regulations. One manufacturer, Framatone, designs, engineers, and manufactures the entire steam system, including the reactor cores. Alsthom-Atlantique, a single contractor, builds all of the systems except the nuclear steam system. Finally, one firm, Cogema, is responsible for designing and building the fuel assemblies.

What would it mean to U.S. nuclear power development if such a standardization system could be set up here? Of course, this would entail major institutional changes that might be difficult and costly to implement. To standardize, some governmental or industrywide organization would have to be set up and given broad powers to control design, construction, and operation of all nuclear power facilities. France already has a single government-controlled utility; the United States has numerous companies that might be unwilling to cede authority to any other body to decide the design and character of their generating units.

But for the sake of this discussion, let us assume that such institutional problems are solvable. Would nuclear power then regain its cost advantage? The best answer is that it is unlikely to happen any other way but still it is uncertain. Important questions remain. Above all, standardization must be able to lower capital costs to the 1.4:1 ratio or better, and this is not guaranteed. Still, it appears very likely that standardization would eventually bring down capital costs considerably.

It is of course possible to cite the French experience for application to U.S. industry. One U.S. study of French nuclear power construction (see EPRI study 1986) suggested that for late versions of French plants the capital cost (in constant 1983 dollars) was approximately $1,000/KW. That would be highly favorable to nuclear power if we could duplicate such cost experience in the United States, because it would put the nuclear/coal ratio at about 1:1, far better than break-even.

But comparing capital costs between the United States and France is difficult and not because of the institutional differences alone. For example, costs of labor in the United States and France differ. So do the costs of materials; to note one difference, oil, which is often used to run construction-site diesel generators, is more expensive in France than in the United States. But perhaps the most significant problem in comparing the two countries is to account for exchange rates and real interest costs. Exchange rates, especially in the last decade, have fluctuated quite widely from year to year. So

real costs, deflated into a given number of French francs (FFr), may have vastly different dollar values depending on the year used. To illustrate the problem, consider a French nuclear plant that cost FFr11 billion in constant 1985 francs. That would have corresponded to about one billion U.S. dollars. But FFr11 billion (1987) would have corresponded to more than $1.8 billion because of changes in exchange values.

As a result, it is misleading to use figures for French construction and U.S. industry probably cannot expect identical results. Even proponents of nuclear power do not envision them, but a 1986 study by the AIF does argue that standardization here could bring the capital cost to under $1,200/KW, a nuclear/coal ratio to about 1.2:1. This result, it should be kept in mind, would be for the *nth* plant. In other words, the cost of the first plant would be far greater, but as construction and engineering experience grew each plant would become increasingly less expensive. The tenth standard plant would, for example, be considerably less costly than the first.

There is logic as well as empirical evidence to support this analysis. Repetition gets the bugs out and improves productivity and performance throughout the construction process and beyond. The French needed approximately 7.25 million man-hours of labor to build the first standard 900-megawatt units, but 30 repetitions later EDF built the same reactor in 6.5 million hours, a productivity gain of 11 percent and a definite cost savings. Design and engineering costs should fall even more markedly; after building a few units, design and engineering become primarily site specific. And the cost of the reactor components – and those for the electric plant as well – may fall because the demand for identical units can permit efficient modular construction methods.

The AIF standardization study tried to put all potential savings into context, attempting to calculate the total savings standardization might bring. The study used a 1200-megawatt reactor and tried to calculate savings in an *nth* standard plant against a "best [U.S.] construction cost experience" for a reactor of that size.[9]

**Table 9.3.** Impact of standardization on the total capital cost of a future nuclear plant

*Authorization 1986*
- Plant size = 1200 MWe
- Construction period (from construction start to commercial operation date):
  —best experience plant model = 11 years
  —Standardized plant model = 6 years
- Cost escalation = 4.1% per year
- Interest expenses (AFUDC)$^a$ = 11% per year

| Nuclear plant components of cost | Best-cost experience plant costs | | Standardized plant costs | | Cost savings ($/KW) |
|---|---|---|---|---|---|
| | ($/KW)$^b$ | (% total) | ($/KW) | (% total) | |
| Permanent equipment and material | 288 | 11 | 273 | 23 | 15 |
| Construction and construction overhead | 354 | 13 | 319 | 27 | 35 |
| Design and management services | 316 | 12 | 108 | 9 | 208 |
| Owner's costs | 117 | 4 | 113 | 10 | 4 |
| *Total of direct and indirect costs* | 1,075 | 40 | 813 | 69 | 262 |
| Escalation | 474 | 18 | 110 | 9 | 364 |
| Financing (AFUDC)$^a$ | 1,101 | 41 | 263 | 22 | 838 |
| *Total of time-related costs* | 1,575 | 59 | 373 | 31 | 1,202 |
| *Total plant costs* | 2,650 | 100 | 1,186 | 100 | 1,464 |

$^a$Allowance for funds used during construction.
$^b$\$/KW expressed in current (as spent) dollars.

*Source*: Adapted from Atomic Industrial Forum (1986).

As we can see in Table 9.3, the AIF study gave a best cost of \$2,650/KW – in mixed dollars – and projected savings for later standardized plants in every aspect of capital expenditures. The savings projected in direct and indirect costs, a total of over \$262/KW, seem logical. Note the largest savings for direct and indirect costs are projected in design and management, which is the most obvious category for savings under standardization.

The remainder of the savings, however, comes under the category of time-related costs. But these costs are not so evident and the reductions under standardization are not entirely convincing. Some nuclear proponents have argued that time delays are the main reason for cost overruns, and that were it not for those delays, often imposed by regulators, plants could be built more quickly and more cheaply. It is true that standardization should shorten construction times. A standard reactor and design

should face shorter periods of authorization, less regulatory uncertainty, and, because of experience, a shortened construction period. But as already noted, the use of mixed-dollar accounting often obscures the real economic costs. While delays have at times led to increased real costs in the nuclear industry in the United States, faster has not always equalled cheaper. The St. Lucie 2 plant in Florida, for instance, was built in only six years and completed in 1983, but in constant-dollar terms was more expensive than other plants built around the same time. The Red River plant was another example of a plant that was quickly built, but at a higher constant-dollar cost than other contemporaneous plants (see Komanoff 1985). In considering the time issue, it is important to keep a certain skepticism. As we saw earlier, quick construction benefits the utility, which can pass along the costs to the ratepayer after completion. That may benefit the utility's finances, but does not necessarily reduce the real capital costs.

While this issue may raise questions about the particular formulation of the AIF conclusions, it does not negate the potential cost benefits of standardization. Indeed, in focusing on time-related costs, the AIF may be understating the benefits. Even industry critics such as Komanoff have noted that U.S. electric utilities have reduced costs through increased experience with nuclear construction and technology. Komanoff found that a utility saves 7 percent in construction costs in real terms for each succeeding reactor it builds. He noted also a 28 percent savings per reactor when a utility builds multiple units instead of only one. He pointed out, too, that certain reactor vendors, particularly Westinghouse, have been more successful than others in keeping costs under control (see Komanoff 1984).

Until a standardization program is implemented, it remains unclear whether these savings would bring capital costs to the 1.4:1 break-even ratio. However, it should be noted that other benefits may result that could give nuclear the advantage even if the ratio remained slightly above that level. First, standardization of components should lower O&M costs or certainly bring them under control. More important, standardization should improve reliability of performance and so raise the availability factor of the reactors. The French found that the AF of their 900-megawatt units rose from 62 percent in 1979 to an impressive 83 percent in 1984 (see Tanguy 1985).[10]

Although standardization holds out some hope of giving nuclear power a cost advantage again, there are two other factors that have never been fully addressed and may make cost effectiveness of nuclear power impossible to achieve. Those are the cost of decommissioning and the cost of waste disposal. Because of the closure of several small plants built early in the nuclear era, there actually have been some attempts to estimate decommissioning costs in the United States and other developed countries. The estimates range in constant 1985 dollars from $0.45 million per megawatt to almost $2 million per megawatt (see Table 9.4). If these cost estimates prove accurate and applicable to current plants, typically in the range of 1,000 MW, utilities would have to earmark as much as $2 billion for decommissioning. Utilities in fact face large costs to decommission existing plants no matter which end of the range proves closer to the mark. However, the tremendous range itself has to concern utilities, especially since the high end might well make even the best-built plant undesirable from a cost standpoint. Alternatively, it is possible that experience will reduce the costs per megawatt for decommissioning. But the industry has too little experience to know if current estimates will prove too high or too low.[11]

If anything, waste disposal costs pose greater uncertainty. Although studies have tried to make allowances for it, any projected number remains largely a guess. In his 1980 study,

**Table 9.4.** Estimated decommissioning costs

|  | MW capacity | Total costs ($M)[a] | Cost per MW ($ M) | Years operated |
|---|---|---|---|---|
| U.S. Atomic Energy Commission, Elk River | 24 | 14[b] | 0.58 | 1962–1968 |
| UK Atomic Energy Authority, Sellafield/Windscale | 33 | 64 | 1.94 | 1963–1981 |
| Pacific Gas & Electric, Humboldt Bay–Unit 3 | 65 | 55 | 0.85 | 1963–1976 |
| U.S. Department of Energy, Shippingport | 72 | 98 | 1.36 | 1957–1982 |
| Commonwealth Edison, Dresden-1 | 210 | 95 | 0.45 | 1960–1978 |

[a]Total costs are in $ 1985.
[b]Actual cost – decommissioning completed in 1974.

*Source*: Adapted from Worldwatch Institute 1986.

Komanoff assumed a cost of 0.15 cents/KW-hr for disposal, which was far more than nuclear industry proponents were allowing. But the cost could be a good deal higher than Komanoff suggested. Industrywide, the cost of disposal will be in the billions of dollars, although just how many billions will depend on the method employed and the problems encountered. Estimates based on likely alternatives were, in the mid-1980s, put as high as $35 billion (constant dollars) for all nuclear power plants.[12] But this figure has little empirical basis, because there is no current U.S. cost experience with permanent waste disposal. As matters now stand, it will be the end of the century or later before the economics of nuclear waste disposal become clear.

## Notes

1. Komanoff Energy Associates, the National Resources Defense Council, and the Environmental Defense Fund are among those who advocate and engage in *alternative* electric power planning, weighing conventional generation against conservation and other nontraditional alternatives.
2. Some of the interest in expanding gas-fired electric generation stems from concerns over pollution and the greenhouse effect (see Chapter 6). The combustion of natural gas produces about half as much carbon dioxide and nitrogen oxide as coal, and almost no sulphur dioxide. Initially, gas turbine technology suffered from reliability problems limiting its application, but in recent years, its availability factor has improved significantly (see Burnett & Ban 1989).
3. Three-hundred-megawatt coal plants and 900-MW nuclear plants are simply representative sizes. The important issue is cost per KW of capacity, not overall size.
4. Debt instruments include bonds, debentures, and notes and are sold generally in the bond markets of the New York and American stock exchanges. Some debt instruments are backed by collateral, others by revenue alone. Public power authorities sell debt at lower interest rates because investors do not pay federal income tax on interest earned from debt issued by government-owned utilities. Investor-owned utilities (IOUs) can also raise money by selling additional stock – equity. This dilutes stock ownership and if done to excess can have a negative impact on a stock price. But

many utilities are so large that the dilution will be small. Pacific Gas & Electric, to cite one example, had over 400 million shares outstanding by 1987. Ten million new shares would have meant a dilution of less than 2.5 percent, but would have raised close to $200 million.
5. In most instances, longer-term debt is preferred because the present discounted value of the principal shrinks the further out in time the instrument matures. Also, a utility will know exactly what its capital fixed charge is if it can borrow for the lifespan of a plant. The utility will know the *peak* cost this way because it can often refinance and reduce costs if interest rates fall. It is a measure of a utility's financial health if it can borrow a great deal of long-term debt because it implies market confidence in its ability to meet interest payments well into the future.
6. This amount is determined by dividing the present value and what is often called the capital recovery factor (CRF). The CRF is essentially the annual payment ($R$) required to amortize the present value ($P$) over $n$ years at a given interest rate ($i$). Or:

$$R = P \frac{i(1+i)^n}{(1+i)^n - 1}$$

For a given $P$ and $n = 30$, we get $P/R = 11.25$ when $i = .08$ and

$$\frac{P}{R} = 6.56, \quad \text{when } i = 0.15.$$

The ratio of the two results is 1.71, that is, a 71 percent increase in total costs where the rate of interest is 15 percent rather than 8 percent.
7. The Perl analysis suggested that costs would be affected by regional differences. It noted a probable advantage for coal in the Northern Great Plains, where abundant surface-mined coal is available, although it gave a substantial advantage to nuclear power in places like the Northeast where there is no local coal supply. There is undoubtedly some value in analyzing costs by region. Other materials and labor charges may also be affected by regional differences. As of the late 1980s, however, nuclear power had become so much less cost effective than coal that regional differences were not sufficient to give nuclear power a clear advantage in any part of the United States.
8. The accounting system is called AFUDC or allowance for funds used during construction. For a critique of this system see Komanoff 1984.

9. The best cost experience model was developed from a Department of Energy economic data base (EEDB VII) and involved confidential data from utilities having nuclear power plants currently under construction and over 65 percent complete.

10. Plant management can have as much to do with its AF as the quality of its construction. The Pilgrim plant, operated by Boston Edison, is almost identical to the Millstone I plant run by Northeast Utilities; both are General Electric-designed BWRs of about 650 $MW_e$ and both opened in the early 1970s. But performance at the two plants has widely diverged. The Pilgrim plant was shut down by the NRC in 1986 because of grave deficiencies in management and control, and as of 1986, its lifetime CF – which, as a base-load nuclear plant, is close to its AF – was a poor 52.9 percent. On the other hand, the Millstone facility had maintained a lifetime CF of over 68 percent. Standardization, its proponents hope, will lead to better, more consistent management of nuclear facilities nationwide. This assumes, of course, institutional changes that would require common operational practices by the nation's utilities – practices that would be monitored, if not controlled, by a central authority.

11. Decommissioning may give nonnuclear technologies a double cost advantage. Nonnuclear plants, of course, avoid the expense of handling and disposing of radioactive materials. But the absence of radioactivity also means that such plants can be more readily refurbished to provide useful service longer than nuclear plants. Old coal, hydropower, or gas facilities can often be upgraded and refitted, whereas comparable nuclear facilities must be carefully dismantled at great expense. Retrofitting can permit the introduction of new technology at a fraction of the cost of building a new plant. Indeed, some existing plants (particularly hydrofacilities) were built more than half a century ago and may still last many more decades. Because an extended life means a longer and larger stream of revenue, a facility's longevity gives it an added cost advantage.

12. Current estimates of disposal costs might change conclusions about future power plant costs. Spreading $35 billion over the lifetime of 100 U.S. nuclear plants on a per KW-hr basis yields a cost of over 0.4 cents at a CF of around 60 percent. Note in Table 9.1 the advantage given for nuclear power is only 0.34 cents, meaning that waste disposal, even from the perspective of a nuclear proponent, might eliminate any economic advantage of nuclear power.

## Bibliography

"Are Utilities Obsolete. 1984." *Business Week*, pp. 116–129, May 21.

Atomic Industrial Forum (AIF). 1984. "A Comparison of Future Costs of Nuclear and Coal-Fired Electricity." Report prepared by the Study Group on Financial Considerations. Bethesda, MD, November.

1986. "Standardization of Nuclear Power Plants in the U.S." A Report by the Study Group on the Practical Application of Standardized Nuclear Power Plants in the United States. Bethesda, MD, November.

Burnett, M. W., and S. D. Ban. 1989. "Changing Prospects for Natural Gas in the United States." *Science*, Vol. 244, pp. 305–310, April 21.

Electric Power Research Institute (EPRI). 1986. "A Comparative Discussion of U.S. and French Nuclear Power Plant Construction Projects." Report by United Engineers and Constructors, September.

"The Fallout From 'WHOOPS.' 1983. *Business Week*, pp. 80–87, July 11.

International Atomic Energy Agency (IAEA). 1986. "Nuclear Power: Status and Trends." Vienna.

Komanoff, C. 1980. "Power Propaganda: A Critique of the Atomic Industrial Forum's Nuclear and Coal Power Cost Data for 1978." Environmental Action Foundation, Washington, DC.

1984. "Assessing the High Costs of New U.S. Nuclear Power Plants." *Public Utilities Fortnightly*, pp. 33–38, Oct. 11.

1985. "Dismal Science Meets Dismal Subject: The (Mal)practice of Nuclear Power Economics." *New England Journal of Public Policy*, Vol. 1, No. 2, pp. 47–59.

Komanoff, C., and E. E. Van Loon. 1982. "'Too Cheap to Meter' or 'Too Costly to Build'?: How Nuclear Power Has Priced Itself out of the Market." *Nucleus*, Vol. 4, No. 1, pp. 3–7, Spring.

Parisi, A. J. 1981. "Hard Times For Nuclear Power." *The New York Times Magazine*, April 12.

Perl, L. J. 1982. "The Economics of Nuclear Power," paper for seminar in *The Economics of Nuclear Power: Should Construction Proceed?* National Economic Research Associates, New York, June 3.

Tanguy, P. 1985. "Safety and Nuclear Power Plant Standardization: The French Experience." *Public Utilities Fortnightly*, pp. 20–25, Oct. 31.

# 10

# Establishing criteria for safety and health standards

A society can choose which energy technologies it adopts. But to choose the benefit of any technology is also to assume certain risks. We have discussed some specific risks in connection with fossil-fired and nuclear-generating technologies. But the issue of risk raises two more general, indeed philosophical, questions. First, how do we assess the risks? That is, how are risks to be determined and then factored into an analysis of costs and benefits? What must be kept in mind is that not only is risk a potential cost, but that avoidance of risk has a cost that we are sometimes unwilling to pay. Second, who – government officials, technologists, the general public, or some combination – decides what society pays and according to what standards? The responses to these questions may be crucial influences in energy-policy decision making. Without some conscious analysis of the method of risk determination and some informed choice among alternative procedures, the exercise of choice on energy technologies may be less than rational or democratic.

## Assessing the risks

What is a risk and what is a benefit? Clearly, some actions and activities, for example those that pose a danger to life and limb, such as handling nuclear waste, contain some element of risk. Acts that better people's lives – such as rural electrification – offer some benefit. But individuals will not always agree that one action is risky or that another action provides benefit. Indeed, any event may present a perception of risk to one and benefit to another. And if it is difficult to determine what is a risk, it is harder still to determine how much a risk costs or how much a benefit is worth. Yet these kinds of evaluations are, and some could argue must be, attempts in the formulation of energy policy. Society needs to be able to analyze the potential consequences of its choices. As a result, policymakers have sought ways to assess and evaluate risk and benefit.

This effort at assessment and quantification originated in utilitarian concepts of the nineteenth century (see Appendix B for a discussion of utilitarian ethical theories). Early utilitarian thinkers believed that happiness itself could be quantified by some objective measure. If it were possible to make such a determination, as some economists of the period believed, then decision makers would be able to count the happiness produced and pick actions that produced the most happiness. Decision making would be simple – a matter of counting. Of course, this was demonstrated to be simplistic, not simple, but the desire to find measures for quantification remained for good reason, without them, policy making is based on random and arbitrary criteria.

Whatever approach policymakers use, they must start from two realities of all cost-benefit decisions:

Perfect safety is unattainable.
Societal resources – money, labor, capital, raw materials, and so on – are finite.

The result is that society must specify the level of safety to be achieved, and allocations must be made between competing claims. This means in some instances that society must choose to expend resources to promote safety in one area and not in another. In making decisions, U.S. government officials have adopted an approach that can be outlined in the following fashion (from Hiskes & Hiskes 1986):

Identify the problem and formulate policy objectives.
Consider alternative courses of action.
Determine consequences (positive and negative) of each alternative.
Analyze the probability of each consequence.
Assign costs and benefits to each consequence.
Select the best alternative.

But how are the costs and benefits to be analyzed?

## Risk-cost-benefit analysis

To evaluate competing claims in a systematic manner, policymakers have come to rely on a quantitative method called cost-benefit analysis (CBA) or risk-cost-benefit analysis (RCBA), a similar though distinct approach to be applied when policy choices put the public at risk. In CBA, engineers or economists weigh the costs of creating and/or controlling a particular technology against its benefits. As the degree of control increases the risks fall, but the costs of control rise. Thus, the objective is to arrive at the optimal (minimum) total cost. In CBA, benefits and costs are figured on a market basis and are measured in standard units of currency.

In many cases, RCBA uses the same initial calculations as CBA, but weighs nonmarket variables as well, often attempting to include the costs of hazards to human life, health, and safety. These risk variables are quantified either in market terms or on some other basis depending on the variation of RCBA methodology the analyst chooses, and risk factors are typically considered on a probabilistic basis to arrive at the likely total cost.

Although, inevitably, any technological project involves a degree of uncertainty in its initial stages, CBA is straightforward when we consider, to use a classic example, the costs and benefits of building a new highway. There, analysts can estimate the construction cost and the value of total benefits, and they can compare those costs and benefits against other alternative uses for the financial resources. Dollar estimates make perfect sense. The costs of building the road, financing it, and maintaining it are estimated in a straightforward manner. It is also realistic to place a dollar value on the benefits, including the economic improvement possible through a better transportation system. Analysts use empirical data; the effects of such projects have been observed before and are relatively clearcut.

The difficulty in RCBA is to weigh all the costs. In many instances, the costs and the benefits of a technology do not lend themselves easily to specific dollar values; indeed, some do not lend themselves easily to quantification at all. Consider the issue of pollution from fossil-fired electric power generation. The potential cost of property damage may be calculable on a dollar basis, although the amount an analyst will choose may vary widely, depending on whether controversial elements, such as the prevention and cleanup of acid rain, are added. Those who see acid rain as a direct consequence of fossil-fired generation may want to increase the potential value of property damage many orders of magnitude higher than someone less convinced of the cause of acid rain. The estimated cost is likely to vary, as well, depending on the extent of control considered necessary.

But the issue of property damage is far less thorny than other aspects of air pollution. What, for example, is the cost if we feel that air pollution might destroy an outdoor stone sculpture? Perhaps we can make a tenuous valuation on the basis of comparative art auction prices, although that value cannot completely compensate for the loss of a unique aesthetic object.

More important, how do we account for the impact on our health and lives? To deal fully with the impacts of pollution, we have to add health and safety to the analysis. But what is the dollar value of human life and suffering? Some economists and engineers have tried to establish such values. What's more, the courts routinely make cash awards to those who have suffered from the impacts of technology – a practice we generally regard as acceptable.[1] Yet such calculations are subject to enormous variation and disagreement by the very people who perform RCBAs (see Rhoads 1980). Indeed, they use different units of measure. While some place a dollar value on human life, others measure the cost on its own terms, for instance, as excess deaths per 100,000 people. But since other costs and benefits have distinct monetary values, this leads to difficult comparisons where deaths are weighed against dollars.

To simplify the comparison, some analysts attempt to arrive at an appropriate dollar value of life, which economists call the *human capital* approach to RCBA. But, of course, to do so raises serious ethical questions. Is a person's value strictly economic? Is he or she to be measured only as the sum of expected earnings – to give an example of one type of measure now used? Can we argue that a project that could kill 100 people and injure, 1,000 has a human cost that is only the sum of lost work days, hospital bills, and other financial losses? Few would say this is the case; in fact, any such measurement of human life, at first glance, might seem contrary to ethical norms.

The human cost inherent in any energy technology is so problematic that it is sometimes simply left out by those decision analysts who would be most expected to quantify it. Some have argued that giving any value for nonmarket, subjective costs is wrong in principle, and their solution is to perform RCBAs that only include readily quantifiable

elements. They maintain that only by avoiding human costs can an unbiased, objective, value-neutral study be achieved.[2] But is this really a better, fairer approach than a human capital valuation?

There are crucial questions raised by value-neutral studies. Implicit in the concept of such a study is the idea that facts and values can be clearly separated and also that an RCBA can and should deal only with objective facts. Many philosophers have strongly questioned whether a fact-value dichotomy can be truly said to exist in RCBA or even in science (see Michalos 1980 and Shrader-Frechette 1985). Science does deal in facts, but these *facts* are subject to change and *evaluation* as the data change. If indeed science dealt only in immutable facts, then how could there be new theories that attempt to evaluate data and explain causality? (See Shrader-Frechette 1985.)

The facts of RCBA are even more problematic. First, the very act of choosing details deemed objective over those deemed subjective is itself evaluative and seems to contradict the premise of a value-neutral study. Second, the objective facts usually included in RCBA involve market valuations, which may be observable and factual, but are at the same time extremely variable and subject to fashion, taste, and many other highly subjective factors.

Even the prices of the necessities of modern life such as energy resources – presumably less affected by taste – truly reflect value only at the instant they are given. Consider the price of a commodity such as oil. On a market basis, it went from a few dollars per barrel to over $30 in less than a decade, and then within a couple years slipped to half that peak amount. Since RCBA is intended to evaluate projects at the planning stage, if the price of oil is to be included, it must be given at some estimated future price – that is, at a time when the plans will have been put into effect. What is the objective valuation of the future price of oil? Clearly, it may depend on when the sudy is performed, and the views and values of the person performing the study. Indeed, there are probabilistic estimates for decisions under uncertain futures, called *Bayesian estimates*, that are acknowledged to be subjective (see Raiffa 1970).

Since even the most objective facts may be inseparable from the decision analyst's values, it seems especially difficult to justify ignoring nonmarket elements because they are too value-laden. But even if facts could be separated from values, it would still be questionable to leave out human costs, where they exist. Doing so might produce a study that is in some sense more objective, but far from complete. Such an approach appears if anything more biased than the one that gives an arbitrary value of human life. At least the latter acknowledges human costs. To avoid confronting them may buy objectivity at the cost of irrelevance.

Some theorists who argue for the elimination of nonmarket variables from RCBA do so not to ignore them, but rather to avoid quantifying factors they believe cannot be quantified. These theorists have suggested that to quantify nonmarket variables gives inordinate power to the decision analysts, since their cost-benefit numbers would then seem to be all-inclusive (see Mishan 1972). According to such a theory, RCBA should focus on what is quantifiable and analysts should simply note those factors that cannot be quantified. Then, it is left to the policymakers (not the decision analyst) to decide which factors – market or nonmarket – should weigh more heavily. In such a policy-making process, it could then be decided that even though an RCBA demonstrated cost effec-tiveness, a project simply was not viable for ethical or social reasons. This approach keeps RCBA in its place as a tool for policymaking, rather than having it determine policy.

But an equally good case can be made for quantifying all the important variables, however difficult such a task may be. The issue may hinge on the answer to this question: If qualitative costs and benefits are not quantified, exactly how are they to be weighed in the political process against costs and benefits that *are* quantified? In other words, although we can say that pollution is bad for our health, how do we weigh that statement in order to determine the required amount of pollution control, which

may range in cost from a few dollars to $100 million for a single power plant.

Philosopher K. S. Shrader-Frechette has noted that quantifying nonmarket variables at least provides a basis for discussion and argument. She notes, for example, that by giving explicit numerical parameters, the Rasmussen study on nuclear plant safety (see Chapter 7) allowed for explicit challenge; vague generalities would have been harder to confront (Shrader-Frechette 1985). The issue for those who believe in complete quantification is how best to do it. Clearly, just assigning arbitrary values for human life and other nonmarket variables is not adequate. But, as we will see, various ways have been suggested to perfect the valuations of RCBA.

## U.S. government decision making

Policymakers in the United States rely widely on CBA, which has been mandated for all federal technological projects.[3] Many of the analyses have proceeded with the assumption of value neutrality. There are numerous examples of government-mandated cost-benefit studies that rigorously avoid quantification of nonmarket elements. For instance, in 1975 an RCBA on oil tanker safety by the Congressional Office of Technology Assessment ignored the costs of oil spills. In particular it omitted such costs as aesthetic damage to beaches, the cost to other vessels of avoiding contaminated waters, and the potential damage to the fishing industry. All of these costs were difficult to quantify and were thus left out (critiqued in Shrader-Frechette 1985).

It should be noted that value neutrality seems to contradict the basic spirit of government policymaking. In the very act of setting policy objectives, the government begins with a set of values, not facts. Policies are adopted (at least in part) because they are beneficial and will improve the general welfare of society. In making such a choice, policymakers use their judgment and are motivated by notions of what is right and good, and of how a government, an economy, and a social structure should function. Often, value neutrality is simply impossible. Government policymakers have had no choice at times but to confront nonmarket issues, in particular the potential human costs of technology. However, at such times, rather than giving exact directions for the establishment of standards for health and safety, they have tended to create vague criteria. In each of the electric-generating technologies we have covered, regulatory criteria avoid explicit calculation of dollar costs and benefits to human life and suffering. Officials have mandated that the projects must have an *acceptable level of risk*, or, equivalently, they have used the principle of a *threshold level* below which no measurable risk presumably exists.[4]

A clear example of how this basic approach works in practice is the Nuclear Regulatory Commission's policy on nuclear power plant safety. Following a review of plant safety in the aftermath of the Three Mile Island accident (see Chapter 7), the NRC issued a document entitled *Safety Goals for Nuclear Power Plants*. These goals are stated qualitatively and in nondollar quantitative terms (paraphrased here):

Members of the public should bear no significant additional risk to life and health, beyond an established baseline level, because of the operation of a nuclear plant.

The risks of generating electricity by nuclear plants should be no greater than those due to viable competing technologies (mainly coal-fired generation).[5]

Phrases such as "acceptable level of risk" or "significant additional risk" are vague and unhelpful as part of the analytical input to policymaking. But in this case, does the quantifiable standard – comparing risks –really provide a better measure? Only if it is clear just what the probability of an accident is. The NRC established a standard of what that probability should be. It stated that each nuclear reactor should be built so that there is only 1 chance in 10,000 that it will have a core meltdown.[6] This formulation stems from the Rasmussen study, which estimated the probability of a

meltdown and of a worst possible accident. As related in Chapter 7, the results showed that such probabilities were very low and were intended to demonstrate that the hazards could be ignored[7] in everyday life by the average citizen. As we noted, however, there has been considerable question about the accuracy of the Rasmussen probability estimates and the standard has evidently not been reassuring to the population at large. In addition, despite the attempted quantification, there has been no widely accepted means of establishing that a reactor design actually meets the safety goal. Thus, the NRC's probabilistic standard remains subject to much dispute.

## Alternative methods of establishing RCBA

Although government regulators have tended to use vague or untestable standards for levels of public risk, that does not mean that stronger, clearer standards cannot be developed and that RCBA cannot play a role in that development. Here are three important alternative methods.

### Revealed preference

One method that has had a number of proponents, in the technical community especially, is called the revealed-preference approach. This method says that there is a level of acceptable risk, and it offers a statistical procedure on how to determine it. The idea is that we accept certain risks in our daily lives. We drive cars, walk the streets, use electrical appliances, ride in elevators, take medications, and so on; all of these can, and have, caused injuries and deaths. If a technology can be shown statistically to have little risk, indeed less risk than most of the above, and if it has substantial benefits as well, then we should have no hesitation in adopting it.

This argument was employed by Dr. Rasmussen, as well as by Nobel Prize-winning physicist Hans Bethe, in supporting nuclear power development in the United States.[8] However reasonable such a position might seem, it in fact contains several problematic elements when applied to nuclear power. First, are the risks of power plants – and the waste they produce – fully known? Second, is the assessment of the need for nuclear (or any other specific kind of) power plants necessarily correct? Third, is it qualitatively the same kind of risk when we choose, say, to drive a car and when a nuclear power plant is built near our homes?

In other parts of this book, we have considered the first two of these questions. The third, however, bears discussion here. The crux of the issue lies in the nature of the risks being compared in the revealed-preference analysis. They are qualitatively different – some are voluntary, others involuntary. We may indeed be willing to assume a risk in order to drive a car, recognizing that it benefits us directly to have this method of transportation at our disposal. But the decision to build a power plant is often governmental and/or corporate, and those potentially at risk may have little or no influence on the decision. In other words, driving a car would be a voluntary risk, while the power plant would be distinctly involuntary. It can be argued that there is a certain involuntary element in driving; in order to function in our car-oriented society we must drive. By the same token, living near a power plant is, in a sense, voluntary. Someone who lives near the site of a proposed nuclear power plant can move. However, the cases are not equal. The former is a life choice that can be freely planned throughout adulthood and allows a considerable number of alternatives. The latter is an either/or choice that is unforeseeable and imposed at a particular time with no intermediate steps available. The two risks are, thus, qualitatively quite different. To accept one voluntarily is not a sufficient condition for allowing the second to be imposed on you. To argue that a person ought to accept one risk because he or she accepts the other is an example of the naturalistic fallacy in ethics.[9] Statistics about relative dangers, even if they are absolutely correct assessments, are far less important to the argument.

Actually, some advocates of the revealed-preference method acknowledge the problem and have attempted to introduce a compensating factor into their calculations. For example, involuntary risks may be assigned a higher value and therefore require a greater amount of benefit to offset them (see Starr 1976). However, attempts at finding an appropriate level of compensation have produced widely disparate results.

But such calculations tend to obscure larger issues, not of reasoning, but of equity in a democratic society. Most important, when governmental authorities decide ultimately whether to build a power plant, we find that their decision requires some individuals to bear special risks. Although the authorities do so, presumably, in order to provide benefits for the many, the decision makes the burden of risk not only involuntary but also unequal. So whenever a power plant is built, some are put at greater risk than others by mere proximity to the plant, if nothing else. In fact, it may well be that some are imposing risks that they themselves would be unwilling to assume – a proposition that seems incompatible with our notions of equity and justice.

## Expressed preference

One creative solution to the ethical and evaluative dilemmas of RCBA methodology has been proposed by E. J. Mishan, an economist specializing in RCBA theory. Mishan proposes a transformation of RCBA, which eliminates some of the ethical problems of the human capital and revealed preference methods. The Mishan approach takes the cost-benefit evaluation, in principle, away from the remote expert analyst (an engineer or an economist) and places it with the population at risk (see Mishan 1976, 1981).

The Mishan transformation is accomplished by:

Moving from an attempt to calculate the value of life in a community affected by a change in risk to calculating the community's own valuation of that change in risk.
Considering, in the process, the valuation that each individual in the population at risk

places on an improvement (or degradation) of his or her own level of risk.

We can readily see the meaning of the transformation. First, as in revealed preference, the emphasis is on *risk*, not lives; the cost of reducing a risk is far more easily and ethically quantified than the value of a life. But this method improves on revealed preference by referring to those affected. It calculates what a change in risk means using the underlying principle that each person knows best his or her own interests. It also gathers data on the preferences of those affected for the specific risks.[10]

Through Mishan's approach, a representative sample of those at risk would be questioned as to how much a risk meant to them and how much they would be willing to pay to reduce or eliminate it. In other words, if a project meant near-certain death or injury to some members of the community, they probably would be unwilling to accept it at any cost or would pay a substantial amount to reduce that risk. If, on the other hand, they were persuaded that the risks were negligible, they might not be willing to pay anything. The statistical facts, although important, would not be as important as the meaning or perception of risks, including the consequences of the hazard in question. Some projects, say nuclear power, might seem nearly riskless to an expert considering only the probabilities of an accident, but would seem daunting to those at risk of the consequences. That perception, which probably upsets the sense of well-being (and is hardly quantifiable by an outside expert), would be more crucial to the peopple involved and would therefore have more weight than more objective estimates of the mere chance of occurrence. No better examples of these perceived consequences are found than with nuclear power. The fear of radiation is deeply felt, as Lifton and Weart have documented (see the Bibliography in Chapter 7).

Although this method, called expressed preference, RCBA, provides data on public perceptions and feelings, it is designed only to create a better means of quantifying risk and is not intended as a referendum of public opinion on a technological project. Nevertheless, because it would focus on the choices of the population instead of the experts, it does at least appear more democratic than either the human capital or revealed-preference approaches. Some supporters see it as part of a general movement to incorporate public perceptions in decisions about energy and other technological issues.

Although this method appears to be an improvement over other methods of RCBA, it nonetheless raises questions concerning its technique and underlying philosophy. For example, the way questions are phrased can bias survey results. Also, since the people at risk generally lack technical expertise, there is a worry that researchers are apt to so oversimplify the issues that they no longer reflect the reality of the problem. There is a drawback to the method, too, when we consider energy projects that are meant to last (or whose effects will last) over several generations. In such cases, we will have some people judging the acceptability of risks today for people in the future whose values and belief systems may well be different, and whose choices about risk are impossible to predict (see Grossman & Cassedy 1985). Indeed, one generation acting for its self-interest could irreversibly damage the world for the next.

## Ethically weighted RCBA

An alternative approach to overcoming the problems of RCBA is to adopt an ethical weighting in calculating costs and benefits. Decision analysts here would have to use ethical values as the overriding criteria in their RCBA. In other words, any technological proposal that would lead to a violation of fundamental ethical values would require significant revision or would be rejected outright – regardless of its market cost effectiveness.

One such concept would require the adoption of a single precept of distributive justice (see Appendix B) to provide the framework for analysis (see Kneese Ben-David & Schulze in MacLean & Brown 1983). For example, one could use the utilitarian concept of the greatest good for the greatest number or a simplified egalitarian idea that actions must improve the lot of those who are worst off.[11] Under such constraints, decision analysts would have to determine, to use the second concept, the group worst off and see whether it would benefit from the project under study. If those who were most disadvantaged became still worse off, the project would be eliminated even if there would be a general improvement of society's welfare.

Through such a system of ethical weight, we might also be better able to make judgments about intergenerational issues. For example, using the utilitarian formulation, we could say that we must include all those in the future who are likely to be affected. If they are likely to be put at great risk or to be burdened with greater costs while we in the present generation reap the benefits, then the project would be unacceptable. Of course, using any given rule might leave unclear the ethical validity of one or another course of action. In that case, policy could be set by the usual factors of standard cost-benefit calculations. However, at least the problem then would have been scrutinized from an ethical perspective and could be cleared of violations of ethical norms.

Of course there is an obvious problem with this approach. It is apt to reduce all ethics to one easily phrased precept. In a pluralistic society such as ours, there is not likely to be a single norm of distributive justice that all would endorse. One way to overcome that problem might be to specify two or three, or even a dozen, different precepts that are widely held and then perform, in essence, several different RCBAs. This, however, could lead to confusion for policymakers. A. V. Kneese and his colleagues have demonstrated that as few as four different rules of distributive justice can lead to four different conclusions – sometimes diametrically opposite ones.

Yet it might be useful for policymakers to have several different positions to consider.[12] One problem with a conventional RCBA is that by providing one set of figures, it tends to disallow ambiguity. Several ethically weighted analyses, in contrast, might point up ambiguities and thereby stimulate a fuller discussion of the social and moral implications of a given technology than might otherwise have taken place. In this way, too, major variables are quantified without any one set of numbers being deterministic. Policy cannot then be decided solely on the basis of a single RCBA. Decision analysts would only provide a range of alternatives from which the policymakers could choose. What is left unclear is just how many RCBAs would be enough to represent adequately the span of values in society. There is no simple answer to this problem.

RCBA might indeed be improved, but what seems evident is that no one method is likely to give us definitive answers as to how best to decide criteria for health and safety. And if RCBA cannot provide definitive answers, then we must consider who should decide. Who should decide the form of RCBA, and the way safety standards are formulated? Indeed, since standards are inseparable from basic technology, the question must be broadened: Who should decide policy over technology?

## Who should decide?

Some have argued that only engineers and scientists have the expertise to understand new technology and that, therefore, only they are competent to make decisions about it. But this position necessarily makes assumptions that are highly debatable. It assumes either that the issues are strictly technological, or that technical experts have the widest insights into all the ramifications of a technology, or both. Of course, this has been disputed for many years (see Laski 1931). And none of these assumptions are likely to be the case as long as people's health, wealth, and value systems are to be considered. In fact, as we have seen, experts have often adopted dubious assumptions in their analyses of technological issues.

Another reason experts do not make ideal arbiters of public policy is that they often disagree, and society is left with the dilemma of which expert to believe. The ordinary citizen cannot trust any given expert simply because he or she appears to use the facts; all sides will employ scientific facts in their arguments. As a result, the public and government officials are often left confused and uncertain by debates on technical matters. This can be seen, for example, in the debate over federal government air quality standards.

The Clean Air Act of 1970 (and its amendments in 1977) established a primary ambient air quality standard that, as we saw in Table 6.13, was closely correlated with the best estimate judgments for an effects threshold. In other words, the standards adopted were related to an estimate of the point below which no measurable risk presumably exists. The standards were apparently based on widely held beliefs of the scientific community.

Despite the assumption of a scientific consensus, the threshold concept in relation to air pollution was soon challenged. First, it was established that certain groups within the population have markedly lower thresholds for health distress due to air pollution. Infants and the elderly are known to be especially vulnerable. But laboratory tests also began to show distinct physiological effects on the general public at pollutant concentrations considerably below the established thresholds. These examples have led some experts to the conclusions that (a) there may be no single threshold standard for air quality that is appropriate for the entire population; (b) the present standard is too lax; or (c) the methods that regulatory bodies use to establish standards are inadequate (see Wilson et al. 1980 in Chapter 6, Bibliography). The debate, however, did not change the views of all the experts and had little impact on the federal regulators; they left the standard unchanged.

Similarly, disagreement and uncertainty have prevailed over thresholds for exposure to magnetic and electric fields. Extremely strong fields, such as those directly beneath very-high-voltage (hundreds of kilovolts) transmission lines have measurable biologic impacts. However, the intensities of these fields drop rapidly wtih distance from the line. For fields of lower intensity (due to lines of lower power as well as distance from high-power lines), cause and effect relationships of harm to living organisms are difficult to prove.

Yet some epidemiological studies have indicated hazards from fields that are far weaker than engineers have thought could be significant. Magnetic fields, oscillating at only the 60 Hz, the frequency of household alternating current (see Chapter 3), have been implicated in several studies that showed a *statistically* significant rise in childhood leukemia. The disease is suspected to be promoted by the disruption of biochemical reactions in the body that themselves rely on small fields to function. But thus far harm has been established only statistically and the implication of cause and effect is not universally accepted. Although some experts would like to see standards established for exposures to low-power magnetic and electric fields, none exist and many experts still do not see the need for any.

Not only do experts disagree, they sometimes do not even see problems the same way. For example, geologists and engineers were asked to estimate the risk of an earthquake at a nuclear reactor site in California. Although we would generally view both engineers and geologists as members of the scientific community, they produced very different results. The latter said that the risk could not be quantified; the former reported that the risk was "so low that it need not be included in the design bases" (Hund 1986). The disagreement lay not in the facts but in the way each group saw the world. The geologists tried to understand the nature of the site. While they could identify a fault structure, they could not determine whether or not it would produce a significant quake during the lifetime of the nuclear facility. The engineers, on the other hand, used standard risk assessment methodology. They could not predict whether or not there would be an earthquake either, but the method allowed them to set odds that were extremely low (see Hund 1986).

An important idea has been advanced for resolving the disagreements among scientists or between groups of people within the technological community. It is the idea of the science court. This concept, as advanced particularly by Arthur Kantrowitz (1975, see also Michalos 1980), calls for the establishment of a system of adversarial proceedings on major technological issues. In this court, disputants would have a chance to present evidence, cross-examine one another, and argue their case just as in a court of law. The decision would be rendered by a panel of scientists acting as a sort of jury. These individuals would not be in fields directly connected with the matter at hand. So, for example, no nuclear engineers or physicists would decide questions about nuclear facilities, since they might directly benefit from the outcome. Kantrowitz did not mean for this body to determine policy per se. The court would determine what could be done as a strictly technical matter, not what ought to be done finally; that decision would be left to policymakers. But at least the court would put to rest technical disputes and allow the policy process to proceed from there.

However, the neatness of this formulation belies several potential problems. First, the concept implies the fact-value dichotomy, which, as we have already seen, may be untenable. Second, there is a danger that court findings may be taken as truth rather than what, at best, would be an opinion about which side produced the best available evidence at the time of proceedings. Philosophers such as Shrader-Frechette have also raised the question of whether scientists in one field are the best judges of the claims of those in another. It can be asked, for example, whether scientists will tend to view the advance of technology as a good in itself, biasing them to favor a project and tending to require the onus of proof to be much greater on opponents of any given project.

Shrader-Frechette (1985) has argued instead for a technology tribunal. It would preserve the opportunity for an open hearing of conflicting views, but the tribunal would render a judgment that would be less final than the judgment of a court. In this concept, the jurors would include nonscientists who could view the project in a broader context. This would help avoid the bias of the expert who, as Harold Laski (1931) noted more than half a century ago, may tend to confuse the facts with "what he proposes to do about them." Of course, this idea is challenged by those who believe that nonscientists cannot understand the complexities of scientific concepts. But there is good reason to reject the notion that a sufficient understanding of a technological project requires years of study of science and mathematics and that technology is inevitably a mystery to all but an elite.

The issue of a science court or tribunal and beliefs about its composition are tied up with the larger question of who should make policy. The view we adopt probably will depend on what we see as the proper role of government officials, experts, and the people. Many would ascribe to government the right to make rules for the public good, even when that conflicts with the wishes of some citizens. The United States has laws against child labor and free use of some drugs and others that regulate the use of toxic substances – all of which are opposed by some, but are generally accepted. Although some government restrictions are legislated, others are determined by regulatory agencies; the Food and Drug Administration, for example, permits or denies the public access to drugs. The agency determines what is right for citizens. Government in these cases is considered to know best and to be acting paternalistically to protect its citizens. As part of the paternalistic view, expert opinion plays a key part of the decision-making process.

Along these lines, we can imagine how government might formulate energy policy decisions. Experts would provide information and analyses (including RCBAs), and then government officials would mandate binding rules about the use and development of energy technology and for the standards for health and safety – presumably for the benefit of society as a whole. Of course, few in our society would attribute to government absolute authority, and most would expect that the government would attempt to learn the public's views and weigh them in any analyses. But in a paternalistic framework, the executive (or legislative) branch of the government could establish authorities to act where they perceived the need to do so. These authorities would not have to consult with the public or its representatives before taking any action (though they might be held accountable after the fact). Essentially, they would have a free hand to make and establish standards and implement policies.

Others, however, see paternalism as authoritarian and argue along the lines espoused by John Stuart Mill. In this view, government has virtually no right to impose policies upon a citizen against his or her will. Mill wrote:

The only purpose for which power can be rightfully exercised over any member of a civilized community, against his will, is to prevent harm to others. He cannot rightfully be compelled to do or forebear because it will be better for him to do so, because it will make him happier, because in the opinion of others, to do so would be wise, even right.

Thus the fact that an energy policy would benefit most people in a society – unless it could be shown to be a matter of life or death – is not, in this view, sufficient reason for imposing it on others, particularly if it placed anyone at risk. All those affected would have to consent to a policy before it could be adopted.

Although allowing individuals to restrict technological development of society as a whole may seem extreme to some people, many would still argue for a greater democratization of the process. They would argue against government-imposed policy, especially where there is risk to life and health. In this view, policies would be arrived at through democratic means including local or even national referenda. It could be required that voters be supplied, perhaps through television and radio, with all the relevant information, including technological information. But the idea of referenda to decide technological development, while democratic, does require an enlightened public. In reality, it presents the possibility of questions being decided by better advertising campaigns rather than on their merits.

Also, if we use democratic methods, we pay a price in efficiency. Even with our current and somewhat limited right to challenge policy in the United States, projects are held up for years, whereas in countries with more authoritarian forms of government projects are planned and implemented much more quickly. Speed may even mean greater social equity. For example,

completion of a nuclear power plant may mean less hardship for the poor as a result of greater economic development. However, the more democratic we make the process, the slower the wheels grind and the more this may lead to hardship.

Perhaps the best choice would be in some middle ground. For example, government officials would begin by determining the basic social goals of technological development. These would be based on widely held views of distributive justice and would perhaps be formulated with the help of ethical philosophers, social scientists, and the public. Decision analysts would then conduct RCBA studies (expressed preference, ethically weighted, or some combination of studies) to lay out the consequences of a specific project. Assuming there was no clear violation of society's ethical norms, some form of science tribunal could hear any arguments as to the feasibility and impact of the proposed technology. The tribunal's report would be used along with the RCBA by policymakers. The latter would also consult with advisory panels, perhaps one entirely composed of citizens and another of experts in different fields – from the social as well as physical sciences. The citizens' panel, according to one suggestion (see Hiskes & Hiskes 1986), could even refine the policy options and help the policymakers determine the weighting of all the evidence. At that point, government officials would make policy decisions.

Although such a system might work in a relatively efficient and equitable manner, given the constraints and realities of a democratic society, it seems unlikely that we will see it implemented any time soon in the United States or other industrial democracies. Indeed, the undeniable complexities surrounding technological decision making can leave one cautious about adopting a set policy-making system. It perhaps explains why we have not set up a definitive system to date. Instead, we continue to try in a somewhat haphazard fashion to balance efficiency and democracy, government authority and individual rights, and expert opinion and public perception.

## Notes

1. The courts and insurance companies award money for injuries and death, but that is different from the matter we are discussing here. Court awards represent compensation *after the fact*; human capital RCBA is an attempt to place some market or intrinsic value on human life and suffering *ex ante*, or before it can occur.

2. For a critique of the value-neutral study see especially Shrader-Frechette 1985, Chapters 3 to 6.

3. In 1981, President Reagan signed Executive Order 12291, effectively requiring all government agencies to perform cost-benefit analyses for all their major regulatory decisions. For discussion of the implications see D. Whittington and W. N. Grubb (1984), "Economic Analyses in Regulatory Decisions: The Implications of Executive Order 12291," *Science, Technology & Human Values*, Vol. 9, No. 1, pp. 63–71.

4. U.S. courts have rules that risks may be too slight (*de minimus*) to be considered. See Ricci and Molton, 1981.

5. In adopting these goals, the NRC was criticized by its own Advisory Committee, which charged that a more stringent principle had been abandoned. This principle stated that risks should be *as low as reasonably achievable* (ALARA) and was based on the idea of utilizing the best available technology.

6. This refers to chances per reactor year of operation (see Chapter 7).

7. Such a notion has an interesting intellectual history. Around the turn of the century, the mathematician Edouard Emile Borel (1871–1956) speculated that there were probability levels sufficiently low that the average person would conclude he could not observe the event in his lifetime.

8. Other notable proponents include Chauncy Starr of the Electric Power Research Institute and Bernard I. Cohen. Cohen and I. Lee developed an extensive table of risk probabilities for decision analysts to use. See Cohen and Lee, "A Catalog of Risks," *Health Physics* 36, June 1979, pp. 708–721.

9. The *naturalistic fallacy* can be any one of three different errors: (1) the confusion of ethical and factual statements; (2) the derivation of ethical statements from nonethical statements – or the confusion of *is* and *ought*; and (3) the definition of ethical concepts in nonethical terms. For a discussion of the naturalistic fallacy with regard to nuclear power see Shrader-Frechette 1983, Chapter 6.

10. Specificity of risk points up another problem with the revealed-preference method. The assumption in compensation efforts for involuntary risks is that the public regards one involuntary risk in the same way that it regards another. However, there is no reason to believe that this will be the case. The expressed-preference approach, by focusing on specific risks, avoids that problem.

11. The formulation used by Kneese and his colleagues for the egalitarian view is that we must try to "maximize the utility of the individual with the minimum utility... until he catches up with the next worse off individual." Or more to the point, until "all utilities are identical." This is taken from the theories of John Rawls (1971). Inevitably, to create an easily phrased precept of distributive justice, Kneese and his colleagues have had to simplify Rawls's ideas greatly.

12. Shrader-Frechette (1985) has proposed a different system of ethically weighted RCBA based on the idea of lexicographically ordered ethical claims. Although her system is too complex to describe here, it is worth noting that it requires a weighting scheme for claims that may also lead to several alternative RCBAs. She defends the idea of alternative RCBAs, noting that this would mean that "citizens could select the *values* by which they would choose social policy as well as the policy itself."

## Bibliography

Grossman, P. Z., and E. S. Cassedy. 1985. "Cost-Benefit Analysis of Nuclear Waste Disposal: Accounting for Safeguards." *Science, Technology & Human Values*, Vol. 10, No. 4, pp. 47–54.

Hapgood, F. 1979. "Risk-Benefit Analysis – Putting a Price on Life." *The Atlantic*, pp. 33–38 Jan.

Hiskes, A. L., and R. P. Hiskes. 1986. *Science, Technology, and Policy Decisions*. Westview Press, Boulder, CO.

Hund, G. E. 1986. "The Fault of Uncertainty." *Science, Technology & Human Values*, Vol. 11, No. 4, pp. 45–54.

Kantrowitz, A. 1975. "Controlling Technology Democratically." *American Scientist*, Vol. 63, pp. 505–509, Sept.–Oct.

Laski, H. 1931. *The Limitations of the Expert*. Fabian Tract No. 235, The Fabian Society, London.

MacLean, D., and P. G. Brown (eds). 1983. *Energy and the Future*. Rowan and Littlefield, Totowa, NJ.

Michalos, A. C. 1980. "A Reconsideration of the Idea of a Science Court." In P. Durbin, (ed.), *Research in the Philosophy of Technology*, Vol. 3, pp. 10–28. JAI Press, Greenwich, CT.

Mill, J. S. 1947. *On Liberty*. Bobbs-Merrill, Indianapolis.

Mishan, E. J. 1972. *Economics for Social Decisions*. Praeger, New York.
   1976. *Cost-Benefit Analysis*. Praeger, New York.
   1981. *Introduction to Normative Economics*. Oxford University Press, New York.

Okrent, D. 1980. "Comment on Societal Risk." *Science*, Vol. 208, April 25, pp. 372–375.

Raiffa, H. 1970. *Decision Analysis – Introductory Lectures on Choice Under Uncertainty*. Addison-Wesley, Reading, MA.

Rawls, J. 1971. *A Theory of Justice*. Belknap Press, Cambridge, MA.

Rhoads, S. E. (ed.). 1980. *Valuing Life: Public Policy Dilemmas*. Westview Press, Boulder, CO.

Ricci, P. F., and L. S. Molton. 1981. "Risk Benefit in Environmental Law." *Science*, Vol. 214, pp. 1096–1100, Dec. 4.

Shrader-Frechette, K. S. 1983. *Nuclear Power and Public Policy*. D. Reidel, Dordrecht, Holland.
   1985. *Science Policy, Ethics, and Economic Methodology*. D. Reidel, Dordrecht, Holland.

Starr, C. 1976. "General Philosophy of Risk-Benefit Analysis." In *Energy and the Environment*, Pergamon, New York.

Wilson, R., and E. Crouch. 1982. *Risk/Benefit Analysis*. Ballinger, Cambridge, MA.

# III

# Energy technology in the future

# 11

# Alternative technologies

## Introduction

*In the future, we will need alternative techno-
logies to replace our present conventional sources
of energy.*

Few would disagree with this statement.
Indeed, most people would say that alternative
technologies are desirable right now. If we are
able to develop inexhaustible energy sources,
our concerns over the limits of fossil fuels can
be forgotten (Chapter 2). If we can find a new
source that does not involve combustion then
our concerns over carbon dioxide – the greenhouse effect – will diminish (Chapter 6). And if
we develop a new alternative to nuclear (fission)
power, we can then leave the controversies of
that technology behind (Chapter 7).

Yet even though people generally agree on
the usefulness and desirability of alternatives,
they may well have different opinions about the
extent to which societal resources should be
expended now in the development of such
technologies. Promoters of alternative technologies express few doubts. Typically, they
represent alternatives as technically feasible,
economically viable, and environmentally safe –
if not immediately then in the near future. Such
promotion, often by the technical innovators
themselves, is not necessarily intentional mis-
representation, but more likely reflects the
innovators' hopes and aspirations for their
projects. Also, promoters of alternatives may be

motivated by a belief that a technical fix must
exist for every problem besetting society (see
Bury 1932).

As we will see in this chapter, however, the
development of new energy technologies (or
any technology) is an uncertain process. From
the conception of the project to its final
adoption by society, the uncertainties may
encompass not just *whether* the concept will
work, but just as importantly, *when* and *if*
widespread adoption can be feasible. We will
see, in fact, that the time scale (*term*) of expected adoption of an evolving technology
is an important consideration in judging its
chances of success; the longer term the prospects, the greater the uncertainty of its ultimate
success.

Finally, we should note that alternative
technologies frequently promise great advances:
freedom for the future from dependence on un-
reliable sources of energy, the reduction in the
harmful impacts of current technologies, and the
elimination of the risk of catastrophic accidents.
These claims should be regarded with some
skepticism. Not only must a technology function, but, we must ask, at what cost – both
financial and environmental? In some cases, the
cost of development and adoption of an alternative technology could be so great that public
welfare would be harmed. And environmental
costs may be lower than those of conventional
technologies, but that does not mean that

alternatives necessarily will be environmentally benign.

## The evolution of new technologies

### Stages

The development of a new technology, from its initial conception to its widespread adoption, is both a technological and a social process. There are many historical examples of the evolution of technologies. (Examples can be found in analyses of technology and the social process; see Kranzberg & Pursell 1967 and Linstone & Sahal 1976.) For the most part, they have evolved through stages – from invention to adoption – without developmental subsidies from government or big business. Development essentially was determined by the market. So, for example, starting in the nineteenth century, steel makers gradually discarded the Bessemer process in favor of the open-hearth process of steel making because it was in their economic interest to do so. The new process worked better and gave its users and producers advantages over those who still relied on the Bessemer process of steel making. The same kind of market-driven evolution – sometimes called the *natural* pattern of evolution – has given us many other familiar technologies, including many energy technologies. Fossil-fired electric generation and industrial process heating are two examples of energy technologies that have followed the natural pattern.

The natural pattern generally proceeds as follows:

1. *New knowledge* – the establishment of basic scientific principles.
2. *Development of technical capability* – proof of the principle for technological application.
3. *Prototype construction* – the first working model designed for practical operation.
4. *Commercial introduction* – low-cost design and manufacture for mass markets.

5. *Widespread adoption* – wide utilization by society.

Although many present-day energy technologies have followed the natural pattern, others have not. Rather they have developed with the help of programmatic government sponsorship. The most notable example is nuclear power. It had government assistance through the process of commercialization and may not have been developed at all were it not for government support (see Camilleri 1984). Some alternative energy technologies may also require significant government sponsorship. The costs and risks of underwriting their development may be prohibitive for individual inventors and entrepreneurs.

While government-sponsored technology development may not be market driven, it nonetheless parallels the natural innovation process. The stages of government-sponsored programs are:

1. *Research* – proof of the scientific principle for technological potential.
2. *Development* – engineering design for technical feasibility.
3. *Demonstration* – the proof of operational feasibility and economic competitiveness for the technology.
4. *Commercialization* – the process of dissemination of information, building of experience, and overcoming various barriers to widespread adoption of technology; the stage often leading, in our economy, to the privatization of the production process.

These stages of programmatic development are called *RD&D*, standing for *research, development, and demonstration*. Commercialization, as a programmatic stage, is relatively recent and has been given different emphasis by various presidential administrations. For these reasons, this last stage is not included in the jargon of government-sponsored development.

### Time scales

There are currently many alternative energy technologies under development – some at every stage of the natural and programmatic stages of evolution. Accordingly, each technology has a different time scale of expected adoption. These time scales are commonly grouped into the following categories:

*Near term* – essentially at the commercial stage, but presently underutilized (impact expected within 5 to 10 years).

*Medium term* – in the engineering–development stage; proof of technical feasibility and prototype demonstration achieved; proof of economic viability to come (10 to 15 years before expected impact).

*Long term* – in the research stage; scientific feasibility yet to be proven; development of technical workability or economic viability awaiting the outcome of this stage (15 to 25 years before expected impact).

Examples of each category are given in the following lists. Each time-scale category carries a different implication of uncertainty. Technologies with near-term prospects, for example, are the most certain to proceed to wide-scale use, whereas the outcome for those in the medium term is significantly less certain.

*Near term*
1. *Industrial conservation* and *technical efficiency* – reduction of heat losses, materials recycling, combustion efficiency, heat recuperation, cogeneration.
2. *Solar thermal* – domestic hot water.
3. *Solar-direct conversion (small-scale)* – remote site use of photovoltaic conversion for, rural and wilderness sites, and small electric loads.
4. *Small-scale, remote-site wind generation* – rural and wilderness sites, and small electric loads.
5. *Geothermal power generation* – particular sites of availability.
6. *Small hydroelectric generation.*
7. *Enhanced oil recovery* – extending recoverable resources.
8. *Fluid-bed combustion* – cleaner, more efficient coal combustion.

9. *Resource recovery* – heat recovery from municipal solid waste.
10. *Heat pumps* – domestic heating.

*Medium term*
1. *Synfuels* – coal-derived and biomass-derived.
2. *Alternative fossil fuels* – tar sands, oil shale, nonconventional natural gas.
3. *Solar heating and cooling* – space conditioning of residential and commercial buildings.
4. *Solar industrial process heat* – processing industries, low to medium temperatures.
5. *Solar-direct conversion (medium-scale)* – domestic and commercial uses of photovoltaic conversion, moderate electric loads.
6. *Dispersed-site generation* – solar or wind electric generation feeding utility grid.
7. *Resource recovery* – reduction of solid waste to synfuels.
8. *Advanced storage* – advanced batteries, hydrogen storage, thermal storage.
9. *Alternative conversion* – fuel cells, magnetohydrodynamics.
10. *Breeder reactors* – uranium-cycle fission.
11. *Superconducting transmission* – more efficient electric transmission.

*Long term*
1. *Nuclear fusion* – power reactors.
2. *Solar (large-scale)* – large-scale plants; widespread (dispersed-site) use; low-cost, direct conversion.
3. *Breeder reactors* – thorium-cycle fission.
4. *Hydrogen economy* – large-scale transmission and distribution of energy using the medium of hydrogen.
5. *Universal, dispersed-site generation* – widespread use of small-scale and/or renewable source generation in autonomous or electric grid usage.

Any technology that reaches the commercialization phase will have passed from stage to stage, going from long- to medium- to near-term prospects as the result of scientific or engineering breakthroughs or changes in market conditions. In the late 1980s, we saw an example of breakthroughs leading to a change in the time scale of development of a technology. Superconductors, materials that offer no electrical resistance (to direct currents) have been known for decades. Yet despite the fact that superconductivity could mean, among other advances, improved energy production and transmission (see Appendix C), it had virtually no technological potential, because the previously known superconductive compounds had to be operated at such low temperatures that the capital and operating costs were prohibitive for commercial use. In the late 1980s, however, researchers in the United States and Switzerland found compounds that proved to be superconducting at temperatures dramatically higher than ever before. Suddenly many practical uses could then be considered and engineering development for those uses started. Thus superconductors could be considered as moved from the *long-term* to the *medium-term* time scales for expected adoption.

The case of superconductors also demonstrates the way sponsorship of development changes. When it was strictly a long-term prospect, superconductivity research was relegated to the laboratory under the sponsorship of government and, in a few cases, big business. Once the breakthrough was achieved, some venture capital was invested by entrepreneurs in product development companies. The major work remained in the laboratory, but with improved prospects, both government and big business stepped up funding of research. Because of a relatively high degree of uncertainty, even medium-term technologies often require programmatic sponsorship to succeed. For the long term all efforts require sponsorship usually from the government,[1] because the great uncertainty of a payoff makes the risk too high to investors. But as a technology draws closer to commercialization, it probably will draw (as the case of superconductors illustrates) increasing interest from private sector investors as well.

## Near-term technologies

The near-term technologies have reached the stage of commercialization. Technical feasibility

has been proven, demonstrations (or pilot units) have operated successfully, and costs are at least close to those of conventional technologies. Still these technologies have not penetrated the relevant market to their full potential. In many cases, penetration is nowhere near that market potential. We here explore some of the reasons for this lack of penetration.

THE COST BARRIER TO ADOPTION. The viability of a new energy technology depends first on its cost effectiveness. Because it will substitute for a conventional source (see Chapter 4), it must be comparable or superior in terms of costs before it can have wide success in the market. Although the near-term alternatives approach cost competitiveness, they may fall just short. In other words, there may have to be some change in the energy market or a small cost-reducing advance in the technology itself to make the alternative market competitive. So, for example, an increase in the price of oil might suddenly make enhanced oil recovery technology economically viable. Alternatively, the government might step in and offer incentives that will at once make an alternative cost effective. It has been argued that incentives would make solar domestic hot water heating (see Chapter 12 for a technical discussion) competitive with conventional means of hot water heating.

The general economic situation can also be a deterrent to investment in alternatives. During the early 1980s, there were high interest rates and a recession. Both discouraged any major capital investments, especially by businesses. And so while heat recuperators, fuel-efficient boilers, and cogeneration units were available and could at least in some cases have been cost effective, adoption proceeded at a slower rate than it might have in a period of economic growth and lower interest rates.

Finally, there may be a problem for consumers simply in assessing the relative costs of technologies. As we discussed in Chapter 4, the cost of an alternative (or of conservation) may be an added capital cost, whereas the savings may be in terms of factor costs, particularly fuel. It may be difficult, however, to persuade an individual to spend a large amount now to save small increments over time, especially when the savings are not entirely apparent because the future prices of factors are unclear. For example, we may project that the cost of oil will average $20/bbl over the next decade and then we can calculate how much oil (and how much money) a solar hot-water system will save. But the savings may depend on oil prices, and those prices will be increasingly uncertain the further into the future the analysis is taken. In some cases, it may take several years for an investment in an alternative to payoff – that is, for the buyer to make up the extra capital cost in saved fuel costs. But consumers probably will be wary if there is a long projected payback period.

THE SOCIAL PROCESS OF ADOPTION. Cost effectiveness is not the only reason some near-term alternatives have failed to gain widespread adoption. Even with financial incentives, adoption can be slow. The U.S. government for a time offered tax incentives to make solar hot-water heating a viable alternative and the technology made some market headway before the government withdrew its support. But solar hot-water heating still did not achieve wide market penetration even in the Sun Belt states, where it had demonstrable financial advantages. The problem lay mainly in the basic social process by which a new technology is adopted.

The behavioral patterns of people in adopting a new technology have been studied by sociologists (see Rogers & Shoemaker 1971). (A *new technology*, incidentally, includes *any* innovation in the methods of accomplishing practical ends, whether it involves a working mechanism or not.) The process leading to widespread adoption is called the *innovation process*. When tracked in time, it can be shown typically to follow the S-shaped pattern in Figure 2.1. The S-shaped curve is familiar, of course, from our discussion of natural resource exploitation and turns out to be applicable to human technological behavior in general. An S-curve for the innovation process would show the cumulative number of adoptions of a new

technology versus time. In the beginning phase, the increase in adoptions would be exponential, where the rate of rise at a given point in time would be proportional to the number of adoptions at that point. The last phase would represent market saturation, indicating the limits of adoption, such as has occurred with many household electrical appliances.

In each phase of this time pattern of innovation, individual adopters have distinct characteristics. In the very earliest phase, for example, the adopters tend to be risk takers who are eager to be seen as trend setters. The *early adopters* in the domestic solar market were most often wealthy and well-educated people who could risk a financial failure of investment in order to be the first with a new technology. The decision to be an early adopter is not, however, strictly a matter of economic class, because many people in the same socioeconomic group are not so adventuresome. Early adopters were then both of a social class and of a particular personality type.

After the early adopters, we next see the *early followers*. These people are innovative, but not quite as willing to take the initial risk on an unknown technology. They may be influenced merely by the fact that their more innovative neighbors have already purchased solar systems, but they would not necessarily wait to see what that neighbor's experience would be. They would, however, tend to be influenced more by their neighbors than by advertising.

This means of building confidence in a new technology is an example of what sociologists call *networking*. The term networking here refers to interpersonal communications and influence, where the participants are most probably unaware of their contributions to the process of innovation and the diffusion of a new technology into society.

As the process continues, we encounter the *late followers*. Here, we are definitely out of the exponentially rising part of the S-curve of adoption. If we were to move into the future to even later adoption, we would begin to move into the saturation range of the curve. Solar hot-water technology has not yet approached the late follower stage, much less the saturation point. But other technologies have in recent years more quickly scaled the upper portions of the S-curve. The personal computer represents a notable example. In the late 1970s, it was still in the early adopter phase, but within the next decade millions of personal computers were purchased.

Part of the success of the personal computer depended on the expansion of applications (its usefulness) and a reduction in its cost. But other factors were important in overcoming resistance to adoption. Communications – first interpersonal networks and later the general media – helped overcome *psychological barriers*, particularly fears about the new technology's complexities. The personal computer market also began to exhibit what social scientists call *trialability*, which may be required for the late follower stage. Trialability is, in essence, the estimation of whether the technology is going to be workable and convenient for an individual. Late followers want to know: Will there be organizations and facilities to repair the equipment and will repair services be readily available? Will repair costs be reasonable? Is the product guaranteed against malfunction? And is there a dependable performance rating of normal operation?

Repairs, guarantees, and ratings are *institutional* barriers to adoption. In other words, because no organizations preexist in society to facilitate the use of the technology, they have to be established before widespread adoption is likely to take place. Not only must repair businesses and a credible system of performance testing and ratings be set up (by government or industry), there must also be some institutional system (public or private) to provide consumer protection to identify and end dishonest business practices. A solar hot-water system needs the equivalent to the nameplate ratings and warranties of a hot-water boiler. The failure of the manufacturers of domestic solar equipment to provide such institutional guarantees has, in the opinions of observers, contributed to the low penetration of the potential market in the late 1970s and early 1980s (see issues of *Solar Engineering & Contracting* magazine from that period).

LEGAL BARRIERS. Legal questions may also produce barriers to adoption of new technologies. In the 1970s, entrepreneurs who saw the potential for dispersed-site generation of electricity found legal barriers to the marketing of their energy. They proposed to use a number of different alternative technologies: cogeneration, small hydroelectric and small-scale solar or wind generation (see Appendix C). But regardless of the kind of technology proposed, local electric utilities refused to purchase the electricity. And because the utilities controlled the power distribution grid, and because the dispersed-site innovators had to tie into that grid, the utilities effectively blocked this form of alternative energy development. They were supported by public utility laws, which gave them monopoly control. In other words, the problem for the dispersed-site innovators was not technical, financial, or sociological; it was strictly legal. The utilities had the right by law to refuse to open the market.

The utilities were acting primarily in the interests of their stockholders; however, legislators, who considered the perspective of the public interest overall, took a different view. At both the state and federal levels, lawmakers recognized that public utility statutes, drawn up in an earlier era to enhance the expansion of the utility grids, were now being used to prevent adoption of alternative energy sources – alternatives that in an era of shrinking domestic resources had social as well as economic value. The result was legislation at the federal level,[2] and in some states, that required the utilities to make such purchases at a fair price. The legislation has not entirely settled the matter, however, as several utilities have gone to court to oppose the required purchases.[3]

Old laws have also affected the adoption of modern solar energy systems. The issue revolves around the rights of an investor in a solar energy system to have an unobstructed flow of sunlight. This question of solar access deals with disputes that can arise between owners of adjacent properties, when buildings or vegetation on one property cause a shading of the solar collector of the other.

Several approaches have been proposed to

resolve such conflicts. One would grant the solar-collector owner unobstructed access to the sunlight in perpetuity. Legislation of this type, proposed in California, harks back to the doctrine of ancient lights in British common law. In the old common law, the householder was simply concerned with his or her level of illumination rather than collection of solar energy. This absolute doctrine, however, conflicts with other constitutional rights because it would limit the right of the owner of the other property to develop it. Therefore, this type of law has not been widely passed in the United States. Some, in fact, have argued for the other extreme: that the only absolute right to unobstructed sunlight is from directly overhead.

Workable legal doctrines on solar access have been explored to enhance the adoption of this technology, while at the same time preserving constitutional rights (see *A Forum on Solar Access* 1977). Toward these goals, perhaps the most workable doctrine is called *solar easement*. A solar easement is analogous to a land easement in which a strip of land owned by one party may be used by another (usually to gain access to another piece of land) under an agreement or contract. The agreement usually involves an exchange of money in return for use of the easement. In this case, the solar collector owner would negotiate an easement of sunlight whereby his neighbor would forbear obstruction of his or her access to it. The legal issues have not been resolved and it may well take experience with various statutes and many court precedents before there is general agreement on what legal formula works best. In the meantime, the legal uncertainty may pose an institutional barrier for solar development.

## Medium-term technologies

Medium-term technologies are those that are scientifically feasible but still a long time away from widespread adoption. None have been developed to the point where commercialization is possible. At present, working models can be used for limited applications and for the most part are simply prototypes. Most important,

production or application (or both) of these working models are not economically viable.

But medium-term technologies have the potential to substitute on a vast scale for the conventional fuels used by society. Some, like synfuels and alternative fossil fuels, would provide *direct substitutes* for petroleum and natural gas. That is, the physical and chemical properties of the substitutes would be indistinguishable from the conventional fuels (see Chapter 12). There would be no need to adopt new capital equipment in order to distribute and use these fuels, and they would be sold in the same markets as the fuels they replaced. Solar technologies, on the other hand, offer what might be termed an *indirect* substitute. They would generate heat or electric power through the conversion of sunlight and, to a greater or lesser extent, supplant the markets for conventional fuels; in this case, however, new capital equipment will be required.

The available energy that medium-term technologies could provide ranges from the huge to the inexhaustible. Synfuels can be derived either from coal or from biomass: the former, as we saw in Chapters 2 and 5, is an immense resource, while the latter, if properly managed (Chapter 5), would be renewable and so effectively inexhaustible. Others are comparably large. Tar sands in the western hemisphere alone, particularly in Venezuela and Canada, contain resources well in excess of known reserves of petroleum, and shale deposits in the western United States contain the equivalent of as much as two trillion bbl of oil or about 400 years of domestic supply at the current rates of use. Unconventional natural gas is known to have resources *in place* that, by some estimates, would increase conventional reserves fivefold.

The solar technologies in the medium term would provide vast amounts of industrial heat and add significantly to electric generation. Indeed, since they use sunlight, they offer the possibility of tapping a *superabundant*, inexhaustible energy source. But the medium-term solar technologies, unlike near-term examples such as solar hot-water heating, are nowhere near commercialization. With all the medium-term technologies, in fact, there is need for further development before economic and/or technical viability can be achieved and adoption can take place. Even tax incentives would not, at present, induce societal adoption.

FEASIBILITY OF MASS PRODUCTION.   Medium term technologies have to be cost effective to users and producers. To become widely adopted, any technology must be capable of cost effective *mass* production. Coal-derived synfuel plants, for example, have to be able to process hundreds of thousands of tons of coal per day cost effectively to become a significant technology. In the late 1970s, during the energy crisis, the United States embarked on a synfuel development program.[4] The original goal of that program was to produce the energy equivalent of 2 million barrels of oil per day, which would require processing over half a million tons of coal per day nationally. The largest of the U.S. prototype plants in the program were designed to handle only 15,000 tons/day, and synfuel processing has not advanced beyond this capacity for a single plant (see Chapter 12).

Alternative fossil fuels present similar problems. Over 20 tons of tar sands – deposits of heavy hydrocarbons mixed with sand (see Appendix C) – must be mined and then refined to produce an equivalent barrel of oil. Thus a daily national aggregate production of 2 million barrels per day would require processing over 40 million tons of tar sands per day. The largest single tar sand operation to date is at the Athabasca deposits in the Canadian province of Alberta. This operation is able to process a little over 100,000 tons per day in a harsh semiarctic environment.

Solar industrial process heat presents a different kind of obstacle to mass production. In order for the technology to be cost competitive,[5] the solar equipment itself must be inexpensively manufactured using mass production techniques (see Chapter 12). Until such production techniques are developed, solar industrial process heat will not be able to move from the medium term toward the commercialization phase. Still, the technical developments required for such mass production do not seem great. Major breakthroughs do not appear

necessary. Rather, a series of practical engineering innovations and design improvements are needed.

To say that no major breakthroughs are required does not, however, guarantee early adoption or even a smooth path toward eventual adoption. Ordinary engineering development requires effort and time to progress. One synfuel demonstration project, for instance, took over five years from the start of design to complete construction of the prototype plant. Once built, prototypes need years of operation and testing to demonstrate no major flaws in the basic design. Since flaws are not uncommon in prototype development, it could be years before commercialization becomes possible.

TECHNICAL BREAKTHROUGHS.   There is uncertainty in the most straightforward technology development programs, if not for the ultimate success of the final designs, at the least for the amount of time it will take for the program to succeed. When some sort of breakthrough is required, the uncertainties are magnified and the time frame made increasingly uncertain.

Photovoltaics – the direct conversion of sunlight into electricity – represent a technology where breakthroughs have been needed, and throughout the 1970s and into the 1980s they seemed near but never really occurred. Only in 1988 was there a sign that significant progress was finally being made, although further advances were still needed (see Appendix C).

Photovoltaic (PV) conversion is a working technology and has had certain limited current applications, for example, in powering space satellites. But its promise has always been greater; some have envisioned large arrays of solar conversion cells powering industry and providing major additions to electricity supplies, especially in Sunbelt areas. This has not happened because the capital costs have been too high. Engineering efforts have focused both on improving efficiency to get more power from each solar cell, and on reducing the cost of producing cells. Improvements in either would lower the cost per watt of peak power output, but major improvements have been needed to bring the costs to competitive levels.

Photovoltaic cell manufacturers have aimed for improvements in basic cell materials (semiconductors) and in the techniques for handling them. The original technique, like that for all semiconductors, was to grow perfect single crystals of silicon, gallium arsenide, or some other crystalline-ordered solid. Since only a very thin layer near the surface is effective in photovoltaic conversion (Appendix C), and since the single-crystal material, as grown, is very expensive, attempts have been made to slice thin layers of it. These attempts, however, still have resulted in very high fabrication costs, considering the large surface areas of solar-cell arrays required for significant electric output.

More recently, attempts have been made to develop an amorphous[6] semiconductor that could be deposited in thin films on flat sheets of less expensive materials. Despite optimistic predictions of success (and occasionally some unwarranted claims by private entrepreneurs), no breakthroughs have been achieved in reducing the costs of production.

However, an announcement in early 1988 by the Electric Power Research Institute gave some hope for the future. Laboratory work had created new cells with efficiencies of around 30 percent, close to the theoretical limit of conversion and double or more the efficiency of cells in commercial production at that time (see Appendix C). Of course, greater power output per cell decreases the need to reduce manufacturing costs and the new cells may indeed bring PV technology closer to large-scale application. But this new development is only a step in that direction. Although these new cells promise a cost reduction, it has not been realized in commercial production. Furthermore, these cells, though efficient, will require intricate production engineering. And even then, estimates are that without further refinements the cells will not be quite cost competitive with conventional sources (for instance, peak load electric supply in the Sunbelt states.[7] In all likelihood, further refinement of the cells, development of manufacturing techniques, and site testing will be needed before they can be widely utilized.

ALTERNATIVE TECHNOLOGIES AND ENERGY MARKETS. To say that photovoltaics or any new technology is economically viable is to say that it can supply its product at a price that is competitive in the market for that product. If the product is an exact substitute, such as synthetic fuel oil is for petroleum, then that price must be equal or less than the product it is to displace from the market. If some properties of the new product differ somewhat from the existing product, then the price must be such that the consumer would still want to purchase it instead of the existing product. This holds, incidentally, regardless of whether the new product is a direct – or an indirect – substitute.

The market price of any product must, of course, be greater than the cost of production. Otherwise, the producer could only then sell at a loss. The production of any product – and hence its market price – must include costs for investment, operation, and for the input factors of production (see Chapter 4). In the case of coal-derived synfuels, the cost of the input coal will be low, and therefore the remaining problem is that the initial capital costs and the cost of processing must be low enough so that the output – synfuel – can be priced comparably to the oil or natural gas it is meant to replace. For the user, there is no other measure than price, since it is presumed that synfuels will directly substitute to operate existing capital equipment. However, experience to date shows that either the price of conventional oil will have to rise greatly or the technology of synfuel production will have to improve before these substitutions can approach cost competitiveness.

With solar technology, on the other hand, economic viability is determined when the market price of new equipment (including the interest cost to borrow the money) results in a net savings over conventional energy sources for the user. A solar user has limited operating costs – no fuel and small operating expenses. But he or she has to be convinced that the initial capital cost represents a savings over conventional sources where the ongoing expense of fuel and maintenance may be greater. Solar technologies, and any indirect substitutes, have a greater hurdle to cross because users must make a large outlay of money in the present for an estimated, but ultimately uncertain, amount of future savings.

Market uncertainty, of course, also will affect the willingness of investors or entrepreneurs to back development ventures that are often needed to make a technology viable. Even a huge rise in oil prices might not encourage investment in synfuels for several years, until it would become clear that the rise was permanent, not temporary. This has been especially true in recent years, when prices have fluctuated and investors, who had invested quickly, lost money. The drop in world oil prices of the early 1980s caused several alternative technologies thought then to be nearly viable to become uneconomic under the changed market conditions. Investments in solar collector production for domestic hot water and various projects for synfuels and alternative fossil fuels depended to one degree or another on private financial backing, and many of those investments failed.

Actually, shale oil and liquid, coal-derived synfuel development have been undercut a couple of times by market conditions in the past. Shale oil research, conducted by the U.S. Bureau of Mines, was initiated in 1916. World War II accelerated this research and by 1944, the government passed the Synthetic Fuels Act (SFA) to develop shale along with coal-derived and other alternative fuels. By the late 1940s some small pilot projects had been completed and the construction of larger prototypes was being considered. But this was a period of increasing availability of large amounts of low-priced oil, principally from the Middle East. For that reason (and, some have claimed, political pressures from the oil industry – see Blair 1978 and Commoner 1979 in the Bibliography for Chapter 2), the SFA projects were stopped by the early 1950s.

Of course this start-stop process was repeated in the 1970s and 1980s. First, concerns about energy supplies led to government-sponsored development, as well as investment by private industry, in alternative fuel pilot projects. Then in the early 1980s, with the falling world price

of oil as a result of the oil glut, the U.S. government dropped its support and private backers for the most part did as well. Only South Africa continued to develop synfuels in the decades following World War II, so that by 1983 a prototype synfuel oil plant was in operation producing 58,000 bbℓ/day. However, the South African government has had to provide ongoing subsidies because the output has not been cost competitive under current conditions in the oil market.

ENVIRONMENTAL UNKNOWNS. The uncertainties of eventual mass use of some medium-term technologies include unforeseen or unclear risks to the environment and safety. Such concerns have tended to be ignored by proponents in their enthusiasm over the potential benefits of these technologies. But environmental questions must be addressed before societal resources – perhaps in the billions of dollars – are expended in development. Depending on the technology in question, the environmental impacts may include air pollution, waste disposal, occupational hazard, and water pollution. Environmental issues are especially important with respect to synfuels (see Chapter 12) and alternative fossil fuels.

A good example can be found in shale-oil processing. Shale oil is essentially a solid form of oil – called *kerogen* – locked up in shale rock deposits. To get the oil out of the rock, the shale must be mined, crushed, and heated (called the *retorting process*, see Appendix C), and then the oil must be refined. There may be many environmental risks throughout the process. Mining may pose occupational hazards similar to coal mining, including pneumoconiosis and possible gas explosions. The sheer volume of material in the process might also prove environmentally detrimental. In order to produce 25 gallons of oil – less than one barrel – a ton of shale must be processed; 1 million bbℓ would leave almost 2 million tons of waste rock to be disposed of. The processing operation would also require large amounts of water – although most of the shale is in relatively arid country. Not only does that leave a problem of water supply and probably conflict over water resource rights, the shale is extremely alkaline and water runoff could irreparably damage rivers and aquifers. Clearly, such impacts would have to be minimized before shale development could proceed. The potential environmental hazard not only adds safety questions, it adds economic ones as well. The environmental cost, when added to the cost of production, could keep shale oil from economic viability even in the event of a dramatic increase in the price of oil.

PUBLIC CONFLICTS. The period leading to adoption of massive production of synthetic or alternative fuels, if and when it comes, will be stressful. As the discussion of shale-oil indicates, substitute fuels – whether they are gaseous or liquid, fossil derived or biomass derived – probably will extract an environmental cost. As a result, a policy decision by government to adopt one or another of these fuels would certainly engender public opposition, regardless of the availability or price of conventional fuels at that time.

Shale-oil projects of the late 1970s provide examples of the kind of opposition such projects probably will face in the future. Critics leveled two charges in particular against the shale operations. They claimed that shale disposal was leaching harmful chemicals into water supplies, and they raised concerns over the safety of shale workers. The dispute pitted the oil companies, such as Union Oil and Occidental Petroleum, who operated the projects, against environmental groups, public interest groups, labor unions, and federal regulatory agencies, including the National Institute for Occupational Safety and Health (NIOSH) and the Mine Safety and Health Administration (MSHA).[8]

Out of this dispute came many charges, but little by way of resolution of differences. MSHA did fix blame for a mining accident on the haste of one company to meet a project objective. But some had hoped that there would be general agreement on occupational and environmental safety standards to provide guidelines in the event of a great expansion of the industry. This did not occur; in fact, one environmentalist complained that the situation resembled that of nuclear power (see *Newsday*, December 12, 1983), where disputes have continued for years without resolution. It can be expected that any decision to develop shale or any other synfuel or alternative fossil fuel will result in intense disputes whose solutions will be difficult to obtain. These only add to the uncertainty of success of such medium-term technologies.

## Long-term technologies

PROOF OF PRINCIPLE. Long-term technologies have the greatest uncertainty of all. Not only do we have the uncertainties of commercial potential and wide adoption, as we do in the medium-term and near-term time scales, but also we have to answer a more basic question: Will the proposed technology work?

The proof required goes beyond *technical feasibility*, which should be understood in the sense of engineering and design. Synfuels, for example, can be produced by the processes described in the next chapter and so we can say the technology is scientifically feasible. Synfuels, however, are not yet viable or cost competitive, and they will not become either until they are proven on a mass scale. Prototype plants alone cannot fully determine feasibility since the problems of large-scale operations may not appear in demonstration projects. Nuclear fusion, in contrast, has not yet been proven scientifically feasible and is therefore classed as a long-term prospect. With such prospects, we have not yet left the initial research stage of technological evolution. In the research stage, a *proof of scientific principle* is needed in order to know that the technology has the potential to work *at all*.

Nuclear fusion research provides the clearest illustration of the distinction between the proof of principle and technical feasibility. In the fusion program, the specific proof of principle has to be that a *net yield of energy* can be achieved from nuclear fusion reactions. This net yield, furthermore, must be in a form that would produce significant amounts of useful energy for power production. This proof of principle has yet to be achieved, as we will discuss in detail in Chapter 12.

The first major objective in the fusion research program overall has always been to conduct a laboratory experiment that shows a *breakeven* of energy, that is, the energy output from the fusion reactions should at least equal the energy input required to ignite and control a fusion reaction. If the breakeven condition can be demonstrated, then the next research objective would be to achieve a net energy yield. In over 35 years of nuclear fusion research in the United States and elsewhere, no breakeven experiment has been successful. It therefore has not yet been proven that nuclear fusion can be a useful source of energy.

This lack of success should not be attributed to the lack of funding of fusion research programs. With the possible exception of nuclear fission, fusion research funding has had more continuity than other energy research and development programs in the United States. Also, the scientific talent working in national laboratories, corporate research laboratories, and universities has been of top quality – both in the United States and elsewhere.

The underlying reason for the failure of fusion experiments to reach the breakeven point appears to be in the intractability of the problems in physics (discussed in the next chapter). Such .an assessment is not only accepted within the fusion research community, but is the focal point of critics of the program's direction. The critics would contend that from the beginning the program should have concentrated more on basic physics before attempting to construct breakeven experiments and programmatic steps beyond (see Furth 1984, Lidsky 1983, and Rose 1986, in the Bibliography for Chapter 12).

Similar debates have taken place at the Joint European Torus (JET), an internationally sponsored fusion research project,[9] over where the proper emphasis should be. These debates indicate the institutional difficulties in running research programs. There are strong pressures on those in charge of such programs to demonstrate short-term results to maintain public funding.

Not all technologies in the research stage have the large public support that fusion has had. Superconductors, for example, went from the long-term to the medium-term category with a small fraction of the research funding supplied to fusion. Photovoltaics research, too, has comparably low public funding. Both superconductors and photovoltaics have research funding from the U.S. private sector, but the investments are a tiny fraction of those the government has made in fusion.

SUPPORT OF RESEARCH. As is apparent, a technology at the research phase of evolution needs financial support, but has highly uncertain prospects. There are no assurances as to *when* or *if* the basic principles will be proven. Few private investors are willing to put their money into such speculative ventures and none are willing to risk large sums. Consequently, large research projects, such as fusion, are funded *only* by governments; but governments, too, will often decline to expend their resources on the basis of mere promises of success from researchers.

The problem with government research funding, when results are slow in coming, lies in assuring its continuity on a long-term basis to allow the possibility of ultimate success. The ongoing operation of research projects requires career commitments from the scientists in them, starting with years of graduate training in universities and continuing into professional specialization within the projects themselves. Large experimental projects, such as those related to fusion, also require cadres of highly trained technicians and design engineers. The construction of experimental devices demands contracting with specialized suppliers for custom-built equipment that often requires years to complete. Disrupted or unpredictable funding in this complex management of large research efforts has impacts far exceeding those of simpler organizations. Thus, we can say that research projects have virtually no chances for success without long-term continuity of support.

The long-term continuity of government funding is a political issue in representative democracies, where government expenditures are determined annually by legislation. Long-term commitments of funding are very difficult to legislate, because legislators are usually elected on a short-term basis and their perspectives are thus molded. Also, there are problems of accountability on the part of large projects that are given long-term commitments of funds for purposes that cannot always be clearly understood by the legislators.

These difficulties were evident in the attempt by the U.S. Congress to set up a long-term fusion research program. The McCormick Act (passed in 1980 and named for its author Representative M. McCormick of Washington) laid out $20 billion for a fast-paced program, but one geared to tangible results rather than scientific inquiry. It aimed first for breakeven experiments and ended with a demonstration prototype of a fusion-power reactor, all within 20 years. Although the McCormick legislation had apparently achieved a step toward assured long-term funding, it did not start at what the critics felt was the necessary beginning – the basic science questions.

The legislative success was, in any case, short lived. Following soon after its passage in 1980, the oil glut ended congressional concerns over energy supplies. The incoming Reagan administration was also strongly interested in budget cuts on all nonmilitary federal funding. As a result, the fusion program, although still able to command far more in funds than photovoltaics or any other energy technology, was curtailed and lost the assurance of vast sums for decades to come.

Interestingly, a few large research projects in the United States and in Europe have had more success in maintaining continuity of funding; some have even been undertaken since the failure of the McCormick Act. The prime examples are the huge accelerators used in particle-physics research. Indeed, Congress considered huge new projects to develop super accelerators even as it was reducing money for fusion research. Some critics have charged that such priorities were just a matter of politics, while others have contended that the accelerator projects have simply had better lobbyists. These

scientist-lobbyists are, it is said, more effective at selling the importance of their proposals to politicians using the mystique of high-energy-particle[10] physics, which they portray as the frontiers of knowledge.

But there is an important distinction between the accelerator research and fusion research. Unlike fusion researchers, particle physicists are not trying to prove an established principle in physics for application to a new technology; rather they are attempting to discover new principles. Such new principles may then be tried and proven for new technologies, but what the new technologies will be is not predictable. Basic research, such as particle physics, precedes the research stage of a new technology such as fusion. It corresponds to the establishment of new knowledge, an effort with no other definite end in itself.

A possible conclusion to draw is that national governments recognize the need to support *basic* research better than they do the need to support *applied* research. If this conclusion is valid, it indicates a political will to make open-ended commitments to efforts with grand, yet vague, prospects, rather than to efforts with specific objectives when both have uncertain chances of successful outcomes. Of course, we should not make too much of this conclusion; fusion has received hundreds of millions of dollars in the United States alone and other similar research efforts have also received support. But what seems to be at work is a difference in criteria for maintenance of funding. Although basic research can continue to provide grand theory alone, applied research needs to show practical results as quickly as possible or run the risk of a cutoff of money.

## Long-term choices

A long-term technology like fusion presents us with an almost unimaginable hope. If it works, it promises not just an improvement in our energy technology, nor merely a reprieve of a few more decades or centuries from the problems of resource depletion. Instead, it promises a *solution* to our energy dilemmas, perhaps for the rest of human history. As such, it would represent a revolution in science, transforming society in a fundamental way. But at the same time, it is uncertain that the technology will ever work, even if billions of dollars are expended in development. Indeed, the uncertainties are so encompassing that even if the technology proves technically feasible, we cannot be sure now that it will not beget a new set of unforeseeable problems; in other words, we may find only after decades of effort that it is not *the* solution after all.

Such promise and uncertainty leaves a society such as ours with large questions: Should we invest massively in a hope or in more probable – though limited – realities? Should we make a great effort to see if we can obtain a solution to our energy problems, or should we focus our effort on medium-term and short-term technologies, even though they may only help us forestall the need for a solution?

Of course, there is no need to make either/or decisions. Society may, and probably will, continue to seek ultimate answers even as it pursues stopgap measures. The mix of choices and the emphasis in energy policy may finally depend less on grand questions and more on immediate concerns. For example:

How can we manage our exhaustible fossil fuel resources?

Which new technologies will assist in creating a more equitable world distribution of energy resources?

Are there attainable alternative technologies that will reduce the spillover effects of energy conversion, including pollution and radioactive hazards?

Which technologies will best help eliminate the possibilities of major ecological disasters, such as the greenhouse effect, deforestation, and widespread radioactive contamination?

Which energy technologies can best sustain worldwide economic growth?

New energy-conversion technologies offer different answers to these questions. If we consider the exhaustible resource question, for instance, we can readily see that if a policy focused on development of synfuels and alternative fossil fuels, it would only extend the life on fossil resources. A more direct solution would be the adoption of *renewable resource* technologies such as biomass fuels (see Appendix C). Biomass fuels could provide energy on a truly renewable basis,[11] if systems of sustainable harvesting of energy crops or forests were adopted. If biomass management could be accomplished on a mass basis, fossil fuels might nearly be displaced and the dilemma of exhaustible fossil resources eliminated.

At the same time, we could focus more on solar or fusion technologies to overcome the problem of exhaustible resources. Both technologies offer the hope of tapping inexhuastible sources of energy. Of course, the former can at present only be utilized on a limited basis and the latter not at all. The question becomes: What route do we choose to deal with the problem of exhaustible resources? Biomass, solar, and fusion all have great potential, all could be pursued, and all are to a greater or lesser degree uncertain.

Any renewable or superabundant source could also solve the present problems of world resource distribution (see Chapter 5). Thus, if fusion technology could be successfully developed and made available, it would eliminate the problem of fuel deficiencies in resource-poor regions of the world. In the event of improvement in solar technology, those regions with highly available solar energy could alleviate local resource deficiencies and at least lessen the worldwide demand for fossil resources (see Chapter 12).

Also the solution of energy resource deficiencies would be a major step toward the development of Third World countries and the reduction of poverty. Either solar or fusion offers the potential of powering development in a *sustainable* way (see Repetto 1986; see also Brown 1986, in Chapter 5, Bibliography). The present (and unsustainable) trends of massive deforestation, erosion, and pollution in many areas of the developing world could be reversed and development could move forward.

In the industrialized world, solar, fusion, and, to some extent, biomass[12] energy technologies could reduce spillovers. Solar energy would appear to be the most environmentally safe. The worst known spillovers in solar conversion relate to the mass production of semiconductors. There are no apparent spillovers in solar operation. On the other hand, many industrial countries such as Great Britain are geographically and climatically poor candidates for widespread use of solar power.

It should be noted that spillovers connected to fusion technology at this point are unclear. Fusion reactors almost certainly will produce radioactive wastes. The reactors will most probably use (radioactive) tritium. In addition, neutron bombardment will induce radioactivity in reactor structures. It is virtually certain, however, that the radioactive by-products will be neither as long-lived nor as potentially lethal as those of a fission reactor, but how much less lethal is not entirely clear. Estimates on the amount of radioactivity of a fusion reactor vary widely by as much as three orders of magnitude – from as little as one-hundred-thousandth to as much as one-hundredth the radioactivity of a comparable fission plant. But the fusion program cannot have a better idea of which estimate is closer to the mark until net energy is achieved and prototype fusion reactors have run for some time.

Eliminating the possibility of major ecological devastation may require the adoption not just of renewables or superabundant sources, but only those alternatives that do not involve combustion. Biomass, for example, though renewable, poses a threat in two ways: potential deforestation (already a concern in developing industrial countries such as Brazil, see Chapter 5) and the carbon dioxide greenhouse effect. Though the latter effect is uncertain, the way to reduce the chance of human causation is to reduce the production of carbon dioxide through the mass adoption of noncombustive energy technologies. This would, however, require not just the adoption of fusion and solar technologies to replace all present fuel-burning means of supplying electricity, industrial heat, and space heating, but also the development of transportation vehicles that do not have internal combustion engines. For that to occur,

there would have to be breakthroughs not only in fusion and solar technologies, but in electric vehicle technology (see Appendix C on energy storage).

It may seem at this point that the arguments favor development of renewable and super-abundant sources, particularly solar and fusion, over all others, since these seem to be most potentially beneficial and cause the least damage to the environment. But we must also consider the economic costs and benefits. It is not clear that superabundant and renewable energy technologies will provide the most economic benefit. Indeed, given the uncertainties, some may provide no benefit at all. Policy choices must consider the uncertainty of the outcome. Should we spend vast financial resources now, taking them from other worthwhile areas such as health and welfare, when the outcome may mean much higher energy costs and lower economic growth? Doing so may mean hardship for some people now; if the alternatives do not work out as hoped, hardship for many may result.

Of course, as we noted in Chapter 4, economic growth does not depend solely on the cost of energy. In the event of somewhat higher energy costs, improved efficiency and productivity can still maintain a high standard of living. Then, the benefits of these alternatives may make them worthwhile even at a slightly higher cost. Much will depend on how much greater the cost will be and how much of that cost increase can be overcome by conservation.

Yet it is the potential for rising prices of conventional sources of energy that gives the strongest economic argument in favor of the development of alternative technologies sooner rather than later. Expectations of very high prices, of course, spurred development of alternatives in the late 1970s and early 1980s. The oil glut and the drop in prices changed expectations of the consuming and investing public, and the impetus toward alternatives faded. But, in the view of many economists and technologists, the long-term outlook is for prices of conventional energy sources – particularly oil and natural gas – to rise dramatically in the 1990s, or after the year 2000, because of increasing demand and decreasing supplies. If these analysts are correct, market conditions probably will lead to renewed conservation and substitution of alternative energy sources for fossil fuels. Substitution will begin in earnest when the costs of energy production by the new methods are lower than prevailing prices of the conventional sources, and when consumers are convinced there will be no reversal of that trend.

## The paths

To this point, we have focused on the particular costs and benefits to society of new energy technologies. But in recent years, philosophers have looked at the evolution of new technology itself as a choice between two paths: *hard* and *soft*. Each path has social and ethical implications as well as more strictly technological implications.

The hard path requires large mass production facilities; in the case of energy, this means the construction of large-scale power plants or fuel-conversion plants, which take advantage of *economies of scale* to produce at the lowest cost. However, the hard path – as viewed by detractors – is not just concerned with ecological impact or hazards, but rather with production for a mass demand, which is unfettered by any conservation ethic. Less critically, the hard path would see mass production as a way to generally improve the condition of all society and would regard spillovers as solvable through further applications of technology. Large coal-fired or nuclear power plants and oil refineries fit into this category, and fusion and synfuel plants would be the corresponding new technologies that follow the hard approach.

Soft technology advocates, in contrast, envision production that is small scale and decentralized. Energy would be provided by on-site solar collectors or wind generators, which have low environmental impacts. The soft path also emphasizes the use of renewables, which include not only solar and wind energy, but also biomass and water resources. Soft technology proponents also advocate conser-

vation of resources, including voluntary reduction in energy consumption.

Principal promoters of the soft path are Schumacher (1973) and Lovins (1977), who in fact coined the term *soft path*. They typically see the decentralized soft technologies in a decentralized sociopolitical context of "integration with nature, democratic politics.... compatible with local culture.... dependent on [the] well-being of other species, innovation regulated by need, steady state economy, labor intensive... operating modes [of technology and society] understandable to all." To its adherents, the soft path would be less stressful and society would function better because, as another advocate has put it, "general efficiency increases with smallness."[13]

Opponents of the soft path have pointed both to the benefits society has historically received from the innovative hard path – electric appliances, automotive transportation – as well as to problems and possible contradictions in the soft path concept itself. Opponents wonder, for example, how this transformation of industrial society will take place. It has been suggested that the massive reordering involved would require a massive and *centralized* industrial effort to build thousands of solar collectors and wind generators as well as the scattered site shops and factories they would power (see Florman 1977). Indeed, it might be argued that hard-path technical breakthroughs in production will be needed to make the soft path realistic. By implication, any attempt to introduce the soft path now into Western industrial society might be socially disruptive and cause economic hardship.

The soft versus hard path debate parallels the debate between the limitationists and the expansionists discussed in Chapter 2, where we focused on the exploitation of exhaustible resources. That debate weighed the values of preserving natural resources versus promoting economic growth. However, both limitationists and expansionists also (at least tacitly) considered the prospects of new energy technologies, with the expansionists taking the optimistic view that new technologies would come along in time to replace declining fossil resources, whereas the limitationists were pessimistic. On the need for conservation of energy, limitationists and soft path advocates favor a voluntary curtailing of consumption and practicing an ethic of doing without to conserve resources and reduce spillovers.

Proponents of hard and soft technologies have debated the question of which path is best, not solely for the industrialized countries but for the less-developed countries as well. Though both sides agree on the need for economic development in poorer nations, soft path advocates are unwilling to concede that the LDCs need to adopt hard technologies. The LDCs, soft path advocates maintain, should adopt renewable energy sources to avoid the kind of spillovers that industrial countries have had to face. Or to put it simply, the LDCs should not make the same mistakes the industrial countries have made. But to hard path advocates, this position seems a prescription for continuing poverty, and they would claim that hard technology is essential if the LDCs are to raise their standards of living; as for spillovers, the argument might be that a temporary problem with air pollution is better than persistent poverty.

In evaluating these arguments we should keep time scales in mind. For example, when we consider renewable energy sources for the LDCs, we must remember that any new technology must be viewed in terms of its present technical feasibility and economic viability. Both of these factors have to be taken into account regardless of the state of economic development in the country where a technology is to be applied. How feasible and viable are renewable energy technologies? Arguably, no such technology is both feasible and viable and so, some have claimed, adoption of renewables as appropriate technologies of LDCs is out of the question for the *short term* (see Cassedy & Meier 1987, Chapter 5, Bibliography). LDCs have the choice either to adopt technologies that might slow growth or to invest to some extent in fossil-fired power plants and other conventional technologies to spur development.

# Intergenerational issues

Underlying much of the discussion on alternative technologies, are ethical questions. Soft path and hard path proponents make implicit and explicit assumptions of what is just and equitable. Policymakers must weigh the relative benefits from funding one program over another or one kind of technology over another or, even, technological development itself versus alternative uses of scarce financial resources. By their votes and personal choices individuals within society may help direct those policies. The choices are ethical as well as technical.

Above all, it is evident that our choices today on energy alternatives will have implications far into the future, affecting the welfare of generations to come. If we invest in one alternative over another, that choice will be a given for future generations; they must either accept it or undo it, where the cost of doing either may not now be known. If we choose no investment in alternatives at all, that too becomes a reality for the future. No present commitment on energy technology exists in isolation. We may leave the future with secure resources or not, cheap resources or expensive, growing spillover problems or diminishing ones, a permanent solution for our energy needs or only a greater need to find one.

It would be simpler if we knew for certain that one superabundant resource was feasible and widely applicable – what is sometimes called a *backstop technology* – for us to develop now and bequeath to the generations to come (see Nordhaus 1979). Hopes exist that either fusion or solar technology will provide that backstop. But no backstop can be promised now or soon.

This leaves contemporary society in a quandary. We can adopt the expansionists' view – that a backstop *will* be found or at least that stopgaps will carry us and our descendants far into the future. Meanwhile, policy should be oriented to exploit resources for the present good. It is true that outcome of investment in alternatives is uncertain, but then so is the decision not to invest. Investment may fail and there may be hardship, but the supposition that no investment is necessary may be wrong. Then generations of the future will pay the price. Because they have had no voice in the decision, the decision may be both wrong and unjust.

Limitationists, on the other hand, see ecological and/or social disaster approaching because of resource exploitation. In this view, we must pursue a range of alternatives now, and because a backstop or anything close to it is at best a mere possibility, we must also decrease our use of energy resources. This may seem morally preferable; after all, it means saving resources and aims to protect the world from potential harm. But if we are wrong, the choice may be unjust nonetheless. We may cause a needless reduction in economic growth now affecting those most in need and actually bequeath an economically poorer world to future generations.

As we noted in Chapter 8, U.S. government agencies have for some policies adopted the basic principle of intergenerational equity. The EPA, for example, has stated that future generations should bear no greater risks than those acceptable to the current generation. In the case of nuclear waste, the risk is clear, although the acceptability of the risk in terms of present policy is disputed. But what of alternative energy technologies? We risk failure, but the outcome is unknown. We risk financial resources, but the outcome is unknown. We risk spillovers, but they are unknown. But we also take a risk if we do nothing, even though once again, the outcome is unknown.

Perhaps further refinement of the policy direction is needed. But because uncertainty can never be removed from the prospects of alternatives, it seems unlikely that any standard can be definitive. In the end, the decision on how or whether to develop alternative energy technologies will come down to a judgment by society of whether the promise and hope are worth the investment – for us today and for those who will follow.

## Notes

1. Much energy research is done for the federal government at a group of national laboratories located at Argonne, Illinois; Brookhaven, New York; Los Alamos, New Mexico; Oak Ridge Tennessee, and Sandia, New Mexico.

2. In 1978 the Public Utilities Regulatory Policy Act (PURPA) was passed as part of President Carter's National Energy Plan II. The act, among other provisions, required public utilities to purchase electric power at full "avoided costs." This meant that the utility should pay for a block of power what it would have cost had it generated that block of power itself.

3. See T. Alexander, "The Little Engine that Scares Con Ed," *Fortune*, Dec. 1978.

4. The U.S. government, starting in the late 1970s, sponsored synfuel projects through the U.S. Department of Energy (USDOE) and the specially created Synthetic Fuel Corporation (SFC) (see Chapter 12).

5. The issues of technical feasibility and economic viability are of course intertwined here. The test of feasibility, in this case, is whether cost-effective energy can be produced.

6. An amorphous material is not in crystalline form. Thus, it does not have a distinctive geometric shape, which perfect crystals commonly have, nor will it break along specific cleavage planes as crystals do. These properties of crystals are attributed to a perfect ordering of the atoms in repetitious (periodic) patterns, at the microscopic scale, in what is called the *crystal lattice* (see Appendix C).

7. Researchers believe that once mass production of these new cells begins, costs will be brought to a range of $2 to $3 per peak watt for a massive installation of panels capable of peak output of 100 $MW_e$. The capital cost would be lower than that of most nuclear plants and large hydroplants. However, the cost of such solar energy might still be too high because the availability factor (AF) would only be around 30 percent even in the sunniest regions of the country. As we noted in Chapter 9, the AF is crucial in analyzing the cost of electric energy from a generating facility. At the same time, we should note that solar energy does not use fuel and probably has low operational costs. It is possible that further development will soon make these cells or ones derived from them cost effective for peak-load generation in certain areas of the United States and elsewhere.

8. A series of articles on the synfuels program appeared in Long Island *Newsday*, on November 28 to December 12, 1983.

9. Reports of controversy on the program's

directions appeared in *Physics Today*, November 2, 1984.

10. High-energy particles move at extremely high speeds, even higher than the neutron and proton speeds mentioned in discussions of nuclear fission, with energies of individual particles measured in GeV ($10^9$ eVs, see Appendix A). Extremely high particle energies are required to penetrate the basic nuclear structure of matter.

11. We define *renewable* as a source of fuel or energy that replenishes itself, such as water resources or biomass. The simplest example is in forest management, where the trees harvested annually are held equal to the annual new growth of trees, so as to sustain the constant stock of trees. Other technologies that are often called renewable, such as solar, we will put in another category of *inexhaustible* or superabundant sources.

12. For example, the alcohols, which burn with far fewer pollutants.

13. These quotes are from soft path advocate Robin Clarke of Biotechnic Research and Development in England (cited in Winner 1986).

# Bibliography

Basalla, G. 1988. *The Evolution of Technology*. Cambridge University Press, Cambridge.

Bhatia, R., and A. Pereira (eds.). 1988. *Socioeconomic Aspects of Renewable Energy Technologies*. Praeger, New York.

Bury, J. B. 1932. *The Idea of Progress*. Macmillan, London (reprinted in 1955 by Dover, New York).

Camilleri, J. A. 1984. *The State and Nuclear Power*. University of Washington Press, Seattle.

Cassedy, E. S., and P. Z. Grossman. 1988. "New Technologies – Promise and Uncertainty." International Symposium on Energy Options for The Year 2000, Wilmington, September.

Cohen, I. B. 1985. *Revolution in Science*. Harvard University Press, Cambridge.

Florman, S. C. 1977. "Small is Dubious." *Harper's Magazine*, August.

*A Forum on Solar Access*. 1977. Proceedings of a Forum on Solar Access held by the New York State Legislative Commission on Energy Systems, July 28. The National Solar Heating and Cooling Information Center, Rockville, MD.

Hayes, D. 1977. *Energy: The Solar Prospect*. Worldwatch Paper No. 11, The World Watch Institute, Washington, DC.

Kranzberg, M., and C. W. Pursell (eds.). 1967. *Technology in Western Civilization, Vol. 2, Technology in the Twentieth Century*. Oxford University Press, New York.

Linstone, H. A., and D. Sahal. 1976. *Technological Substitution – Forecasting Techniques and Applications*. Elsevier, New York.

Lovins, A. B. 1977. *Soft Energy Paths: Toward a Durable Peace*. Ballinger, Cambridge, MA.

Mansfield, E. 1968. *The Economics of Technological Change*. Norton, New York.

Nordhaus, W. D. 1979. *The Efficient Use of Energy Resources*. Yale University Press, New Haven.

Partridge, E. (ed.). 1981. *Responsibilities to Future Generations*. Prometheus, Buffalo.

Repetto, R. 1986. *World Enough and Time – Successful Strategies for Resource Management*. Yale University Press, New Haven.

Rogers, E. M., and F. F. Shoemaker. 1971. *Communication of Innovation: A Cross-Cultural Approach, Second Edition*. The Free Press, New York.

Schumacher, E. M. 1973. *Small is Beautiful*. Harper & Row, New York.

Winner, L. 1986. *The Whale and the Reactor*. University of Chicago Press, Chicago.

# 12

# Paradigms of new technologies

In Chapter 11, we discussed the general obstacles to the adoption of new technologies. Here we will focus on the specific issues connected with three alternative energy technologies – solar-thermal, synfuels, and nuclear fusion, one from each time scale described in the previous chapter. An appreciation of the specifics, particularly of the technical specifics, is a necessary complement to the more general points we have already raised. Only by having a grasp of both can we understand and judge the prospects of future technological innovation in the energy field.

## Solar-thermal technology

By solar-thermal technology we mean any method of collecting *heat* from the sun's rays and finding useful purposes for that thermal energy. These purposes can include space heat for homes and buildings, hot water heat for home use, industrial process heat, or steam for electricity generation. Although we place solar-thermal technology in the near-term time scale for adoption, the technology for applications such as large-scale industrial heat and electricity generation has not been so well proven technically and may therefore be placed in the medium-term category.

To be useful, solar energy must be collected and the simplest form of solar-thermal collector

is a window facing toward the sun on a cloudless day. This, however, is not solar energy technology, because normal windows are not designed to optimize heat collection, retention, or use; they are a primary source of illumination. At the same time, window design and placement can be a component of modern solar-thermal technology. In recent years, architectural designs have been developed to enhance the collection and storage of solar heat, purely through the basic structure and components (walls, windows, and so on) of the building. These designs are classified as *passive-solar* technology, because they operate basically without any moving parts or controls.

Other designs, termed *active solar*, attempt to capture and utilize a greater amount of available solar energy through the use of moving parts and controls. Most important, all active systems have a circulating coolant. The coolant (which plays the same role as the coolant in a thermal power plant, see Chapter 3) transfers the heat energy from the collector to its point of use. Such a circulation system requires a separate source of power (usually electric) as well as instrumentation to monitor and control its operation.

Solar-thermal technology has been proven technically feasible, and passive systems as well as active ones for home hot water heat in particular have been mass produced and are commercially available. But solar-thermal technology, particularly active solar systems, has had limited economical viability. It may not be cost effective to use active solar thermal technology today in any given site or application. One of the most important considerations in determining whether or not solar thermal will make economic sense is, of course, the intensity and fluctuations of incident solar energy at the desired building site. This starts with the specific building and its immediate surroundings. In some cases, local obstructions may reduce incident sunlight throughout the year, at various times of day, or during certain seasons. Such obstructions, if they exist, are usually other buildings or vegetation and pose the legal problem of solar access if they are located on a neighbor's property (see Chapter 11).

The availability of incident solar energy is also a regional consideration. The latitude of a site, for example, determines the seasonal angles of the sun above the horizon *and* the number of hours of sunlight per day. Nearly as important, however, is the regional climate, especially the amount of cloud cover. Figure 12.1 shows contours of the average solar flux density[1] across the United States for the entire year and for the month of December only. If we examine an east-west line (Fig. 12.1) on either side of the Great Lakes, we observe that regional variations in average solar flux may be different even at the same latitude. This is, of course, due to differences in the weather patterns. Cloud cover and precipitation are much heavier east of the Great Lakes than west. Moreover, at the same latitude in the Pacific

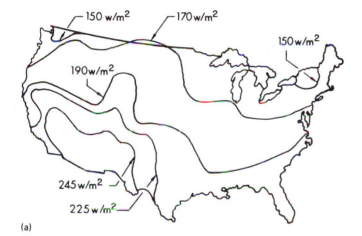

(a)

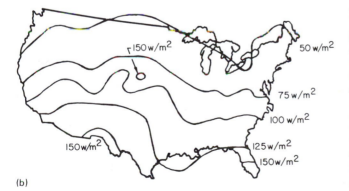

(b)

**FIG. 12.1.** Average solar flux densities for the United States. (a) Yearly average of solar flux densities in W/m². The numbers represent applicable 24-hour average values reaching the surface of the earth. (b) December average of solar flux densities in W/m². The numbers represent applicable 24-hour average values reaching the surface of the earth. *Source:* F. Bennett, "Monthly Mass of Mean Daily Insolation for the United States," *Solar Energy*, Vol. 9, Pergamon Press, 1969.

Northwest, cloud cover again reduces annual insolation.

Weather affects not only average solar performance, but also the *durations* of solar interruption and therefore the relative importance of energy storage (see Appendix C). It is important to realize that any annual or monthly average solar performance includes 24-hour diurnal variations over night and day. Also, an annual average will obscure the fluctuations in sunlight that occur day to day as well as seasonally (Figure 12.1b displays the lower average for the winter season).

The variations in incident solar flux are an important consideration, both technically and economically, because they determine what is required to make energy available continuously to the user. Active solar systems usually include two features to ensure continuity of supply: energy storage and a *backup* source. Solar-thermal storage systems usually store heat in the form of hot water. Although storing hot water is straightforward in concept, the investment cost rises rapidly with increased storage capacity chiefly because the collector size must also be increased to take advantage of it. In most cases, it is prohibitively expensive to attempt to store more than a couple of day's worth of heat energy.

The backup source is merely a conventional heat source – for example, a fossil-fired boiler – which supplies the demand when there is insufficient incoming solar energy or available stored energy. Generally speaking, the economic viability of a solar thermal system is determined by how much conventional fuel is saved. Of course, the economic viability of a system rises the less often the backup needs to be employed. Or more to the point, we can say that viability is at least partly determined by the trade-off between the investment cost of the solar-thermal system and the fuel costs of the backup.

We can get a better sense of the economics of solar thermal by calculating the dollar value of solar energy using the solar flux data in Figure 12.1. Essentially, we must consider how much energy can be captured by a collector, given a certain average annual flux density in a given region. The result will be expressed in a unit of energy per unit of area over which the solar energy falls. Typically, the amount of energy is expressed either in metric units, watts per square meter of area, or in BTUs per hour per square foot. We will use BTUs per hour per square foot here and place the calculation on an annual basis. The resulting conversion factor[2] is:

$$2.75 \times 10^{-3} \text{ MBTU/ft}^2\text{-yr per W/m}^2$$

which would convert an annual average flux density of $100 \text{ W/m}^2$ to $0.275 \text{ MBTU/ft}^2$-yr (that is, 0.275 million BTUs per square foot per year).

This amount of solar energy (assuming that the system is 100 percent efficient) must then be compared with the cost of the same heating requirement using a fossil fuel such as oil. Heating oil at a price of \$1/gallon costs about \$7.25/MBTU. Thus, if solar energy replaced oil, the *annual* cost saving of a square foot of collector absorbing this amount of energy would be $0.275 \times 7.25 = \$2.00$ for $0.275 \text{ MBTU/ft}^2$-yr.

Then we have to consider the initial investment cost in the solar collector and auxiliary equipment such as storage, pumps, and controls. Solar industry experience and prototype projects in the early 1980s indicated system investment costs in the range of \$50 to \$100 per square foot of collector. If we assume such system costs, it would take 25 to 50 years for the user to save enough to pay for the solar system (provided, that is, that there were no interest charges). This would not, of course, be a good investment, even assuming 100 percent collector efficiency and zero interest charges. The costs of the equipment will evidently have to go down dramatically to make a practical system attractive to a cost-conscious user.

This example makes solar-thermal technology appear uneconomical. But actually if we consider specific regions, the economics of solar thermal improve considerably. The area of highest insolation in the United States is in the Southwest (see the $245 \text{ W/m}^2$ contour in Fig. 12.1a), the Sunbelt region where one can expect a great deal of sunshine. The Sunbelt's high ambient temperatures also improve efficiencies of collectors. Certain types of collectors achieve annual average performances in the Sunbelt of around 200 W/m²; or over 0.5 MBtu/ft²-yr. This would improve the payoff by about a factor of two over the 100 W/m² example – a payoff in 12.5 to 25 years, with zero interest. If we add to this higher prices for conventional fuel as well as federal and state incentives typically in the form of tax deductions (which were offered in the late 1970s and early 1980s, but then were discontinued), the payoff would be shortened approaching the point where solar thermal begins to be cost competitive with conventional energy sources.

## Passive solar-thermal systems

The basic concept behind passive solar design is to absorb solar heat in a structure while the sun is shining and then retain the heat when it is not. This is achieved by allowing maximum transmission of the direct sunlight into a building, absorbing the heat in some way inside, and then minimizing heat loss from radiation, conduction, and convection processes.

We can get some idea of how passive solar design works from Figure 12.2. Note in the figure that sunlight heats the room directly through windows and that heat is retained in walls or ceilings. This heat warms the surrounding air, which circulates through the space. The indirect and isolated wall techniques shown in the figure can enhance these storage effects.

Through the use of still other passive technologies, the heat is prevented from being quickly lost. For example, some windows have an interior coating of substances like tin oxide that allow sunlight to enter, but deflect infrared heat from escaping. Indeed, a passive solar design will provide space heating even after the sun is no longer striking the building. Architects will also orient the building's south-facing and overhead windows to enhance the capture of sunlight. At the same time, they will minimize heat loss by heavily insulating north-facing walls and having few if any northern exposures.

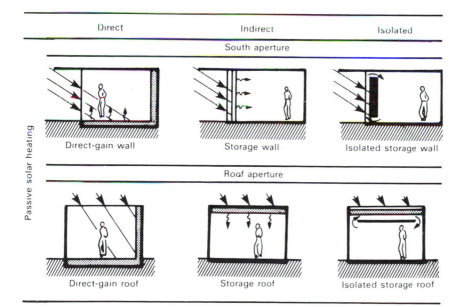

**FIG. 12.2.** Passive solar heating. *Source:* U.S. Department of Energy (1981).

## Active solar-thermal systems

As we have noted, active solar systems in general operate with collectors and a circulating coolant that transfers the heat from the point of collection to a point of use. Figure 12.3 shows schematic diagrams of two active systems operating on the same principle with different coolants. The top diagram depicts a hot water system using a liquid coolant. The coolant, either water or an anti-freeze solution,[3] is heated in the collector and flows to the hot water storage tank, where it circulates through a separate heat exchange coil (not shown) immersed in the household hot water but not mixed with it. The water in the tank is thus ready for use in the building while the coolant is recirculated back to the collector. In the bottom diagram, a hot air system is shown. The air is warmed in the collector and then circulated either to air ducts directly for space heating or to a pebble bed storage unit.[4]

In both of these systems, the coolant is circulated by a pump or blower. Both systems can store the solar heat collected and both have auxiliary heaters (backup sources) to operate in the absence of incoming or stored solar heat. Most important, both systems have to be controlled in order to operate properly. The coolant should flow out of the collector only when the coolant temperature is sufficiently high to exchange heat into the storage tank. (If the coolant is circulated at low temperatures, it will actually take heat *out* of the storage unit.) Similar considerations hold for the circulation of water or air from the storage unit to the point of use. When the storage temperature is too low (all stored heat is exhausted) a bypass path sends the coolant directly to the backup heater.

The three-way valves or dampers shown on the figure operate on command of the control system to ensure these flows. For example, when water is to be heated from the sufficiently hot storage tank, the lower valve allows the supply water to flow into the bottom of the tank and the upper valve allows it to flow out to its use. If, on the other hand, the tank temperature is too low, the lower valve shunts the supply water along the bypass path, thereby skipping the tank, and the upper valve directs it through the auxiliary heater to its point of use. The air dampers on the hot-air system function in a similar way.

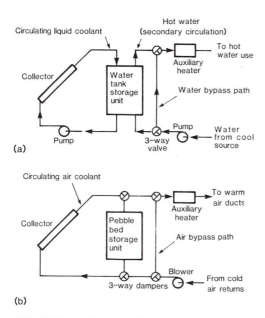

(a)

(b)

**FIG. 12.3.** Active solar-thermal systems. (a) Residential hot water systems. (b) Residential warm-air space heating system. *Source*: U.S. Department of Energy (1981).

## Collector technology

There are several types of active solar collectors. Some focus the sun's rays, whereas others simply absorb the incident solar radiation onto flat plates. We will consider five examples.

In flat-plate collectors, solar radiation falls on an absorbing surface, such as a sheet of blackened metal, that is in close contact with the coolant for heat transfer. In an air heat-transfer collector such as the one depicted on the left in Figure 12.4, there is a simple rectangular box. One side of the box is a double thickness of glass that allows the sun to warm the other side of the box – a blackened flat plate (the collector plate). At the same time, the double glass acts as an insulator to prevent

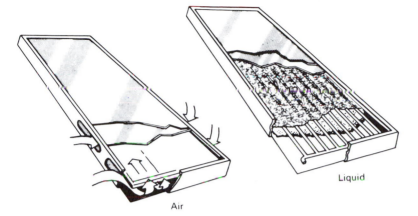

Liquid

Air

**FIG. 12.4.** Flat-plate solar collectors. *Source:* From *Energy Primer: Solar, Water, Wind, and Biofuels.* Updated and Revised Edition. Edited by Richard Merrill and Thomas Gage. Copyright © 1978 by Portola Institute. Reprinted by permission of Dell Publishing, a division of Bantam, Doubleday, Dell Publishing Group, Inc.

the heat from escaping. The coolant (air) flows between the glass and the plate's surface, transferring heat away from the surface and into the system. In the liquid flat-plate collector (Figs. 12.4 right and 12.5), the coolant passes through a series of small tubes, all of which are in contact with the solar-heated surface, again to draw the heat away. The flow rate of the coolant is adjusted to maintain constant temperature. Otherwise, the temperature will increase to the point where the coolant will boil or thermal stresses strain the structure of the collector. Flat-plate collectors can absorb heat from the diffused sunlight of a cloudy day, although they work better with direct, full sunlight.

For the most part, commercial solar thermal systems have utilized flat-plate technology. This is especially true of residential solar systems. But two types of focusing collectors have approached the commercialization point as well, at least for some limited applications. In general, focusing collectors are not as versatile as the flat-plate varieties because they cannot operate in diffused sunlight. They concentrate the sun's direct rays only when the collector is oriented toward the sun. However, focusing collectors operating on a sunny day perform at high efficiencies and, in certain climates, have annual heat-collection averages well above those of flat collectors of equal collecting areas.

Focusing collectors can operate at higher temperatures than flat collectors because there is a smaller heated surface. This means less

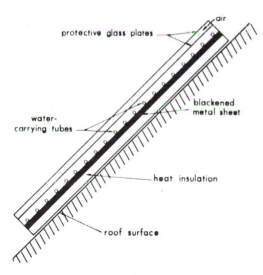

protective glass plates

air

blackened metal sheet

water-carrying tubes

heat insulation

roof surface

**FIG. 12.5.** Hot water flat-plate collector – cross section. *Source:* Penner/Icerman, *Energy,* Vol. II, © 1975, Addison-Wesley Publishing Co., Inc., Reading, Massachusetts, figure on page 380. Reprinted with permission.

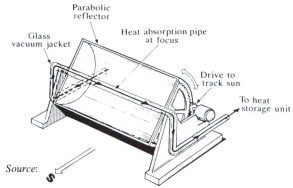

**FIG. 12.6.** Line-focus solar collector. *Source*: Morrow (1973).

**FIG. 12.7.** Point-focus solar collector, parabolic disk type. Photo courtesy of Jet Propulsion Laboratory, California Institute of Technology.

thermal radiation loss at a given temperature, and the focusing process means a higher solar-flux concentration. Thus a line-focus collector (Fig. 12.6) can raise coolant temperatures to the 150°–550°F range, whereas flat-plate collectors typically operate only in the 80°– 200°F range. The coolant in a line-focus collector circulates through a single pipe placed along the (line) focus of the reflector, which is a parabolic cylinder. The pipe has either a blackened surface or is an absorbent-coated rod within a glass tube. In either case, the focused solar rays heat the coolant in the tube or rod, which is then circulated through a system of heat exchange and storage similar in operating principles to a flat-plate system.

Evacuated-tube collectors (not shown) have also demonstrated some promise for widespread use. This variety combines features of flat and focusing collectors. Essentially, they consist of arrays of long tubes that are really tubes within tubes. The innermost tube carries the coolant; the outer is evacuated, providing insulation to reduce heat loss. Behind the tubes are reflectors, which are curved slightly for minor focusing. But a flat array of tubes can gather enough radiation so that, like a flat-plate collector, it can operate on diffuse as well as direct sunlight.

**FIG. 12.8.** Point-focus solar collector, low-cost design. Photo courtesy of Power Kinetics, Inc., Troy, New York.

Like line-focus collectors, though, evacuated tube designs also can generate high temperatures – up to about 350°F.

The highest temperatures of operation are attained by collectors that focus to a *point* rather than along the line. Point focus can be achieved by the use of a parabolic dish-shaped reflector as shown in Figure 12.7. In this type of collector, sun rays are reflected into a chamber where a coolant coil is contained. This chamber is shielded against radiative loses and insulated against conductive losses. Because of the extreme solar intensity at the focus, the system can achieve very high temperatures; demonstration models have produced temperatures in the 400° – 1,500°F range.

Parabolic dish solar reflectors have been demonstrated many times in the course of history (for example, the Paris Exposition of 1878 had an exhibit with a working solar dish). Historically, however, such reflectors were built without regard to cost and were viewed more as scientific instruments than as practical alternative energy devices. Present-day solar demonstration projects, in contrast, emphasize low-cost construction as a prime feature for feasibility. The point-focus collector shown in Figure 12.8 is an especially good example of a prototype, low-cost, point-focus design. This reflector consists of a series of curved mirrored slats whose angle of elevation can be adjusted up and down, while the horizontal angle is adjusted by rotating wheels on a circular track.

Although all focusing-collector systems share some operating similarities with flat-plate systems, their range of application is wider because of the higher temperatures of the focused systems. Line-focus collectors, for example, have been demonstrated for use in

FIG. 12.9. Solar industrial process heat collector field at Johnson & Johnson Company, Sherman, Texas. *Source: Solar Engineering* (1980).

industrial process heat and in replacing boilers to run steam turbines for electric generation. Point-focus collectors have been used in similar industrial demonstration projects, and because of their higher temperatures are more efficient for such tasks. But large-scale application of solar thermal to industrial uses would require development on a vast scale. Single projects may contain hundreds of thousands of square feet of collector area. Such applications are, in fact, medium term in time scale. Full technical feasibility in the sense of being ready for mass production and wide commercialization has not yet been achieved. As we noted in Chapter 11, one of the key engineering problems is in the mass production of collectors. Until such collectors can be fabricated on a mass scale and cheaply, they will be unlikely to provide much of the energy needs of the rest of industry.

At the same time, engineers are contemplating ways to make *low-cost* mass-production possible. One concept for low-cost fabrication of line-focus collectors is to build parabolic-trough reflectors (Fig. 12.9) that employ sheet-metal fabrication techniques such as those used in the mass-production of household appliances. The reflectors would in effect be stamped into the desired shape by large metal-forming machines. They would then be assembled with line-focus tubes using mass-production riveting or welding. One problem with this idea is that reflectors are nearly like optical instruments, requiring more precision than is typically demanded of other items using mass-production metal-fabrication techniques. But it is believed that some small degradation of performance would be acceptable and may be necessary in the interest of cost containment. In any case, such a mass-production effort has yet to be attempted.

## Toward adoption

Passive solar technology has reached the commercialization stage, elements of it are being included in building design every day, and it is often proving economically viable. Still, only when there is a greater concern over energy costs or conservation of resources will it be demanded in new buildings and thus exploited and developed to its fullest. A few other barriers, however, remain to widespread adoption.

Although flat-plate collectors, as well as some line-focus and evacuated-tube collectors, have been demonstrated to be technically feasible and have enjoyed some commercialization, they are not close to widespread adoption for any major application. As we have indicated, the major barrier is economic. Until the cost of active solar-thermal technology is competitive with conventional energy sources, commercialization will be limited. But solar thermal probably will increase its market penetration considerably *if* the prices of oil and gas rise significantly; the government reinstitutes tax incentives; the prices of solar systems fall; and/or any combination of the above. We can, as a matter of fact, expect that at some point prices of conventional fuels will rise, and that event alone might make active systems economically viable. If fuel prices rise enough, they very well might induce government to offer tax-policy-motivated incentives as well. The difference between economic viability (at least for certain regions) and lack of it may well be considered even now the difference between a price of oil of $1/gallon (or less) and $2/gallon (or more), plus a tax deduction on the cost of the solar system. It should be emphasized that only systems for such applications as domestic hot water heating could make a quick leap to commercialization in the event of a large increase in fuel costs. Solar-thermal heat for industry probably would still need more development. However, as solar thermal began to replace conventional sources in domestic use, development would probably begin on a wider scale, accelerating the trend to adoption for all kinds of applications.

On the whole, solar thermal remains a promising technology. It is environmentally benign and superabundant and may well offer a means of saving significant amounts of fossil fuels in the future.

## Synthetic fuels

Synthetic fuels are the creation of fuels, in gaseous or liquid form, that can substitute for conventional fuels, such as natural gas or oil. The two principal sources of raw materials for these alternative fuels are *coal* and *biomass*. Coal, as we learned in Chapter 2, is a huge fossil resource, whereas biomass, a renewable resource, has hardly been utilized as an energy source. We will focus in this chapter on coal-derived fuels and discuss biomass briefly in Appendix C.

What is basically involved in creating a synthetic fuel, or synfuel, is a process of chemical reaction, heat treatment, and sometimes pressure that leads to new chemical compounds having the heat content and physical state (gaseous or liquid) of conventional fossil fuels. As such, synfuels have a different derivation than alternative fossil fuels, such as tar sands, which require extensive physical processing but *not* the synthesis of new chemical compounds.

The question might logically be asked: Why bother with the extra step of synthetic processing when coal is itself usable as a fuel? Why not simply burn it directly? The answer is that coal in its natural state does not lend itself to some purposes, particularly those related to automotive and air transportation. Also, gas and liquid fuels have advantages in storage and portability over coal.

Consider automotive fuels. Conventional automotive fuels, gasoline and diesel fuel, are derived from petroleum. Petroleum (both crude and refined oil) is now transported very efficiently – pipelines, tankers, and trucks. Supertankers can carry as much as 5 million barrels of oil at one time. Huge stocks of crude oil and refined products are routinely stored in commercial operations and even larger amounts of crude oil have been stored in strategic national reserves.[5] The sheer bulk of coal, on the other hand, makes transportation and storage far more difficult. To transport 1 million tons annually, for example, would require the service of 30 railroad cars each day throughout the year. But the energy in 1 million tons of coal equals the energy in the oil stored in one supertanker. Coal also weighs about 70 percent more than the energy equivalent amount of oil.

Liquid automobile fuel also is extremely portable. A small tank holding 10 gallons of gasoline will provide the fuel to power an internal-combustion vehicle for more than 200 miles. Such portability makes it virtually impossible at present to replace current automotive technology with another having a comparable technical performance (see Appendix C, section on advanced storage). Any alternative fuel that will be substituted for petroleum products in automotive applications must then be equally *transportable, storable,* and *portable.*

An alternative substitute fuel should have other equivalent properties. Natural gas, for example, not only lends itself to easy transport and storage, it is clean – providing heat with low levels of polluting emissions. Any substitutes for it, then, must have comparable chemical properties. In order to achieve such a substitute, however, a synthetic coal-derived product must be well purified and contain high fractions of methane and hydrogen as the combustible components. A product like this could be extremely valuable, offering the technological potential for substitution not just for existing gas, but for coal itself. The result could be better air quality in regions that today do not meet national environmental standards. Synthetic gas would need to use the same pipeline distribution network as natural gas. It would, above all, need the same heating content as natural gas so it could be used for the same purposes. Otherwise, partial substitution of synthetic gases within natural gas networks would not be possible.

Synfuel conversion then would have desirable effects, and as we will see, chemists have proven that conversion of coal to liquid and gaseous fuels can be accomplished physically. But synfuel development still has a long way to go before it can be produced commercially. Indeed, what development engineers have shown so far is that there are a number of alternative means of producing liquid and gaseous synfuels. We will discuss several of the more promising synfuel processes, but we should keep in mind that as yet none of them has been conclusively demonstrated as having superior technology for commercial application. Of course, none of them has been proven technically feasible or economically viable for mass substitution.

## Coal-derived synfuels

Coal can be used as the basic ingredient for either synthetic petroleum or synthetic gas, although for both it must be extensively treated and its chemistry altered. Coal is composed primarily of two elements: carbon and hydrogen. In general, it might be said that the main scientific goal in the development of all coal-derived synfuels is to create gaseous or liquid hydrocarbon compounds that have a higher ratio of hydrogen to carbon than exists naturally in coal itself. We will consider both gasification and liquefaction technologies.

## Coal gasification

The production of a synthetic substitute for natural gas is complex and costly. The start of the process, however, is comparatively simple, because coal can be gasified simply by heating it. A more highly combustible gas will be created by reacting it with water and air in addition to heat. These reactions will produce two gases primarily: carbon monoxide (CO) and hydrogen ($H_2$). Both gases are combustible and can be used as energy sources and are known together as *town gas*, a term that dates back to the last century.

But neither alone nor together are carbon monoxide or hydrogen comparable to natural gas in energy content. A simple gasification process using heat, water, and room air produces a low-BTU gas (LBG), about 150 BTU/SCF (BTUs per standard cubic foot). The energy content is limited because the $CO + H_2$ is diluted with noncombustible nitrogen and carbon dioxide ($CO_2$). If pure oxygen is employed instead of air, the coal gas output will be a medium-BTU gas (MBG) of about 350 BTU/SCF. Both LBG and MBG are used for heating, principally in industry; MBG gas is also used in industry for petrochemical feedstocks. But natural gas is a high-BTU product (HBG) on the order of 1000 BTU/SCF and is composed mostly of methane ($CH_4$). Although some methane

is created in simple gasification processes, it is likely to be only a small proportion of the output. In order to turn coal into a comparable product to natural gas, it must be subjected to further reactions and often requires chemical catalysts to raise the proportion of methane.

We can get a clearer sense of coal gasification by considering the Lurgi gasifier depicted in Figure 12.10. The Lurgi process, which has been known since the 1930s, is used commercially to produce gas, albeit low- or medium-BTU gas. However, Lurgi gasifiers have also been employed to create gases from coal that are then processed further into synthetic natural gas (SNG).

As we can see in the figure, coal is fed into the gasifier, and as it is distributed around the top of the chamber, it is subjected to heat from below that dries the coal. But at temperatures of 1,100°–1,500°F, the coal is also devolatized. This reaction leads to a small amount of gasification. As we can see on Table 12.1, the devolatization reaction is a chemical process that produces $CH_4 + C + H_2$. But in the Lurgi system, the gasification process has just begun. As the coal falls through the chamber, it not only undergoes further heating, it reacts with both air and steam. The air and steam, along with continued application of heat from coal burning at the bottom of the gasifier, result in three different gasification reactions, which can occur simultaneously. The *carbon-oxide* reaction (see Table 12.1) – from the effect of heat on carbon and oxygen – results in release of both noncombustible carbon dioxide as well as combustible carbon monoxide. With the addition of steam, we also get *steam-carbon* and *water-gas* reactions. The former produces carbon monoxide and hydrogen, the latter carbon dioxide and hydrogen. Note that the coal ash falls through the grate at the bottom of the chamber while the lighter gas flows to the top and out to a scrubber, where impurities, such as tars and sulphur, are removed.

Despite the multiplicity of reactions, the Lurgi gasifier produces gas with a low percentage of methane (see Fig. 12.12, for instance). Greater amounts of methane can be generated even in variants of the Lurgi process

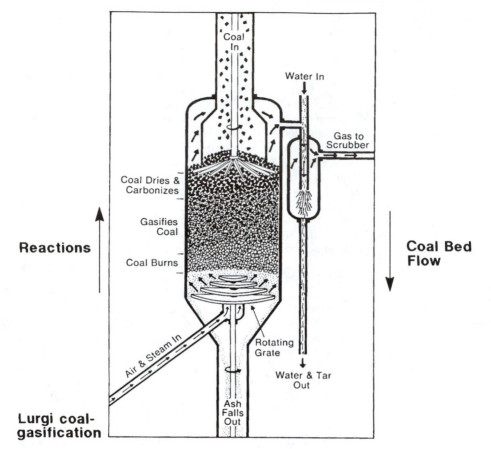

**FIG. 12.10.** A Lurgi coal gasifier. Adapted from
*Popular Science* and Pringle (1975).

**Table 12.1.** Coal gasification reactions

| Chemical reaction | Reaction name | Component of grade |
|---|---|---|
| $1{,}100°-1500°F$ $Coal \rightarrow CH_4 + C + H_2$ | Devolatilization | All |
| $C + O_2 \rightarrow CO_2$ $2C + O_2 \rightarrow 2CO$ | Carbon-oxide | Medium BTU |
| $C + 2H_2 \leftrightarrow CH_4$ | Hydrogasification or methanation | Low or medium BTU |
| $C + H_2O \leftrightarrow CO + H_2$ | Steam-carbon | Low or medium BTU |
| $CO + H_2O \leftrightarrow CO_2 + H_2$ | Water-gas shift Catalytic shift-conversion | Low or medium BTU Feeds high BTU |
| $3H_2O + CO \leftrightarrow CH_4 + H_2O$ | Catalytic methanation | High BTU gas |

*Source*: U.S. Department of Energy (1981).

through the application of both heat and high pressure. This leads to the reaction of carbon and hydrogen gas – called the *hydrogasification* (or *methanation*) process (Table 12.1). But even a methanation phase in a Lurgi-type reactor is insufficient by itself to produce high-BTU SNG.[6]

To create SNG, the low- and medium-BTU gases of the Lurgi (or a similar) process must be purified and reacted further. The final step listed on Table 12.1, catalytic methanation, is typically required. For this reaction, the important issue is not the amount of methane in the gas, but rather the ratio of hydrogen to carbon monoxide. There must be a ratio of three or more hydrogen molecules to every CO molecule (or $H_2/CO > 3$). If it is lower than three, the gas must undergo a procedure prior to catalytic methanation, called the catalytic shift conversion, in which carbon monoxide and steam react with a catalyst to raise the proportion of hydrogen (see Table 12.1).

Catalysts are substances that enhance chemical reactions without themselves entering the resulting compounds. Engineers have experimented with various elements as catalysts in the syngas process. Only by using catalysts can the concentration of methane be raised sufficiently, through the catalytic methanation reaction (Table 12.1), to create a high-BTU output. Often catalysts are metals, such as nickel, or metal oxides. But one system uses sodium carbonate as a catalyst, and another uses potassium chloride. It should be noted that in some cases the exact composition of a catalyst is considered proprietary information and its makeup is not publicly known.

Coal gasifiers were first built in the early nineteenth century and were in relatively heavy use in industry by the later part of the century. The gaseous product was generally a low-BTU gas used for industrial heat and household lighting just prior to electrification. The

**Table 12.2.** Commercial coal gasifiers in operation (1980)

| Gasifier | Product gas | Maximum coal throughput per gasifier (tons/day) | Number of gasifiers | Locations |
|---|---|---|---|---|
| *United States* | | | | |
| Wellman-Galusha | LBG | 85 | 23 | 12 installations in Pennsylvania, Ohio, and Texas |
| Chapman (Wilputte) | LBG | 60 | 12 | Houston, Texas |
| Wellman Incandescent | LBG | 80 | 1 | York, Pennsylvania |
| Foster Wheeler/Stoic | LBG | 80 | 3 | Minnesota, Michigan, and Texas |
| *Foreign* | | | | |
| Wellman-Galusha | LBG | 85 | 14 | South Africa |
| | MBG | | 7 | |
| Wellman Incandescent | LBG | 80 | 23 | South Africa |
| Foster Wheeler/Stoic | LBG | 80 | 2 | South Africa |
| Winkler | MBG | 1,100 | 51 | Worldwide |
| Woodall-Duckham | LBG | 200 | 32 | Worldwide |
| Koopers-Totzek | MBG | 450 | 40 | Worldwide |
| Lurgi | LBG | | 2 | Worldwide |
| | MBG | 1,200 | 70 | |
| Otto-Rummel | MBG | — | 3 | Germany |

*Source*: U.S. Department of Energy (1981).

water-gas reaction was developed in the late nineteenth century, yielding a superior medium-BTU product.

Synthetic gas production was all but abandoned in the United States when natural gas became widely available following World War II. Natural gas could be extracted at a fraction of the cost of synthesizing a fuel, and the construction of long-distance gas pipelines provided low-cost transport of the natural fuel from the gas wells to demand centers. As a consequence, development of new synthetic gas processes virtually ceased. Only with the energy crisis of the 1970s did interest in development and commercialization of syngas revive. Then government and private industry, together and separately, began to assess various processes. Though no one process could be deemed ready for widespread commercial development, pilot projects on the more promising concepts were planned and in a few instances constructed.

Before these new projects started there were a few small commercial gasifiers in the United States and abroad. Table 12.2 lists gasifiers that had achieved a certain level of commercialization by 1980.[7] They were all relatively small-scale operations, providing LBG or MBG with coal input ranging from 100 to around 1,000 tons/day, at most one-tenth the daily coal input of a conventional 1,000 $MW_e$ coal-fired power plant. Many of these plants, as a matter of fact, were serving as the process-fuel source for a single industrial furnace where gas is preferable to coal itself.

Table 12.3 lists various coal-derived, synthetic gas development projects underway around 1980. Several were HBG projects that were intended to bring SNG processes closer to the commercialization phase. All were experimental and heavily subsidized by government and/or industry. Few of these pilot projects were actually operating, and fewer still produced significant quantities of gas as of the late 1980s. Others, like the Great Plains plant, were intended to test prototypes of large-scale operations and indeed the later successful opening of that plant demonstrated that significant quantities of HBG syngas could be produced and marketed, although it would be

at a much greater cost than natural gas itself.

The HBG projects adopted various techniques and relied on different processes to produce SNG. The Great Plains plant, for instance, is based on Lurgi technology, but most of the methane is produced in a final catalytic methanation stage. On the other hand, the HYGAS plant, a small-scale syngas pilot project jointly developed by the Department of Energy and the natural gas industry's Institute of Gas Technology, uses a different approach. In this process, coal is reduced to small particles and mixed with liquid – in this case a light oil residue from the syngas process itself – and then pumped as a slurry into the gasifier (see Figure 12.11). The HYGAS hydrogasifier is unlike a Lurgi unit in several ways. Most important, it uses a two-stage approach. In the first stage, the coal is not only heated but kept under pressure (about 1,000 pounds per square inch, or psi) to create a significant amount of methane through the methanation reaction shown in Table 12.1. In fact, about two-thirds of the methane needed for SNG is produced in this initial stage of gasification; only one-third is produced from a final catalytic methanation.

In the second stage of the HYGAS hydrogasifier, the input of steam and oxygen produces reactions similar to those of the Lurgi process. At a temperature of about 1,850°F, the steam and oxygen cause steam-carbon and water-gas reactions, which produce hydrogen and carbon monoxide gases, adjusted to the proper ratio for the final catalytic methanation. But even at this stage there is an important difference in the technology between the HYGAS and Lurgi gasifiers. In the HYGAS chamber, the coal is processed on a fluidized bed – that is, while it is heated it is suspended on a bed of air (see Appendix C for fluidized bed technology). In contrast, in the Lurgi process the coal is moving downward in a moving-bed system.

The differences between these two types of gasifiers give some indication of the kind of questions that must be resolved before technical feasibility is achieved. Engineers, at present, cannot say with certainty whether a fluidized bed, or a moving bed, or yet another concept,[8] will work most efficiently and cost effectively,

**Table 12.3.** Coal gasifiers under development

| Gasifier | Project | Gas | Status |
|---|---|---|---|
| Slagging Lurgi | Consolidated Coal Company | HGB | Demonstration in design; pilot plant operated in Westfield, Scotland. |
| COGAS | Illinois Coal Gasification | HBG | Demonstration in design; pilot plant operated in Leatherhead, England. |
| HYGAS | Institute of Gas Technology and U.S. Department of Energy | HBG | Pilot plant work near completion. |
| BI-GAS | Bituminous Coal Research, Inc. | HBG | Pilot plant work in progress. |
| $CO_2$ acceptor | Consolidated Coal Company | HBG | Pilot plant work completed. |
| Lurgi, Mark IV (Great Plains) | American National Resources Company | HBG | Completed and operational, Beulah, ND. |
| U-Gas | Memphis Gas and Light | MBG | Demonstration in design; pilot plant in operation at Institute of Gas Technology. |
| Texaco | Grace | MBG | Feasibility of demonstration design. |
| | Texaco | LBG | |
| | | MBG | Pilot plant operated by Texaco. |
| | Dow Chemical | LBG | |
| | | MBG | Pilot plant in operation. |
| | TVA | MBG | Pilot plant in start-up phase. |
| | Southern California Edison, Coolwater Consortium | MBG | Syn gas and electricity; demonstration in design. |
| Shell-Koppers | Shell International (Federal Republic of Germany) | MBG | Pilot plant in operation. |
| Winkler high temperature | Rhienbraun (Federal Republic of Germany) | MBG | Pilot plant in operation. |
| Saaberg Otto | Saarbergwerke, A.G. (Federal Republic of Germany) | MBG | Pilot plant in operation. |
| Combustion Engineering (atmospheric) | Combustion Engineering | LBG | Pilot plant in operation. |
| Westinghouse | Westinghouse | LBG | Plant pilot unit in operation. |
| Hydrogasi- fication | Rockwell International | LBG MBG | Pilot development unit operated. |
| Catalytic gasification | Exxon | HBG | Pilot plant in operation. |

*Source*: U.S. Department of Energy (1981) and the *New York Times*, May 6, 1983.

especially over time on a large scale. These demonstration projects can produce SNG, but they are mainly designed to provide data that may help in deciding the best means for large-scale production. Prototypes will need years of operating experience to develop the technological capability to cope with the numerous aspects of syngas production. Not only will engineers have to answer questions connected directly to the production of gas, they will also have to show whether plants can handle huge amounts of coal and whether

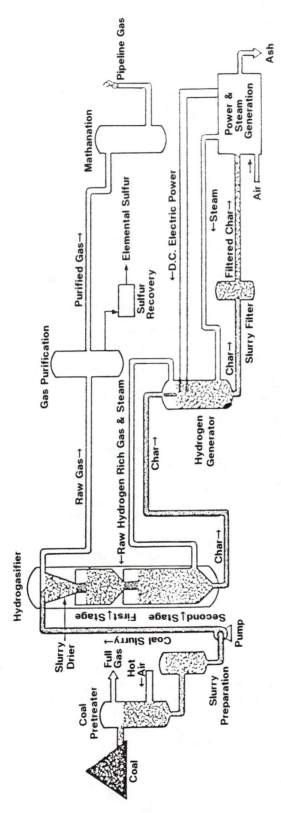

FIG. 12.11. The HYGAS process. Courtesy of the Institute of Gas Technology.

waste products can be handled in an environmentally safe fashion (an issue we will consider later). Only the Great Plains project is big enough in scale to begin to give data about such questions as environmental safety. The HYGAS and many other pilot plants produce too little gas and handle too little coal to provide data applicable to large-scale production.

Of course, syngas is not at present near economic viability either. The Great Plains project processes about 5 million tons of coal annually into 125 million cubic feet of high-BTU synthetic gas daily. The total cost of the project was over $2 billion and its output costs about $6.75 per 1,000 cubic feet to produce; in contrast, in 1985, pipeline natural gas[9] was sold to distributors at a profit for around $2.50/1000 cf.[10] The Great Plains plant could thus not survive without subsidies. This was demonstrated in 1985; when the U.S. government announced it was withdrawing price supports for the output from the Great Plains plant, the consortium of companies that had built it promptly defaulted on $1.5 billion in government-subsidized loans. But the plant was kept in operation by the Department of Energy and by a guaranteed purchase agreement for the output. Without the latter, the gas could not have been sold commercially.

The development work needed to make synthetic HBG feasible and viable is proceeding slowly. As of this writing, only those few projects like the Great Plains plant that were built by the early 1980s still operate. All other U.S.-government sponsored programs, or government incentives for private syngas development, have ceased.

## Coal liquefaction

Coal can also be liquefied to produce substitutes for the products of crude oil. One way is to start with gasification, adjust the composition of the gas to achieve the desired ratio of hydrogen to carbon, purify it, and then react it in a separate process to turn it into a liquid. Since it starts as a gas, this method is called *indirect* liquefaction.

There is one process of indirect liquefaction

that is today producing significant quantities of synthetic fuel. As we noted in Chapter 11, in South Africa over 50,000 bbl/day of synfuels are produced and marketed, albeit with highly subsidized prices. They are produced at three plants that use a similar indirect liquefaction method called SASOL[11] (Fig. 12.12). This process begins with a Lurgi gasification scheme. Once the gas is cleaned and purified, it is treated by a process called the Fischer-Tropsch synthesis. Developed in 1925, this process reacts the gas at 300 psi with an iron catalyst. The reactions release heat. As the heat is removed by coolants, gaseous hydrocarbons are liquefied. The output from the Fischer-Tropsch synthesis includes most of the products of ordinary crude oil, from light to heavy hydrocarbons.

Though the SASOL plants have demonstrated the fairly large-scale operation of indirect liquefaction, the indirect method of liquefaction is relatively inefficient and thus it is believed will ultimately prove more costly than the alternative: *direct liquefaction.* The basic chemical objective in direct liquefaction is the same as in indirect and pure gasification processes: to obtain a higher ratio of hydrogen to carbon than exists naturally in coal. But in gasification (and thus in indirect liquefaction), the molecular structure of coal is broken down into small components. Direct liquefaction requires that the molecular structure of coal be broken down, but into larger units.[12] This process – called a *cracking process* – requires temperatures up to 900°F and pressures up to 10,000 psi. Once cracked, the coal can be hydrogenated directly with pressurized hydrogen or through *donor solvents,* which transfer their hydrogen to the output compounds in solution. However, this output is for the most part already in the form of liquid hydrocarbons that do not require further synthesis. Thus, the direct process is more efficient than the indirect process. But direct processes are more difficult to control. There is a problem simply in breaking down coal's organic structure, which if not properly handled leads to gasification, not liquefaction. Direct processes also may pose problems of purification, for example, the removal of catalysts or compounds found in

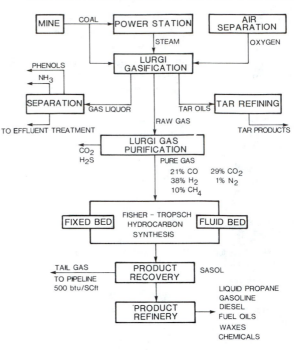

**FIG. 12.12.** The SASOL process – flow
diagram. Adapted from Lee (1982).

coal ash, which may be suspended in the liquefied coal output (see Lee 1982).

There are basically three techniques for direct liquefaction: pyrolysis, solvent extraction, and catalytic hydrogenation. The pyrolysis process, known since the nineteenth century, is the simplest, since it merely reacts the coal at high temperatures in an oxygen-deficient atmosphere. Pyrolysis produces both liquid and gaseous outputs and a low-hydrogen solid waste called *char*. It is not considered promising for large-scale operation, largely because of the char, which needs to be separated from the liquid hydrocarbons and then disposed of. A large operation would produce vast amounts of char.

Solvent extraction is more promising. As Table 12.4 shows, a number of different solvent extraction processes have been developed. Essentially, in all of these processes, coal is mixed with and dissolved in a solvent. They differ, though, on exactly how hydrogenation is accomplished.

In the solvent refined coal (SRC) process, for example, dried coal is mixed with a solvent oil that is distilled and separated from later stages of the liquefaction process itself. This combination of coal and solvent is then pumped into a reactor where it is subjected to heat (about 850°F) and a pressurized atmosphere (1,500 psi) rich in gaseous hydrogen. This process creates liquid and solid hydrocarbons. Typically, the liquids are removed and refined, while the solids can be gasified to provide hydrogen for the liquefaction process. The SRC process (a forerunner of which was first developed in the 1940s) has been tested in prototype plants. Initially, its output more resembled tar than oil and it proved to have some value as an industrial furnace fuel. But the general goal of liquefaction is to create liquid fuels that can replace gasoline and diesel fuel. Toward that end, a variation of an SRC plant, called SRC-II, was subsequently built. It subjected the coal and solvent mixture to an atmosphere that had a greater amount of hydrogen and reacted it for a longer period of time to produce a more liquified output.

Catalytic liquefaction, though far more experimental than SRC, was intended from the outset to give high yields of liquids. In one

**Table 12.4.** Comparison of direct liquefaction techniques

| Technique | Comparative advantages | Comparative disadvantages | Representative processes |
|---|---|---|---|
| Pyrolysis | Low or atmospheric pressure operation | Not attractive for large-scale liquid fuel production: low liquid yields; large quantities of by-product char must be disposed of. | Lurgi-Ruhrgas |
| | Hydrogen, oxygen, synthesis gas not needed | | COED (FMC Corp.) |
| | | | Occidental (Occidental Coal Research) |
| | Relatively short reaction times | Liquids are heavy, difficult to separate from char | Toscoal (TOSCO) |
| | | | Clean-Coke (U.S. Steel) |
| | Simple, low capital cost equipment | Liquids require hydrogenation to produce environmentally acceptable fuels | Coalcon (Union Carbide) |
| | | | Flash Liquefaction (Rockwell International) |
| Solvent extraction | High liquid product yields | Limited knowledge exists for preheating and handling of coal-solvent slurries | Consol (Conoco Coal Development Co.) |
| | Flexibility of product slates | | SRC I and II (PAMCO) |
| | Substantial sulfur removal | | SRL II (University of North Dakota) |
| | | | CO-Steam (DOE) |
| | | | Exxon Donor Solvent (Exxon Corp.) |
| | | | Extractive Coking (A. D. Little, Inc.) |
| | | | Extraction by Supercritical Fluids (Coal Research Establishment – England) |
| | | | UOP Process (Universal Oil Products) |
| Catalytic liquefaction | Low operating pressures | Separation of oil, catalyst, and undissolved coal and ash difficult by filtration | Bergius |
| | Short retention times | | H-Coal (Hydrocarbon Research, Inc.) |
| | High degree of product quality control | Fouling and deactivation of catalyst due to char and trace elements | Synthoil (DOE) |
| | | | Catalytic Coal Liquids (Gulf Oil) |
| | High liquid yields | | Clean Fuel From Coal (C-E Lummus) |
| | Substantial sulfur removal | | University of Utah Process |
| | | | Schroeder Process |
| | | | Liquid Phase Zinc Chloride Process (Continental Oil Co.) |
| | | | Dow Chemical Process |

*Source*: U.S. Department of Energy (1981).

example, H-coal, a slurry of coal and a solvent is again heated under high pressure and mixed with hydrogen gas. But the coal solvent is also reacted with a catalyst of cobalt molybdenate. Although the output is lighter than that from

solvent processes, there are engineering questions relating to the long-term handling and durability of the catalyst itself (see Lee 1982). This process would need to be tested for years to demonstrate how long-lived the catalyst will be and how well and economically it can be used over the long term.

But data on liquefaction processes and engineering ideas will be limited unless interest is revived. As we have noted, the oil glut has diminished support and interest from both industry and the U.S. government for all synthetic fuels. Indirect liquefaction projects were never more than small-scale prototypes; ambitious plans for greater indirect liquefaction development were scrapped. Pilot projects for direct liquefaction have been scrapped as well. For example, a large SRC-II plant (the first prototype handled only 50 tons of coal per day) planned under the auspices of the Department of Energy in the late 1970s was dropped in 1981. In fact at one time, the Department of Energy also sponsored pilot plants in the H-coal process as well as several other processes.[13] The announced goal of the Department of Energy program in 1980 was 2 million bb$\ell$/day of synfuels (over all types, by 1992) – a goal that was shortly abandoned.

## Environmental concerns

One of the most important and still unanswered questions affecting the ultimate feasibility of synfuels in general is the impact they would have on health and the environment. Coal-derived synfuels could have profound effects on both, from the mining process through utilization. In fact, a large synfuel industry is expected to have the same environmental consequences as are caused by industries that presently use coal as an energy source. In addition, it will add problems commonly found in the refining and the burning of conventional liquid and gaseous fuels. And finally, it may have impacts specific to it alone.

For example, in the production process of coal-derived synfuels, heat is applied either in the form of steam or directly as a requirement for the reactions of the process. The heat energy will be supplied by burning coal itself, and indeed more BTUs of coal will need to be expended than are created of gas or liquid. With current processes, about 3 BTUs equivalent of coal are needed to create 2 BTUs heating value

of gas. As a result, a synfuel plant will, like a coal-burning electric power plant, require controls on sulphur and nitrogen oxide emissions and on particulates. In fact, when synfuel development was being pushed around 1980, various government agencies, labor unions, and environmental groups expressed concern about the potential damage to air quality from synfuel plant operations.

As with conventional coal-burning power plants, environmental concerns do not end with air pollution. Coal ash must be disposed of as well. Even though the ash represents only 10 percent of the original coal mass, large-scale synfuel plants will use huge amounts of coal. The Great Plains gasification plant alone consumes about 16,000 tons of coal every day. If the United State met one-third of its demand for natural gas with synthetic gas, the syngas industry would consume over 2 billion tons of low-sulphur Western coal per year. This would be an order of magnitude larger than the highest projected extraction rate of coal for conventional purposes. This would mean, first, that huge quantities of ash as well as scrubber sludge and desulfurization residues would be produced and would require safe disposition; if improperly handled, the waste could cause massive pollution of rivers and aquifers. And since Western coal is strip mined, it would also mean disturbance of tens of thousands of acres of land that would later need reclamation.

These spillover effects, and those typically found in conventional oil and gas refining, are at least predictable and technology exists to cope with them. Others are less predictable and may occur on such a scale that major engineering efforts will be required to solve them. In gasification and indirect liquefaction, there is concern about the environmental impacts of *acid gases*, primarily hydrogen sulfide, hydrogen cyanide, and naphtha. These gases are particularly hazardous; in high concentrations, they can be fatal, in low chronic exposures, they may be carcinogenic – a hazard particularly for synfuel workers. With commercial-scale synfuel operation, plants are expected to produce large quantities of acid gases. These must be eliminated at the gas purification stage, and although promising purification schemes have been proposed, they must be built and tested over a wide range of operating conditions.

By-products of liquefaction could also be extremely toxic both to the environment and to workers. Solvents, such as benzene and alkyl creosols, are known to be carcinogenic. In fact, many of the heavy distillates and solvents found in the various liquefaction processes are believed to have mutagenic characteristics.

Control of these hazards will require special designs of physical systems and will demand industrial hygiene procedures to meet the conditions of an emerging synfuel industry. At present, it is not clear whether current engineering concepts will be adequate. Engineers will need the benefit of years of prototype operation to gather the information and experience to refine controls and procedures to assure environmental safety.

Finally, it should be kept in mind that carbon dioxide is the inevitable product of combustion of any sort and the conversion of coal hydrocarbons to other forms of fuel does not change the final product of combustion. Therefore, synfuels do not provide any solution to the dilemma of the greenhouse effect, as discussed in Chapter 6. In fact, if their widespread use leads to increased and prolonged use of fossil fuels, the dilemma could be made even worse.

## Toward commercialization

Coal-derived synfuel · development is clearly possible, and yet the technology is far from commercialization. The problems are not simply technical, although technical hurdles remain to be overcome. The most important technological questions relate to the scale of the operation. Can such facilities be properly controlled? Can they operate efficiently and safely? Will new equipment and materials be required to maintain long-term operating effectiveness and cost efficiency? How should plants be organized to facilitate mass production?

But if we project a large-scale synfuel industry, we have a massive economic issue that must be considered as well. In solar thermal, we considered the investment cost versus the fuel saving to determine cost efficiency for the buyer of an active solar energy system. Each residential system involves an investment of several thousand dollars and a payback period of a few years to a few decades depending on fuel costs, subsidies, and so on. At the point where fuel costs go up enough, solar saves money and becomes cost effective for the home owner. In the event of a hike in fuel costs, solar thermal could be ready for the market quickly.

But with coal-derived synfuels, the investment becomes so enormous it cannot be undertaken unless the technology is deemed fully feasible and the need overwhelming. Consider on economic terms the replacement of one-third of U.S. natural gas supply by syngas. The actual cost of producing the gas versus the cost of pipeline natural gas would be a factor, but let us assume that market conditions cause gas prices to rise to levels that make syngas competitive. Yet to reach the one-third level (about 6 trillion cubic feet per year) would require building over *100* plants the size of the Great Plains plant. The cost probably would exceed $200 billion (in late-1980s dollars), a capital investment of staggering proportions. The implications of this price tag are important. First, though it is not inconceivable that society will feel the need to invest that much money for energy, it cannot allocate such large financial resources overnight. Second, even if it could, it would be virtually impossible to begin construction of so many plants all at once.

This means, of course, that even if there would be a perceived need for syngas, it would take years – perhaps a decade or more – to get large-scale production going. But given the cost, the desire to begin construction on such a scale would probably result from a supply crisis that a sudden burst of syngas development would not solve for a decade. It would be foolhardy, however, for society to invest $200 billion or even one-tenth that amount unless technical feasibility were assured. In other words, even if society wanted to produce syngas on a large scale, it would have to decide either to do so blindly or first to spend tax money to develop technical feasibility before investing the major amount for the massive production facilities. This could push actual commercialization a decade further into the future even with government and industry research support.

Of course, as of the late 1980s, synfuel research and development were proceeding very slowly. It remained a medium-term technology and even an energy crisis would not have made it a technically feasible or an economically viable alternative in any shorter time. At the same time, it continued to hold out the prospect of mass substitution for liquid and gaseous fuels that could allow present technologies to be utilized for decades more, perhaps until a long-term superabundant energy source is developed.

## Nuclear fusion power reactors

Nuclear fusion represents the ultimate in energy technologies in several different ways. It is the prime example of the *backstop technology* (see Nordhaus 1979 in Chapter 11, Bibliography), in that its fuels[14] are superabundant. Fusion fuels could replace fossil fuels when they become exhausted and would essentially never run out. It is potentially a clean technology in that it does not emit great quantities of gaseous or solid wastes and its nuclear by-products are far less hazardous than those of nuclear fission technologies. Because of its enormous potential, it has become by far the largest long-term research and development effort for a new energy-conversion technology. In fact, it represents one of the largest commitments worldwide to scientific development for civilian purposes over the past quarter century.

### Physical principles

Nuclear reactions are a source of liberated energy at the two extremes of the nuclear mass scale on the periodic table. At the high end (atomic mass number $> 200$) we have noted the nuclear fission reactions of isotopes of uranium and thorium. Nuclear fusion reactions, in contrast, occur at the low end of the mass scale, with the isotopes of hydrogen and helium.

It might seem paradoxical that the splitting of nuclei leads to the liberation of energy in the case of fission reactors and that the fusing together of nuclei in fusion reactions also leads to energy release. The full explanation of this paradox, and why the reactions occur at opposite ends of the nuclear mass scale, lies in the theory of nuclear binding forces (see Sproull 1963). We are able to understand the outcomes of the reactions in each case, however, as a consequence of the equivalence of mass and energy – the famous equation of modern physics ($E = mc^2$) that we encountered earlier in this text.

In examining the mass-energy equivalence (Chapter 7) for nuclear fission reactions, we saw that the aggregate mass of the fission products is less than the combined masses of the $^{235}U$ nucleus and the neutron from which they were created. Similarly in a typical fusion reaction (of two deuterium nuclei[15]), the combined mass of the helium isotope and neutron fusion products is less than the two deuterium nuclei from which they were created. For both fission and fusion reactions, then, the difference in nuclear masses is manifested as kinetic energy of high-speed motion.

Physicists have considered four different possible nuclear fusion reactions as the basis for fusion power reactors of the future (see Fig. 12.13). For each of the four, the liberated energy is shown on the right side of the reaction equation in millions of electron volts (MeV). In the first reaction, deuterium (D) and tritium (T)[16] fuse to yield the helium isotope nucleus ($^4_2He$, or the alpha particle) and a neutron (n); each such reaction also releases 17.6 MeV of energy (see legend, Fig. 12.13). The reaction of two deuterium nuclei, on the other hand, can produce either a tritium nucleus and a proton (p) plus 4.03 MeV in energy, or a $^3_2He$ nucleus and a neutron, plus 3.27 MeV in energy. Finally, deuterium and $^3_2He$ react to give $^4_2He$ and a proton, plus 18.3 MeV in energy.

Fusion reactions occur only when the light nuclei come sufficiently close together (within

# Fusion Energy

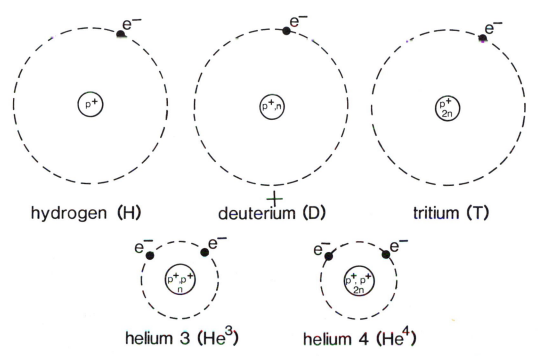

**FIG. 12.13.** Hydrogen isotopes and possible fusion reactions. First-generation fusion reactions (DT); $T \geqslant 4\,KeV$, $40 \times 10^6\,°C$; $D + T \rightarrow \alpha + n + 17.6\,MeV$. Second-generation fusion reactions (DD); $T \geqslant 34\,KeV$, $340 \times 10^6\,°C$; $D + D \rightarrow T + p + 4.03\,MeV$; $D + D \rightarrow He^3 + n + 3.27\,MeV$. Third-generation fusion reaction (DD); $T \geqslant 100\,KeV$, $10^9\,°C$; $D + He^3 \rightarrow \alpha + p + 18.3\,MeV$. Adapted from Penner & Icerman (1975).

$10^{-15}$ meters). The main way that they come into that close proximity, however, is when they collide at high speeds. Otherwise, the repelling force within the nuclei (isolated nuclei are positively charged ions) keeps them separated. The kinetic energy required to overcome those electrostatic repelling forces is shown for each of the four fusion reactions in Figure 12.13, in units of KeV (thousands of electron volts) or equivalent temperature (see Appendix A).

Creating and sustaining nuclear fusion reactions is a fundamentally different problem from that of fission. The fusion reaction is not a chain reaction, nor does it depend on energetic neutrons to initiate it. Rather, the fusion reaction requires the heating of the isotope constituents, in gaseous form, to extremely high temperatures, so as to create high-energy collisions of the fusible nuclei. For this reason the reactions are called *thermo*nuclear reactions.

The nature of the gas as it is heated is suggested by Figure 12.14. It indicates that as the temperature of the gas rises, collisions between neutral atoms become sufficiently energetic to strip the electrons from their nuclei, resulting in a gas of negatively charged electrons and positively charged nuclei (ions) in equal numbers. This gas, which has a net zero charge over all of its volume, is called a *plasma* and much of the research effort for fusion has to do with its dynamics.

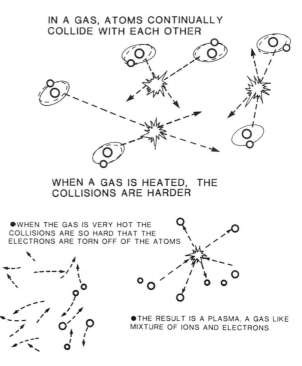

IN A GAS, ATOMS CONTINUALLY
COLLIDE WITH EACH OTHER

WHEN A GAS IS HEATED, THE
COLLISIONS ARE HARDER

● WHEN THE GAS IS VERY HOT THE
COLLISIONS ARE SO HARD THAT THE
ELECTRONS ARE TORN OFF OF THE ATOMS

● THE RESULT IS A PLASMA, A GAS LIKE
MIXTURE OF IONS AND ELECTRONS

**FIG. 12.14.** Principles of plasmas. *Source*: U.S. Energy Research and Development Administration (1976).

● A HOT PLASMA WANTS TO EXPAND QUICKLY. ALSO IF IT IS
ALLOWED TO HIT A WALL IT WILL COOL QUICKLY AND REVERT
BACK TO A GAS AGAIN.

As the gas of hydrogen isotopes is heated, the ionization takes place at temperatures in the range of 100,000°C or, equivalently, when individual atoms have average thermal velocities over 1 million meters per second. In order to achieve fusion reactions by collisions of the ions, however, the temperatures of the gas must rise to over 1 hundred *million* degrees, which corresponds to average thermal velocities about one-tenth the velocity of light (300 million meters per second). Such temperatures create solarlike conditions, and no ordinary material can contain the gas in this state.

Let us assume, however, that containment is possible and consider an idealized fusion reactor. The ideal fusion reactor should provide a way of transferring the thermal energy generated by the nuclear reactions in the confined plasma to a turbine or other means of conversion to useful work. We can see this idealized concept at the top of Figure 12.15.

The fusion plasma is "bottled" and a coolant flows around it, taking heat out for useful work. In other words, the fusion reactor would be like the nuclear fission reactor in one respect: another extremely exotic way to boil water. Although the ideal is unattainable, we can in practice come close enough to contain the plasma and permit controllable fusion reactions.

We must emphasize at the outset that the quest for controlled thermonuclear reactions (CTR) is still very much in the research phase. A successful fusion reactor technology must solve three problems: the temperature must be *hot* enough for the high-energy collisions, the plasma must be confined *long* enough for the nuclear reactions to heat the plasma for further reactions, and the plasma must be *dense* enough for the rate at which reactions take place to overcome the rate at which energy is lost to the outside.

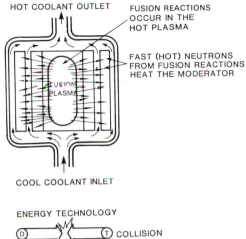

(a)

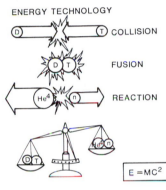

(b)

**FIG. 12.15.** Principles of nuclear fusion. (a) Fusion plasma confinement (b) Nuclear fusion reaction. *Source*: U.S. Energy Research and Development Administration (1976).

## Fusion research

The first scientific goal of fusion research programs throughout the world is to achieve the *breakeven* condition. Breakeven is the point where the energy yield from the fusion reactions equals the energy input required to create the reactions. The required energy input is the amount necessary to confine and heat the plasma to fusion conditions. If the breakeven condition can be achieved, the next goal would be a net gain of energy, that is, where *more* energy is generated by the fusion reactions than has to be invested to initiate them. This condition is also termed *ignition*, with the implication that a thermonuclear *burn* would take place in a manner analogous to combustion. Only after ignition is attained can the fusion reactor be developed on an engineering basis and demonstrated for commercial use.

## Plasma confinement

One of the most crucial problems that the researchers had to solve was how to confine the plasma, because there is no known material that could be used for a containment vessel that could withstand the solar temperatures of the plasma. Confinement of the fusion plasma has been conceived in two major ways: *magnetic* confinement and *inertial* confinement. We will focus on the former in this chapter, because it is the primary thrust of fusion research today, and will consider inertial confinement briefly in Appendix C. Magnetic confinement takes advantage of a well-known property of magnetic fields (Fig. 12.16). These fields deflect the motions of charged particles that are directed across the lines of force, so that they tend to spiral or gyrate around the lines (see Fig. 12.16).

FUSION REQUIREMENTS

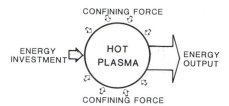

Specifically,
$T_{ION} > 5KEV (50,000,000°)$
$n\tau > 5 \times 10^{13} cm^{-9} seconds$

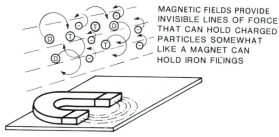

MAGNETIC FIELDS PROVIDE
INVISIBLE LINES OF FORCE
THAT CAN HOLD CHARGED
PARTICLES SOMEWHAT
LIKE A MAGNET CAN
HOLD IRON FILINGS

**FIG. 12.16.** Principles of magnetic confinement. *Source*: U.S. Energy Research and Development Administration (1976).

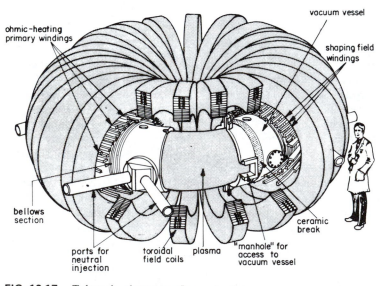

**FIG. 12.17.** Tokamak plasma confinement – the Princeton large torus (PLT). *Source*: U.S. Energy Research and Development Administration (1972).

The leading design configuration for magnetic confinement is the *tokamak*,[17] an example of which is shown in Figure 12.17. In the tokamak configuration, confining magnetic fields bend the plasma into a torus (a donut-shaped ring). These fields are created by a series of coils looped around the plasma toroid (see Figure 12.18, the toroidal-field magnet coils). These magnetic lines of force are concentric with the plasma donut, following its large diameter.

The toroidal magnetic fields of the tokamak deflect (and thus confine) any radial motion of electrons or ions out of the toroid. Thus, the natural force of expansion of the hot plasma is overcome. This confinement field must be maintained to sustain fusion reactions. As we will discuss later in this chapter, the enormous forces for a fusion plasma at ignition temperatures make continuous confinement a major engineering challenge.

Magnetic confinement in practice is imperfect due to physical processes, such as diffusion, within the plasma. Diffusion is a natural process in any body of particles in thermal motion. Although the magnetic field can contain rapidly *moving* particles, it is ineffective against a random displacement of a particle momentarily brought *to rest* by a thermal collision. As a result, the objective of magnetic confinement in the face of diffusion and other losses is simply to sustain the density of the plasma for sufficiently long time periods for fusion reactions to take place. Alternatively, the plasma density would have to be sustained by continuously supplying new fuel to the reactor.

After confinement, the next major task in fusion research is heating the plasma. In a tokamak, the principal method is to induce a strong electric current through the plasma, thus causing electrical losses that in turn heat the gas. This form of heating is called *ohmic heating*, so named after the electrical unit for resistance (ohm). The current is induced magnetically in

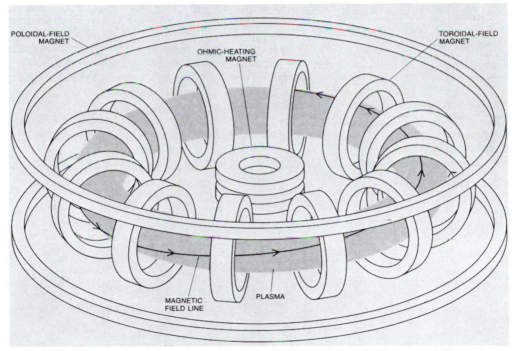

**FIG. 12.18.** Idealized tokamak. *Source*: "The Engineering of Magnetic Fusion Reactors," by Robert W. Conn. Copyright © October 1983, SCIENTIFIC AMERICAN, Inc. All rights reserved. Illustration by Gabor Kiss.

the plasma torus in the manner of an alternating current (or a.c.) electrical transformer (see Appendix A). The primary winding of the ohmic heating coil is concentric to the plasma torus (Fig. 12.18).

The tokamak uses another basic concept of the a.c. transformer. When there is a second winding coupled by the magnetic field of the primary coil, then a current will be induced in the secondary coil. If the secondary coil has fewer turns than the primary, then there is a step down in voltage but a *step up* in current. The plasma torus is, in effect, a single-turn coil, whereas the primary (ohmic heating) coil is designed with 10 to 20 turns. When 50,000 to 100,000 amperes are driven through the primary coil, 1 million amperes or more get induced in the torus.

Ohmic heating, as powerful as it might seem, is insufficient for reaching ignition temperatures in the tokamak. Efforts have been made toward *supplementary heating*. One means of supplementary heating is through the injection of a high-energy beam of neutralized hydrogen isotope atoms into the plasma torus (Fig. 12.19). The injected particles must be neutralized – that is, have neither a positive nor a negative charge – in order not to be deflected by the strong magnetic field in the tokamak. Once into the plasma, the injected atoms collide with the particles of the plasma, become ionized components of the plasma, and add to the thermal energy.

Supplementary heating may also be achieved by irradiating the plasma with high-power radiowaves or microwaves, somewhat in the manner of a microwave oven. This process, called *rf* (for radio frequency) *heating*, is thought to work particularly well when the frequency of the incoming waves is near one that resonates naturally with the magnetized plasma. One such resonance frequency is the *cyclotron frequency*, which is the natural rate of gyration of the electrons around the magnetic field.

In addition to the confining and heating, plasma physicists have faced the challenge of controlling the plasma against *instabilities*. A hot plasma is inherently unstable and can break into violent motion of various types, including strong streaming motions and what are called magnetohydrodynamic (MHD) instabilities. The conditions for controlling plasma are so delicate that some plasma researchers have likened the problems to *trying to suspend water in a bottle upside down*, while others have said they have to control a stellar Jello. The MHD instabilities, in which the whole body of the plasma would twist, bend, or swell under the combined magnetic and fluid forces, were among the earliest problems plasma researchers encountered. They found it could result not just in undesirable fluctuations, but even in an entire breakup of the plasma.

It was found, however, that the addition of a *poloidal* magnetic field tended to stabilize the tokamak plasma against such motions. The poloidal fields are created by a set of separate

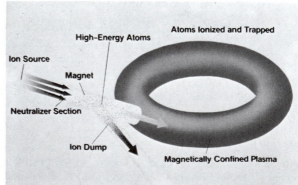

**FIG. 12.19.** Neutral beam concept. Courtesy of the Princeton Plasma Physics Laboratory.

magnetic coils (see Figs. 12.17 and 12.18). The resulting poloidal fields are a set of rings that are roughly concentric with the original *toroidal coils*. They cross the toroidal fields and give combined total field lines that spiral along the plasma torus, as shown in Fig. 12.18. The poloidal-field windings carry a steady current (d.c. or a long, steady pulse) to generate a sustained magnetic field with the toroidal field.

On a microscopic scale, the addition of the poloidal field creates motions of electrons and ions that follow spiraling trajectories around the spiraling lines of the total (toroidal plus poloidal) magnetic fields. Both types of magnetic fields, however, tend to confine the charged particles, because motion toward the boundary of the torus will be deflected by the fields.

Fusion researchers have proposed and tried alternative types of magnetic confinement to the tokamak. Although the tokamak scheme remains the leading contender for the fusion reactor, magnetic *pinches* and magnetic *mirrors* have also been tested. In the magnetic pinch, the plasma is squeezed down in size by rapidly increasing a magnetic field that surrounds it. The pinch, which both increases the plasma density and heats it, can be applied in several different geometric configurations. A mirror, on the other hand, is a magnetic field configuration that causes the particle trajectories to double back on themselves, providing a way to bottle up the ends of a straight column of plasma. The

poloidal fields in a tokamak also cause trajectories to close on themselves and this contributes to the stability of confinement in that configuration.

These alternative concepts, although not as promising overall as the tokamak, have in recent years been utilized to help researchers close in on the conditions needed to achieve breakeven. There has been, however, a sense of competition between alternative confinement projects and the main tokamak project. This competitive element was not discouraged by the research sponsors at the Department of Energy because they knew that the alternatives could serve as insurance against an unexpected collapse of the tokamak concept.

The status of tokamak fusion research as of the mid-1980s is summarized on Figure 12.20. The vertical axis shows the ion temperature ($T_i$), and the horizontal axis depicts the Lawson criterion. This criterion is in the form of a product: plasma density multiplied by the confinement time ($n \cdot t$, where $n$ is the number of particles per cubic centimeter and confinement time $t$) measured in seconds. The Lawson results are thus expressed in numbers of atoms per second per cubic centimeter times the number of seconds of confinement. Past results from tokamak experiments, over the period from 1978 to 1984, are shown in relation to the temperatures and Lawson criteria needed to achieve breakeven and ignition. As we can see, breakeven requires temperatures of 10 KeV or

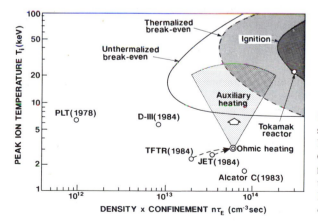

**FIG. 12.20.** Tokamak fusion research status. PLT = Princeton large torus; D-III = Doublet III (General Atomic Company); TFTR = (Princeton) tokamak fusion test reactor; JET = joint European torus; Alcator C = MIT Alcator. *Source:* Furth, *Physics Today*, American Institute of Physics (1985).

more along with an $n \cdot t$ of at least $10^{13}$ sec/cm$^3$. The PLT experiment, for example, achieved the highest temperature of any experiment to date, but had the lowest $n \cdot t$ product and was therefore far from breakeven.[18] At the other extreme was the MIT Alcator project; it was highest in $n \cdot t$, but was lowest in temperature and thus also far from breakeven.

Currently, the emphasis in the United States is on the TFTR, which started experiments in 1984, and for which a breakeven demonstration was planned for – but not achieved by – 1988. Comparable hopes were placed on the Joint European Torus (JET), whose experimental results to date are also shown on Figure 12.20, and on large tokamak experiments in Japan and the Soviet Union (see Furth 1985). The figure indicates that researchers expect the TFTR and the JET to attain breakeven with ohmic and auxiliary heating (neutral beams and/or rf heating). The vertical, cone-shaped area on the figure above ohmic heating, extending into the breakeven region, purports to show the possible range of $(T, n \cdot t)$ conditions attainable above ohmic heating by auxiliary heating.

If the breakeven condition is attained in the current generation of tokamaks, then the next logical step would be for a tokamak (or an offshoot) to achieve ignition. There are questions, however, of what the strategy should be for the next generation of fusion experiments. As we noted in Chapter 11, the fusion research community (in the United States and internationally) has been debating whether the program should be basically science oriented or technology oriented. That is, should the focus be on basic scientific questions, or should researchers continue to pursue a working reactor. This debate on fusion RD&D, although argued largely in scientific terms and on technical grounds, is colored heavily by concern over the future of funding and by uncertainty of the outcome of the present generation of research. The doubts of the late 1980s represent a distinct change from the confident programmatic plans proposed in the late 1970s by the McCormick legislation (see Chapter 11) and internationally by the INTOR reactor project.[19]

Both the McCormick and INTOR programs set out steps leading to a demonstration fusion reactor, with coordinated efforts for the engineering aspects of working fusion power plants to be carried out. The first priority of these efforts was to arrive at the demonstration reactor after about two decades. With funding cutbacks, however, by 1985 the prime objectives had been scaled back to attain a science-oriented fusion ignition experiment; the engineering of a demonstration plant had become secondary (see Furth 1985).

## Toward development: the engineering challenge

Fusion breakeven and ignition experiments will, even if successful, only prove the scientific feasibility of a fusion reactor. After that, there will be a vast engineering task to create an operating reactor that will be safe, reliable, and economical – thus beginning the development stage mentioned in Chapter 11. Some of the engineering tasks that would need to be solved have not even been addressed in the current research phase. For example, no work has been undertaken with respect to handling the hydrogen fuel. Magnetic fusion experiments to date have operated with pure hydrogen, not its fusible isotopes, and therefore the problems of breeding and handling radioactive tritium have yet to be explored.

Figure 12.21 depicts the full range of technological requirements of a working magnetic confinement nuclear fusion reactor. We can list the following:

1. *Magnetic field*: Enormous magnetic field strengths, for example 50,000 gauss,[20] must be maintained over large volumes (on the

---

FIG. 12.21. Technological requirements for a nuclear fusion reactor. *Source*: "The Engineering of Magnetic Fusion Reactors," by Robert W. Conn.

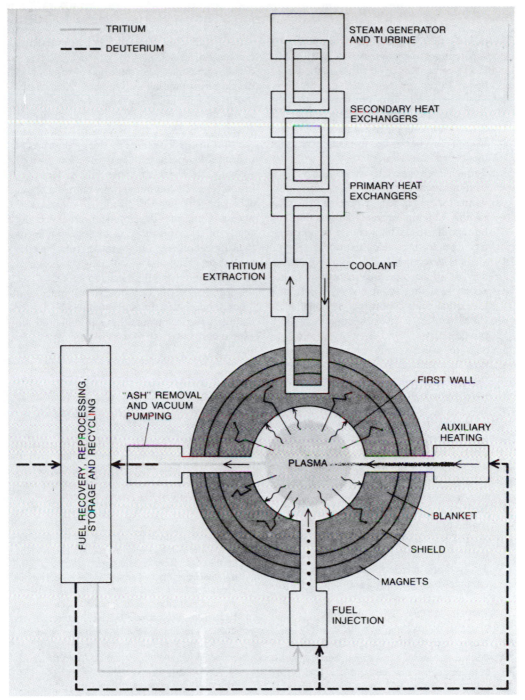

**ENGINEERING PRINCIPLES** common to all magnetic fusion reactors are diagrammed. A magnetic field must confine the fusion plasma; an auxiliary heating system must help to raise its temperature; a fuel-recycling system must keep it pure and well supplied with thermonuclear fuel. The heat the plasma releases must be withstood by the first wall of the reactor. The neutrons the plasma releases must penetrate into the blanket, where the energy the neutrons deposit must be transferred (in the form of heat) to a coolant. In turn the coolant can generate steam, the steam can drive turbines and the turbines can generate electricity. Nuclear reactions in the blanket must also "breed" tritium, which is radioactive and is extremely rare in nature.

order of 10,000 m$^3$). This will require superconducting magnets that will need at least 1 percent of the reactor's useful output power to run.

2. *Fuel recycling*: The reactor must be resupplied as DT is burned up and fuel is lost. Also, the supply of both deuterium and tritium must be purified to exacting requirements.

3. *Tritium breeding*: This is necessary since tritium occurs rarely in nature. It probably will be accomplished by placing a blanket of liquid lithium around the reactor. The bombardment by high-energy fusion neutrons would then create tritium.

4. *Tritium extraction and handling*: This is a delicate process, because tritium is radioactive. And although it is shorter-lived and less lethal than fission fuel, it nonetheless raises concerns over occupational and public safety.[21]

5. *The "first wall"*: The fusion reactor must be able to withstand extreme physical conditions of neutron bombardment and heat deposition. Working power reactors will be many times larger than current experimental units and, as a result, the energy densities will be very large and place enormous stresses on the reactor.

6. *Heat transfer*: Coolant circulation must be reliable and poses a major engineering task.

7. *Radiation shielding and disposal*: This is required against the 18 Mev neutrons of the DT reaction that will penetrate beyond the blanket into the structure of the reactor. This requires development of new neutron-damage-resistant alloys and demands provisions for disposal.

8. *Plasma heating*: The engineering of an efficient source for plasma heating is a major task. Auxiliary tokamak heating, of course, means either neutral beam generators or rf/microwave sources (or both), either of which will require considerable technological development.

The solution to any of these engineering problems will demand a major development project. Taken together, the list suggests that a commercial nuclear fusion technology is decades in time and billions of dollars in development funding away – *after* the achievement of scientific breakeven and ignition.

Harold Furth, director of the Princeton Plasma Physics Lab, wrote in "Reaching Ignition in the Tokamak" (1985): "the success of fusion always seems to be twenty years away." Actually, Dr. Furth, an expert and leader in the fusion research community, is optimistic about achieving that success. But he knows it is uncertain and that development remains many years in the future. His ironic comment indicates, however, that those running the fusion programs have not always been so insightful about the time scale to success. Over the decades, they have promised too much, too quickly – each time to be confounded by nature and the difficult barriers that still remain to be overcome.

## Notes

1. Solar flux, sometimes termed *insolation*, is the density of solar-thermal power per unit of area, thus in MKS units (see Appendix A) – watts/meter.$^2$ In British units it is in BTU/ft$^2$-hr, where BTU/hr is the power (rate of heat-energy transfer) instead of watts (see Appendix A).

2. Using: $1W = 3.41$ BTU/hour, $(24 \times 365) = 8,760$ hours/year, $3.29^2 = 10.85 \text{ ft}^2/\text{m}^2$ and $10^6$ BTU = mBTU.

3. Antifreeze is used in climates where freezing temperatures may occur overnight. The coolant in this case is in a separate circulation loop, isolated from the hot water to be used. Heat is conducted from the coolant into the hot water in the heat exchanger, as in a power plant.

4. Heat from air warmed by a collector may be stored by forcing it through a bed of pebbles and thus warming the pebbles. The heat may later be extracted by circulating cool air (from the space to be heated) through the pebble bed (see Appendix C, on energy storage).

5. The establishment in the United States of a reserve of crude oil, stored in underground caverns, followed the oil embargo crises of the 1970s. As of 1987 about 200 million barrels of oil had been stored by this government program.

6. The Ruhr-100 process is a German-designed, pressurized Lurgi-type reactor. Its output is significantly higher in methane than a standard Lurgi reactor, although catalytic methanation would still be necessary to create HBG.

7. *Commercial* operation means that either a commodity here syngas – was produced and sold to a regular market or that the product was used internally in the manufacture of market commodities.

8. Researchers have also used a fixed bed and have experimented with what is called an entrained bed, in which there is a concurrent flow of coal and reactants. We might also note here the somewhat self-contradictory concept of a moving bed, but indeed the coal descends through the chamber as combustion takes place.

9. Pipeline gas is a specific gas industry term meaning that the gas has a BTU content and a level of purity capable of being mixed with natural gas for pipeline transmission and distribution.

10. In the natural gas market, as in any other market, there is a cost of production and a higher market price for consumers. As of 1988, the average wellhead price was in the range 1.50 to 2.00 per 1,000 scf (standard cubic feet), depending on the age of the well, the depth, the flow rate, and so on.

11. Actually the three SASOL plants – I, II, and III – represent three generations of the same basic process.

12. Coal molecules are very large, with molecular weights over 1,000. Methane molecules, in contrast, have a molecular weight of only 16 and the weight of the compounds in refined petroleum fuels is about 100.

13. Some of the projects for development of direct liquefaction techniques included solvent processes such as the SRL II (University of North Dakota) and the Exxon donor solvent (Exxon Corporation), as well as catalytic processes such as Synthoil (Department of Energy) and the liquid phase zinc chloride process (Continental Oil Company).

14. Nuclear fusion fuels are the isotopes of hydrogen and helium. Deuterium, an isotope of hydrogen, is the principal element and occurs naturally in the world. Approximately one atom in 6,500 in the seas is a deuterium atom.

15. Deuterium ($^2_1$H or D) is the first isotope of hydrogen, composed of an electron, proton, and a single neutron (Fig. 12.3). As we have indicated, deuterium is nonradioactive and is found in abundance in seawater. If all of the deuterium

atoms in a cubic meter of seawater could be extracted and used as nuclear fusion fuel, they would have the energy equivalent of about 2,000 barrels of oil. There are over 1 billion cubic kilometers of seawater in the world, making deuterium effectively inexhaustible.

16. Tritium ($^3_1$H or T) is the third isotope of hydrogen, composed of an electron, proton, and two neutrons. See Figure 12.13.

17. *Tokamak* is an acronym based on Russian words. Successful containment in a tokamak was demonstrated in an experiment by Artsimovich in the USSR in 1970.

18. It should be mentioned that several of these experiments (including PLT and Alcator) were planned without the expectation of achieving breakeven. The PLT effort emphasized high temperatures, whereas Alcator stressed high densities.

19. *INTOR* stands for International Tokamak Reactor, a joint project of the United States, Western Europe, Japan, and the USSR coordinated through the International Atomic Energy Agency in Vienna.

20. Gauss is a measure of magnetic flux density. The earth's magnetic field is a weak magnetic field and has a magnitude from 0.25 to 0.55 gauss at the equator. Magnetic fields found in conventional electric power equipment, such as transformers and generators, run in the range of 10,000 gauss, but are required over much smaller volumes within the apparatus.

21. The scientific possibility exists for clean fusion, wherein radioactive elements are not created. For example, the third and fourth fusion reactions shown in the legend of Figure 12.13 do not create tritium. The fourth reaction, furthermore, has no neutrons resulting from fusion and therefore would eliminate the high-energy bombardment of support structures that is an additional source of radioactivity in the DT and DD reactions. These possibilities, however, are even more remote than the first prospective reaction (DT) because of the vastly higher thermonuclear temperatures that would be required.

## Bibliography

### Solar thermal

Beckman, W. A., S. A. Klein, and J. A. Duffie. 1977. *Solar Heating Design*. Wiley Interscience, New York.

Dorf, R. 1978. *Energy Resources and Policy*, Chapter 17, Solar Energy. Addison-Wesley, Reading, MA.

*Energy Technologies and the Environment*. 1981. U.S. Department of Energy, Report No. DOE/EP0026. Washington, DC, June. Available from National Technical Information Service (NTIS), Springfield, VA.

Penner, S., and L. Icerman. 1975. *Energy*. Vol. 2, *Non-Nuclear Technologies*, Chapter 14, Addison-Wesley, Reading, MA.

## Synthetic fuels

Lee, B. S. 1982. *Synfuels from Coal*. Monograph Series no. 14. American Institute of Chemical Engineers, Vol. 78.

Penner, S. S., and L. Icerman. 1975. *Energy*. Vol. 2, *Non-Nuclear Technologies*, Chapter 10, Addison-Wesley, Reading, MA.

Blake, E. M. 1980. "Fusion in the United States" *Nuclear News* (four-issue series). June, Part 1: "An Overview," pp. 59–67. July, Part 2: "Magnetic Confinement: The Hunt for Heat," pp. 45–8. August, Part 3: Sept., Part 4: "The Issues Yet to Come," pp. 38–41.

Conn, R. W. 1983. "The Engineering of Magnetic Fusion Reactors." *Scientific American*, Vol. 249, Oct., pp. 60–71.

Furth, H. P. 1985. "Reaching Ignition in the Tokamak." *Physics Today*, March, pp. 53–61.

Hagler, M. O., and M. Kristiansen. 1977. *An Introduction to Controlled Thermonuclear Fusion*. Lexington Books, Lexington, MA.

Lidsky, L. M. 1983. "Trouble with Fusion." *Technology Review*, Vol. 87, March.

Penner, S. S. 1976. *Energy*, Vol. 3, *Nuclear Energy and Energy Policies*, Chapter 23, Addison-Wesley, Reading, MA.

Sproull, R. L. 1963. *Modern Physics*. Wiley, NY.

# Epilogue

An event in March 1989 demonstrated anew how difficult and uncertain is the process of scientific advance and technological innovation. At an extraordinary news conference, two researchers in electrochemistry claimed to have discovered a way to create nuclear fusion at room temperature. So-called *cold fusion* required only electrodes of palladium and a heavy-water solution, the researchers claimed. Fusion, they declared, would not require massive magnets or lasers and temperatures in the millions of degrees; with the right combination of materials, we could have fusion in a beaker.

The announcement produced a worldwide sensation. Some scientists rushed to duplicate or explain the phenomenon, others (including some scientists who had labored for decades on *hot* fusion projects) expressed immediate skepticism. Journalists, meanwhile, reported in great detail not only the discovery itself, but also the prospects cold fusion held for the future. Although most reports contained cautionary notes that potential applications were years away, the press on the whole adopted a decidedly optimistic tone that strongly suggested that the *backstop* energy technology had indeed been found and would soon become a part of everyday life.

How is cold fusion possible? A theory based on quantum mechanical probabilities says that when deuterium nuclei are absorbed into the crystal structure of palladium, they can become so densely packed that they fuse – without solar temperatures. If true, it would mean that all of the obstacles of plasma physics described in Chapter 12 would instantly be overcome.

Reality? Or wishful thinking? As this book goes to press, no one is entirely sure. Some additional experiments have appeared to confirm parts of the discovery. But the successes have provided little hope for any practical applications (much less a backstop energy source), and most other attempts at duplication have failed completely.

What the story of cold fusion has confirmed most of all is the slow, uncertain progress of science and technology that we have described in the last two chapters. The process of innovation is one of hope and of disappointment, of ideas that appear fruitful one moment and barren the next. It is often a path that raises more questions than it answers. Because even if all the claims of cold fusion are demonstrated to be true, we will have only proven the scientific principle. Ahead would lie technological development that would have to answer many practical questions. Will cold fusion develop enough energy to be useful? Will it be cost effective enough to be feasible on a large scale? Will there be other social costs that defeat it?

The development of new energy technologies will always involve uncertainty. Unfortunately, there is no reason to believe that any particular technology, be it cold fusion, photovoltaics, superconductivity, or what have you, will arrive on the scene in the nick of time to solve all of our energy problems. No technical shortcut has been found that can overcome the large and varied dilemmas facing society.

The cold fusion story gives us a good reminder: We cannot expect science to provide a panacea in a laboratory beaker.

# APPENDIX A

# Scientific principles

## Energy conversion

The purpose of all energy technologies is to provide energy in useful forms, such as mechanical work, electricity, and useful heat. Indeed, when we discussed society's demand for energy in Chapter 4, we had in mind energy as the equivalent of a material commodity for a consumer or as an input factor in production. That is, we viewed energy as a quantity that is purchased or allocated to provide means of accomplishing useful ends.

In order to be a useful commodity, energy in general must be converted to its end-use form. For example, the potential energy in water stored behind a high dam is converted to a useful form (mechanical work or electricity) once the water has fallen and turned the shaft of a water wheel or turbine (see Chapter 3). The energy tied up in the hydrocarbon compounds of fossil fuels becomes useful when the fuel has been burned and converted into heat. The fossil-fuel heat so derived may also be useful for space heating or a manufacturing process or the heat may itself be converted into mechanical work or electricity.

This last conversion process is of prime interest to us in this appendix. The process of converting heat into useful work energy is said to be the technological underpinning of modern industrial society. The *heat engine* is the historic name for all such machines. Industrial societies today are powered by heat engines of all sorts (steam turbines, gasoline engines, and so on) that are fired almost exclusively by fossil fuels (coal, oil, and natural gas). Such engines universally convert the heat of combustion into a useful form, such as mechanical work or electricity. Engines converting the heat of combustion of biomass alcohols would be in the same category. Even a nuclear reactor is merely a heat source to drive a steam turbine.

The scientific basis for understanding the workings and fundamental limitations of heat engines resides in *thermodynamics*. Thermodynamics involves a general set of principles that apply to energy conversion, heat transfer, or any one of many processes that involve changes in the form of energy. Thermodynamics applies regardless of the medium bearing the energy, the type of physical reaction involved, or the scale of the system in which the process is taking place. It is based, in the best tradition of modern science, upon repeated observations and experiments that fit into a consistent theoretical scheme.

Also of great importance to modern society is the generation of electricity. Electricity provides the means of transmitting and distributing energy that has already been converted to useful form by a thermal process or other means. In Chapter 3, we described a conventional power plant including the electric generator. Here, following our discussion of thermodynamic principles, we will also review the principles underlying electric generation.

## Units of measure

There are several different systems of units in use today for measuring physical quantities such as *energy* and *power*. Each system has a *fundamental* set of units that usually measure length, mass, and time. Thus, the modern metric system is called the *MKS system*, standing for *meters, kilograms*, and *seconds* – specifying the three fundamental units. If we are making electrical measurements, we must add one additional fundamental unit, either *charge* or *current*.[1]

All units of measurement that are not fundamental are called *derived* units, meaning that they can be expressed equivalently in terms of the basic units. Force, for example, is measured in newtons (N) in the MKS system, but can be expressed equivalently as kilograms divided by seconds per second or:

$$[N] = \frac{[kg]}{[sec^2]}$$

which comes from Newton's law for the force required for acceleration of a mass. In this way, the MKS system is made self-consistent and calculations can be carried out with the assurance that the numerical answers will always be in units of the system.

Conversion of units becomes necessary when switching from one system to another. In the energy field, British units such as British thermal units (BTUs), horsepower, and gallons are frequently encountered, due to their common usage and familiarity in the United States and the English-speaking world.

Here is a listing of MKS units as used in our discussions of energy throughout the text. Abbreviations are shown in brackets.

joule [J]: energy
watt [W]: power, time rate of energy delivery or usage
meter [m]: length (basic)

kilogram [kg]: mass (basic)
seconds [s]: time (basic)
newton [N]: force
degree Celsius [°C]: temperature, centigrade scale
degree Kelvin [°K]: absolute temperature, Kelvin scale (absolute zero is −273°C)
kilocalorie [Kcal]: heat energy[2]
liter [l]: volume
ampere [A]: electric current (see note 1)
volt [v]: electric voltage[3]
Electron volt [eV]: the energy ($1.602 \times 10^{-19}$ J) required to move one electron over an electric potential difference (see next section) of 1 volt.

A comparable listing of British units, many of which are used in this text, follows. The British system has been called the *foot-pounds-seconds (FPS) system*, with reference to its fundamental units.

horsepower [HP]: a measure of power, the time rate of energy delivery or usage
foot: a familiar measure of length
slug: mass, not generally familiar in usage[4]
second: time, same as MKS
pounds: force, a familiar measure of weight – the force of gravity
degree Fahrenheit [°F]: temperature
degree Rankine [°R]: absolute temperature on the Rankine scale (absolute zero is −460°F)
Btu: heat energy[5]
gallon: a measure of volume

The following are a number of useful conversions between Metric and British units.

*Energy*
1 Joule = 0.949 $10^{-3}$ Btu
1 KW-hr = 3,412 Btu[6]

*Power*
1 HP = 0.746 KW

*Mass*
1 Kg = 2.2 pounds
1 metric ton ($10^3$ Kg) = 1.1 ton (2,200 lbs)

*Temperature*
$[°F] = 9/5 \cdot [°C] + 32°$

*Temperature/Energy Equivalent*[7]
$1\,KeV \simeq 10^7\,°C$

*Heat Energy*
$1\,Joule = 0.24\cdot10^{-3}\,Kcal$
$\qquad = 0.940\cdot10^{-3}\,Btu$

*Volume*
1 gallon (U.S.) = 3.78 liters
1 barrel (42 U.S. gallons) = 160 liters[8]

# Forms of Energy

The laws of physics tell us that energy exists in the classical forms of *work*, *kinetic energy*, *potential energy*, and *heat*, and in the hidden (nuclear) form of mass-equivalent energy.

## Work

Work is force acting over distance, as expressed in:

$$W = F\cdot d \ \ [J]$$

where $F$ = force [N] and $d$ = distance [m] for translational displacements along straight or curved paths, including purely rotational displacements.

## Kinetic energy

Kinetic energy is the energy of motion, as expressed in:

$$K = \tfrac{1}{2}\cdot mv^2 \ \ [J]$$

where $m$ = mass of an object in motion [Kg] and $v$ = velocity of the mass [m/sec] for translation motion. Kinetic energy exists also for rotational motion.

## Potential energy

Potential energy is energy that can be, but has not yet been, delivered to other forms. The gravitation potential to deliver mechanical energy is expressed as:

$$P = mgh \ \ [J]$$

where $m$ = mass of an object [Kg], $g$ = gravitational constant [N/s²], and $h$ = height of the mass [m] for a mass placed at a height $h$ above some reference plane. Potential energy exists in systems such as electrical and chemical systems.

## Heat

Heat energy is expressed colloquially as *heat content*, that is, the amount of heat energy (such as BTUs) inside a solid body, liquid, or gas.[9] In engineering usage, however, *heat* is usually defined as the energy transferred from one mass to another when a temperature difference exists between the two; thus, the term might better be called *transferred heat*, denoted by the letter Q.

## Mass-equivalent energy

This amazing discovery of modern physics shows that mass and energy are equivalent, leading to the conversion of mass to usable energy in the nuclear reactor.

# Thermodynamic definitions

In order to conduct an orderly review of the laws of thermodynamics, it is first necessary to define certain terms and concepts used throughout.

A *system* is the specific combination of working substance and identified space in which a thermodynamic process takes place or in which certain thermodynamic conditions obtain. The working substance may be gaseous, liquid, or solid.

A *boundary* is the closed surface, real or imagined, enveloping the system and separating it from the rest of the universe.

The *surroundings* include all of the matter and space outside the system boundary.

In a *closed system*, there is no exchange of matter (mass) with the surroundings, but an exchange (transfer) of heat may occur with the surroundings.

In an *open system*, there is an exchange of matter, often a flow of matter, to and from the surroundings that is usually accompanied by an exchange of heat with the surroundings.

An *isolated system* is presumably not influenced by its surroundings in any way thermodynamically, that is, it has no transfer of energy or exchange of mass with its surroundings.

We can use these terms with reference to a modern steam power plant. We can define the thermodynamic system as the entire plant, or we can say that each of its internal parts, such as the boiler, the steam turbine, the electric generator, or the condenser, constitutes a separate system (see Chapter 3). If we elect to define the entire plant as the system, then the system boundary is the building envelope itself (Fig. A.1). This entire plant is certainly not an isolated system and it is not even a closed system, if we consider the mass of fuel, air, and cooling water flowing in and the mass discharge of ash, stack gases, and cooling water necessary for its operation. However, it is possible, if the system boundary is suitably defined, to imagine a closed system for this plant having exchanges of energy with the surroundings but no mass flowing in and out.

Such an imaginary boundary passes through the combustion chamber (furnace), so as to exclude the combustion gases and the mass of the coal and ash. The boundary must also pass through the condenser, so as to exclude the flowing mass of cooling water (see Chapter 3). Through such a boundary, albeit abstract, flows only energy: the heat transferred in the boiler from the furnace into the working substance (water/steam), the condensing heat transferred from the working substance to the cooling water and, of course, the electrical energy transmitted out of the generator toward its load.

The working substance of such a steam generating plant literally flows around a closed loop (see Fig. A.1). It starts as water that is heated into steam in the boiler, passes through the turbine, is condensed back down to water, and then starts circulating all over again. This circulation could be contrasted with the old-fashioned steam engine where the steam (working substance) is discharged from the piston after each power stroke, thus making the engine an open system in thermodynamics.

Finally, isolated thermodynamic systems are idealizations, only approximated in the laboratory or thought of as models of practical operation. The truly isolated system would have no mechanical or electrical means of receiving work energy from the surroundings and would

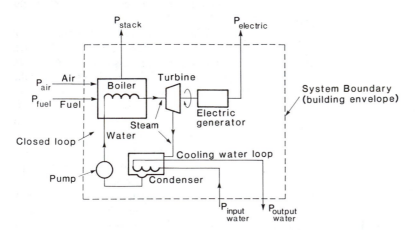

**FIG. A.1.** Schematic diagram of a thermal power plant.

$$P_{input} = P_{fuel} + P_{air} + P_{input\ water}$$

$$P_{output} = P_{electric} + P_{stack} + P_{output\ water}$$

Adapted from R. C. Dorf, 1978, *Energy, Resources, and Policy*. Ballinger, Cambridge, MA.

be totally insulated thermally (that is, no heat transfer could occur through conduction, radiation, or other means) from the surroundings. Its only value is that it can provide a conceptual model to illustrate the workings of thermodynamic laws.

## Conservation of energy: the First Law of Thermodynamics

The First Law of Thermodynamics, most simply stated, is:

*Energy can neither be created nor destroyed.*

It is a law because it has been tested repeatedly in experiments and has never been found violated. It also serves as part of the basis of an entirely self-consistent scheme of physical theory and underlies a cosmic hypothesis that states: "The total amount of energy in the universe is constant."

It is not obvious at the outset how such a broadly stated law can be applied to practical energy technologies. One approach is to consider various thermodynamic systems. The simplest case is the *isolated* system, where the interpretation is easy – the total energy contained within the isolated system cannot change. A more practical case is the closed system, where transfers of energy with the surroundings can occur, but mass transfers do not. Here the First Law can be stated in the form of an energy balance for the system, as follows:[10]

energy in = energy out [joules]

The application of this energy balance to a conventional power plant is shown schematically in Figure A.1. This plant is a closed cycle that recirculates steam and water (see also Figure 3.3). As such, the system boundary excludes the mass flow in of fuel, cooling water (in and out), and combustion gases (air in and stack gases out). But it includes the energy flows in and out of these masses. In the case of a power plant with a constant rate of energy

exchange, the balance equation may be written in terms of power (time rate of energy flow – see Units of Measure section) as follows:

$P_{in} = P_{out}$ [watts], where (referring to Fig. A.1)

$P_{in} = P_{fuel} + P_{air} + P_{input\ water}$

= total rate of energy flow in [watts]

$P_{out} = P_{elect} + P_{stack} + P_{output\ water}$

= total rate of energy flow out [watts]

Thus we see the energy transfers into and out of the closed steam cycle of the power plant. We know from the first law that as much energy goes in as must come out, or for conditions of a steady rate of energy production, the total power in must equal the total power out.

However, not all of the power out of the plant is useful. For an electric power plant generally, the electricity is the only useful output. A measure of the fraction of useful energy out is the *plant efficiency*:

$$\eta = \frac{E_{useful}}{E_{fuel}}$$

$$= \frac{P_{elect}}{P_{fuel}} \quad \text{(for the power plant with steady power out)}$$

The plant efficiency is sometimes called the First Law efficiency, because it is based on that principle. It is also a practical measurement, however, because it only rates the useful output against the energy input from the fuel, and fuel is the only energy input that has a direct operating expense. It might also be noted that the energy content of the incoming air ($P_{in}$) is negligible compared to the fuel input energy and thus the energy balance equation can be written approximately as:

$P_{fuel} + P_{input\ water} \simeq P_{elect} + P_{output\ water}$

which can be solved for the output in the form:

$P_{elect} \simeq P_{fuel} - P_{stack} - P_{net\ output,\ water}$

where the last term has been collected together

into the single term:

$$P_{\text{net output, water}} = P_{\text{input water}} - P_{\text{output, water}}$$

representing the net output of waste heat energy carried away by the cooling water. The importance of this waste energy will be considered during our discussion of the Second Law of Thermodynamics.

For an open system, according to our thermodynamic definitions, we must consider whatever mass flows that take place through the system boundaries. For example, consider the steam turbine alone, that is, define system boundaries as solely enclosing this single component of the steam cycle. Here the general statement (energy in = energy out) must be written specifically in terms of the various forms of energy (work, kinetic energy, potential energy, and heat) in order to account for conversions between energy forms in the open process. In the steam turbine, work is extracted as the mass of steam expands and speeds up, thus producing major changes in pressure, volume, kinetic energy, and internal heat content. These changes are all accounted for in the thermodynamic calculations, which require a conservation of energy – and also of mass – from the input to the output.

## Limits on the conversion of energy: the Second Law of Thermodynamics

No thermal energy conversion process can be entirely complete. The steam of the power plant, for example, cannot be converted entirely into work or electrical energy in any real-life system. Complete conversion would require not only that there be zero losses and zero friction, but also that surroundings be maintained at absolute zero temperature. These requirements are embodied in the Second Law of Thermodynamics. In order to better understand this law, it is first necessary to extend our set of thermodynamic definitions.

First, a *property* is composed of the observable, macroscopic quantities that characterize the working substance within a system. For example, temperature, pressure, and mass are properties of steam in a boiler. Properties are said to be *extensive* if they depend on the size of the system; for example, total volume and total mass are extensive properties of a system of a particular size. On the other hand, properties are often given on an *intensive* basis, reduced to a per unit quantity of volume or mass. Finally, the properties of a working substance are said to be well defined thermodynamically when the system is in an equilibrium state.

A *state* occurs when there is sufficient equilibrium so that the properties of the substance within are uniform and can therefore characterize the system as a whole. A minimum number of such properties for a system are required to define a state. For example, for a system containing an ideal gas, three properties – pressure, temperature, and volume – are necessary to define the system state. The state in this case will be defined only when the temperature and pressure of the gas are uniform throughout the volume enclosed by the system boundary and are not changing in time, that is, they are in *equilibrium*.

In a *process*, one equilibrium state changes to another. A process is usually depicted as a path on a diagram such as Figure A.2, which shows a process of isothermal (constant temperature) expansion of a gas.

In a *reversible process*, the system can return to the original state by return along the same path. This can only take place under certain strict conditions. Ideally, the reversible process takes place slowly enough so that transitional effects smooth out over the system, thereby maintaining a condition of *quasi-equilibrium* while changes in state are taking place. Thus, for example, the process depicted in Figure A.2 would be in quasi-equilibrium if the expansion from state 1 to state 2 were slow enough so that the temperature remained the same throughout the volume and the pressure remained evenly distributed while dropping in value. Reversibility then permits another slow, quasi-equilibrium process to take place, returning the system exactly to state 1.

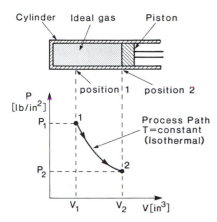

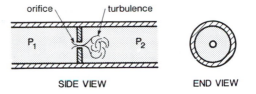

**FIG. A.3.** The throttling process. Gas in a high-pressure $(P_1)$ chamber flows through a small orifice into a chamber at lower pressure $(P_2)$.

**FIG. A.2.** Pressure-volume diagram for an ideal gas process. The system's boundaries are the cylinder walls and the face of the piston, thus defining a cylindrical volume $(V)$ filled with an ideal gas as the working substance. The process is controlled by having the piston move from position 1 to position 2, while maintaining the gas at constant temperature $(T)$. (Note that this system is closed, but not isolated.) The process is portrayed thermodynamically as starting at state 1, characterized by properties $P_1$, $V_1$, and $T$, and proceeding smoothly to state 2, which is characterized by $P_2$, $V_2$, and $T$.

An *irreversible process* does not maintain quasi-equilibrium and therefore cannot return to the original state without input of additional energy. Other sources of irreversibility are friction and energy losses that are not recovered on the reverse path. Perhaps the best example of a nonequilibrium path is the throttling process depicted in Figure A.3, where the turbulence is an excellent example of a nonequilibrium condition.

*Work* applies the basic definition of work $(W = F \cdot d)$ to the substance in a thermodynamic system. Work can be done on the substance (negative work) or it can be delivered by the substance (positive work). An increment of work occurs with an incremental change in volume at a constant pressure as in:

$$dW = p\,dv$$

Thus, an expansion is an increase in volume, with a positive quantity of work delivered by the substance and a compression is a decrease in volume with a negative quantity of work done on the substance. In Figure A.2 the total work delivered by the ideal gas over the path 1–2 is:

$$W = \int_{V_1}^{V_2} p\,dv > 0$$

An integral, taken between two limits as shown here, can be interpreted as the area under the curve of the function being integrated (here $p$).

*Internal energy* is the energy associated with microscopic thermal motion on a molecular or atomic scale in the working substance of a system. It thus includes the kinetic energy of molecules in thermal motion in a hot gas, thermal motion of electrons in liquids and solids, and crystal lattice vibrations in solids. Internal energy, as denoted by $U$, is measured in units of heat energy and is distinct from the quantity of transferred heat $(Q)$. It is an *extensive* property because it depends on the total mass of the system substance; it can be reduced to an intensive property if we measure it on a per unit of mass basis.

An *adiabatic process* is a process in which no transfer of heat takes place into or out of the system. A transfer of heat $(Q)$ must involve changes in the internal energy and/or the work of the system, as expressed by:

$$dQ = dU + dW$$

for an incremental change in heat. An adiabatic process has no transfer of heat and hence $dQ = 0$ everywhere along its path from one state to another. Therefore, increments of work $(dW)$

must be exactly compensated by incremental changes in internal energy ($dU$), with accompanying changes in system temperature. The p-V path for an adiabatic compression is illustrated in Figure A.4.

In an *isothermal process*, the temperature remains constant over the path from one state to another. Such a process can take place by design or occur naturally, as we will see for saturated steam. In the illustration of a piston in Figure A.2, a means of maintaining the temperature constant in the gas is assumed, but not shown, whether it is from the transfer of heat into the system or from the internal energy, during the expansion indicated.

*Other processes* involve a constant property. They include constant volume and constant pressure (isobaric) processes. Each has a characteristic path, dependent on the other properties.

A *cycle* is a sequence of processes for which the end state is the same as the initial state. A hypothetical example of a four-path cycle is shown in Figure A.5, with each path resulting from a different process. In the case of a closed system, the cycle represents all of the changes is property that the same mass of working substance goes through over and over again in continuous energy conversion.

Limits on energy conversion are best discussed in terms of the *Carnot cycle* (after N. L. S. Carnot, 1796–1832, a French engineer),

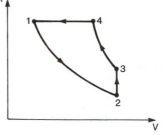

**FIG. A.5.** A thermodynamic cycle. The cycle is composed of four paths: path 1–2: an isothermal expansion process; path 2–3: a constant volume process; path 3–4: an adiabatic compression process; and path 4–1: a constant pressure process. This particular hypothetical cycle would require work into the system substance, that is, it represents a system for conversion of work to heat, rather than heat to work.

an idealization of heat–energy conversion from which is derived the theoretical limit to thermal efficiency. This cycle can be visualized as the physical process shown in Figure A.6, which illustrates reciprocating engine much like the steam engine (Chapter 3). Note in particular the connecting rod from the piston to the rotating flywheel, as in a steam engine, which requires that reciprocation of the piston must accompany rotation of the flywheel. A stroke in each direction is seen to involve two processes; thus, four processes are included in a complete reciprocation up and down. The four processes of the Carnot cycle shown as paths in the figure are:

Process 1–2: isothermal heat input
Process 2–3: adiabatic expansion
Process 3–4: isothermal heat rejection
Process 4–1: adiabatic compression

The ideal means of achieving the conditions required for each process are indicated in the figure at the bottom of the piston. It should be noted in particular that all four processes of the Carnot cycle are reversible processes. The Carnot cycle operation is called an engine because the system delivers net work when operated this way. Net work for a complete

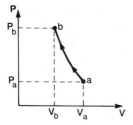

**FIG. A.4.** The adiabatic process. The gas is compressed from volume $V_a$ to $V_b$, with system boundaries that are perfect thermal insulators, thereby allowing no transfer of heat outside. Note that the path of the adiabatic process in the p-V plane is steeper than that of the isothermal process (Fig. A.3). The temperature in this adiabatic compression rises.

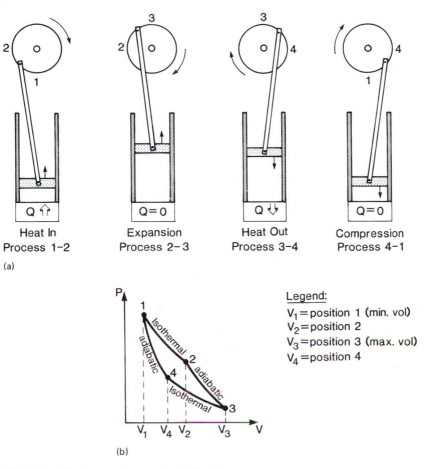

Heat In
Process 1-2

Expansion
Process 2-3

Heat Out
Process 3-4

Compression
Process 4-1

(a)

Legend:
$V_1$ = position 1 (min. vol)
$V_2$ = position 2
$V_3$ = position 3 (max. vol)
$V_4$ = position 4

(b)

**FIG. A.6.** The ideal Carnot cycle. (a) The Carnot cycle in operation. (b) The Carnot cycle p-V diagram.

cycle can be represented using our mathematical definition of work *on* or *by* a substance as:

$$W_{net} = \int_{V_1}^{V_2} p\,dV + \int_{V_2}^{V_3} p\,dV$$

$$+ \int_{V_3}^{V_4} p\,dV + \int_{V_4}^{V_1} p\,dV$$

and can be understood graphically to be the area enclosed by the four paths on the p-V diagram. The energy required for the system to give up net work is supplied in process 1–2 of

each cycle. The action of the Carnot system simply converts heat energy into work.

One of the most useful features of the Carnot conceptualization is that it depicts an ideal (reversible) engine working between two thermal reservoirs – one at a high temperature $(T_H)$ and the other low $(T_L)$. The high-temperature reservoir is the source of heat transfer into the engine and the low-temperature reservoir is the sink for rejected heat. This is shown in Figure A.7a, where heat $(Q_H)$ is transferred from the hot reservoir to the engine, which in turn converts part of that input energy into work $(W)$ and rejects the remainder $(Q_L)$ into the cool

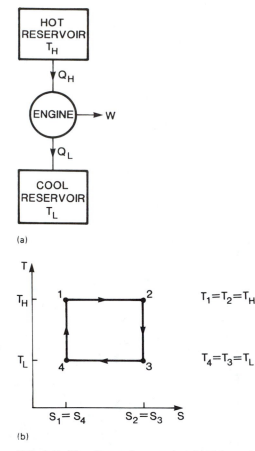

(a)

(b)

**FIG. A.7.** The Carnot heat engine. (a) Schematic diagram. (b) T-S diagram.

reservoir. The Carnot engine is used as an ideal standard to compare any other heat engine that operates between a high-temperature source and a low-temperature sink. Comparisons of thermodynamic cycles are commonly done using the property called *entropy* [S], which is most clearly defined in terms of transferred heat and temperature. Thus, the entropy increment is

$$ds = \frac{dQ}{T} \quad [\text{J/}^\circ\text{K or BTU/}^\circ\text{R}]$$

where
$dQ$ = increment in heat transfer [J or BTU] and
$T$ = absolute temperature [$^\circ$K or $^\circ$R].

Absolute temperature is used because it measures total thermal energy from its reference at 0°K or 0°R. Using this definition, the entropy change for a process going from one state (a) to another (b) would be:

$$S_b - S_a = \int_a^b \frac{dQ}{T} \quad [\text{J/}^\circ\text{K or BTU/}^\circ\text{R}]$$

The Carnot cycle provides particularly simple interpretations of entropy changes. For example, the two isothermal processes in Figure A.7b are simply reduced to summed changes in $Q$:

$$S_2 - S_1 = \int_{Q_1}^{Q_2} \frac{dQ}{T_H} = \frac{(Q_2 - Q_1)}{T_H}$$

with $T_H = T_1 = T_2$ and

$$S_3 - S_4 = \int_{Q_4}^{Q_3} \frac{dQ}{T_L} = \frac{(Q_3 - Q_4)}{T_L}$$

with $T_L = T_4 = T_3$

The other two processes are adiabatic processes, which according to definition:

$$dQ = 0$$

and hence: $S_2 - S_3 = 0$ and $S_4 - S_1 = 0$. These two paths of constant entropy characterize *isentropic* processes, which are adiabatic *and* reversible.

Entropy and absolute temperature provide a useful alternative to displaying the paths of processes and cycles. The T-S diagram for the Carnot cycle is shown on Figure A.7b, using the same state numbers as Figure A.6. On a T-S diagram, area generally represents heat transferred in a process from any state (a) to any other (b) as the quantity:

$$Q_b - Q_a = \int_{S_a}^{S_b} T ds$$

which is the area in the T-S plane under the path from a to b.

In the Carnot cycle, this T-S area representation is particularly simple when area is the rectangle as shown in Figure A.7b. The transferred heat in this case is:

$$Q_2 - Q_1 = \int_{S_1}^{S_2} T dS = T_H(S_2 - S_1)$$

$$= \text{rectangular area under path } 1-2$$

and

$$Q_3 - Q_4 = \int_{S_4}^{S_3} T dS = T_L(S_3 - S_4)$$

$$= \text{rectangular area under path } 4-3$$

The net area of the closed cycle path on the T-S diagram represents that portion of the heat transferred that is *available* for work in the engine. Thus, in the Carnot cycle:

$$W = \int_{S_1}^{S_2} T dS + \int_{S_2}^{S_3} T dS$$

$$+ \int_{S_3}^{S_4} T dS + \int_{S_4}^{S_1} T dS$$

but since:[11]

$$W = \int_{S_1}^{S_2} T dS + \int_{S_3}^{S_4} T dS$$

we therefore have,

$$W = T_H(S_2 - S_1) - T_L(S_3 - S_4)$$

$$= \text{the } T-S \text{ area in the reactangle } 1-2-3-4$$

Our earlier identification of the heat input process (1–2) and the heat rejection process (3–4) then enables us to write for the heat transfer process:

(heat in)  $Q_H = T_H(S_2 - S_1)$

(heat out)  $Q_L = T_L(S_3 - S_4) = T_L(S_2 - S_1)$

and therefore, the ideal available work is:

$$W = Q_H - Q_L.$$

That is to say, theoretically in the Carnot engine all of the available energy goes into work ($W$) and thus we have an equality sign in this equation. It should be noted, moreover, that the heat ($Q_L$) that is rejected into the low-temperature reservoir is all *unavailable* for work, even in the ideal Carnot engine. These results enable us to write down a particularly simple expression for the ideal Carnot efficiency of a heat engine. The Carnot efficiency is defined as:

$$\eta_C = \frac{\text{work out}}{\text{energy in}} = \frac{W}{Q_H}$$

For the Carnot cycle, the simple result is:

$$\eta_C = \frac{(Q_H - Q_L)}{Q_H} = \frac{T_H(S_2 - S_1) - T_L(S_2 - S_1)}{T_H(S_2 - S_1)}$$

$$= \frac{(T_H - T_L)}{T_H}$$

which represents the upper limit for any heat engine working between the temperatures $T_H$ and $T_L$.

There are several reasons why no practical heat engine can achieve the Carnot limit. Heat losses from the system, such as radiant heat loss, would be one obvious way in which the available energy would be reduced. Also, friction causes efficiency loss, and it is readily understood as an irreversible process. Irreversible processes in general, whether from friction, turbulence, or other nonequilibrium conditions, represent the essential ways in which practical engines differ from the ideal. The expansion of steam in a practical steam engine is another way. It is not perfectly isentropic, that is, it is neither completely adiabatic nor entirely reversible in compression. Later in this appendix, we will compare the efficiencies of practical engines with the ideal.

An appreciation of the nature of irreversible processes, and how they turn energy into forms unavailable for useful work, gives us a better understanding of why a machine of perpetual motion could never exist in reality. Such a machine would be capable, for example, of utilizing energy converted by friction into heat.

The Second Law tells us that such energy is unavailable for the work necessary to keep the machine moving. This machine might instead be able to derive net work from a *cool reservoir* (Fig. A.7a) by transferring heat out of the reservoir and then returning it. But here again the energy is unavailable, in this case because it resides in the *lower* temperature reservoir.

The unavailability of energy in a low-temperature reservoir is a form of the Second Law (attributed to Lord Kelvin, 1824–1907, British physicist). The modern heat pump (or refrigerator) illustrates that energy can be transferred from a lower temperature to a higher one only with a net expenditure of work.

## The Rankine cycle: heat engines using steam

Let us look again at the thermodynamic cycle of the reciprocating steam engine (see Fig. 3.1). As we noted in Chapter 3, this engine operates with a piston and therefore a description of it can parallel that of the Carnot cycle. Now, however, we will be considering a real life operation, using a real physical working substance – *steam* – rather than an idealized cycle and substance.

The p-V diagram of the steam engine was given in Figure 3.2. The thermodynamic cycle that the paths in the diagram follow is the *Rankine cycle*. In the Rankine cycle, none of the paths correspond exactly to the ideal processes of the Carnot cycle and two of the paths have mixed processes. Paths 2–3 and 4–1 are, respectively, expansion and compression, but neither can be termed truly adiabatic because of the presence of irreversibilities such as turbulence. Path 1–2 is part compression and part expansion, with the expansion process having a slowly dropping pressure as the cylinder volume starts to increase at the end of the steam input operation. Finally, the exhaust path (3–4), after a short portion in expansion (at the end of the piston displacement), undergoes an almost purely constant pressure process in ejecting the steam into the atmosphere.

Figure A.8 displays the various states of the water/vapor working substance by means of the steam dome lines on a T-S diagram. On the left in Figure A.8a is the *saturated liquid* line, marking the division between pure liquid and a liquid-vapor mixture. On the right is the *saturated-vapor* line, marking the division between the liquid-vapor mixture and the region of pure vapor, called the superheated region. These two lines of division meet on the peak of the steam dome at the *critical point* (Fig. A.8a,b).

If water is heated under pressure, its temperature and entropy will increase along a path just above the saturated liquid line on the T-S diagram as shown on Figure A.8a. When the boiling temperature of the compressed water is reached, the temperature then remains constant as more heat is added and the T-S path becomes a horizontal line. This constant temperature path continues, and only entropy increases as heat is added until the saturated vapor line is reached. On Figure A.8b, we can see the heat input paths of the water/steam mixture at different pressures (measured in pounds per square inch absolute, psia): (a) for $P = 400$ psia, $T = 444.6°F$; (b) for $P = 14.7$ psia, (atmospheric pressure) $T = 212°F$; and (c) for $P = 1$ psia, $T = 101.7°F$.

If the saturated vapors were cooled at a specified pressure, it would trace a return horizontal path at a constant temperature across the steam dome back to the liquid line on the left.

If the steam is heated beyond the saturated vapor state, the temperature would again begin to rise, as shown for the path in Figure A.8a, into the superheated vapor region. Superheated path segments for the pressure 400 and 14.7 psia are also shown in Figure A.8b. We will see that taking the steam into the superheated range can improve the thermal efficiency of a steam plant.

Modern steam power plants (see Chapter 3) use a closed version of the Rankine cycle. Both the p-V and T-S paths for the closed Rankine cycle (shown in Figure A.9) may be compared with those of the Carnot cycle. The heat input path for the Carnot cycle, for example, is entirely isothermal (see Figure A.7), whereas for

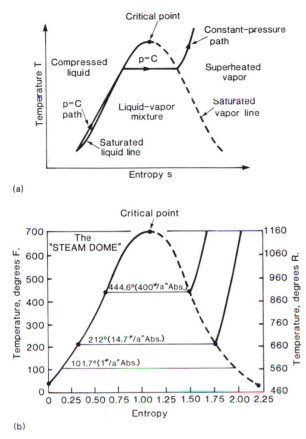

(a)

(b)

**FIG. A.8.** The T-S diagram for water and steam. (a) A single constant-pressure path. Adapted from Burghardt (1978). (b) The steam dome. Adapted from E. B. Norris and E. Therkelson, 1939, *Heat Power*. McGraw-Hill, NY.

the Rankine cycle, this path (1–2) starts by rising in the T-S plane just above the saturated liquid line (see Figure A.9). The remainder of the heat input process becomes isothermal only after the boiling temperature is reached. As we have noted, this constant temperature path is determined by the vaporization properties of steam for the pressure existing in the boiler.

Point 2 on the saturated vapor line in Figure A.9 marks the end of the T-S heat input path and the beginning of the Rankine expansion process. This expansion process is nearly isentropic (constant S) and therefore drops in the T-S diagram in a nearly vertical path, similar to the ideal Carnot expansion, to point 3. Path 3–4 is the steam exhaust process, which we can now see is a constant pressure *and* constant temperature path, due to the

properties of steam condensation. This isothermal path therefore appears similar to the Carnot heat rejection path in the T-S plane. Finally, the Rankine closed-cycle compression process shows as a short *isentropic* (vertical) path (4–1) on the T-S diagram, spanning only a small part of the temperature change achieved in the boiler. Unlike the Carnot compression path, the temperature does not rise over the full range. This compression path for the Rankine cycle is best performed when the working substance is a liquid rather than a vapor or a vapor–liquid mixture.[12]

The Rankine cycle can be, and usually is, extended into the superheated region by continuing heat input, at the same pressure beyond the saturated vapor state (point 2 on Figure A.9a). When this is done in superheater

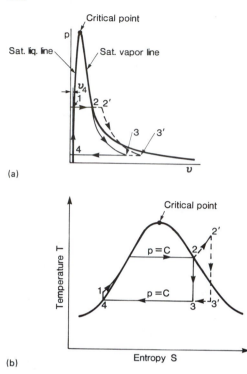

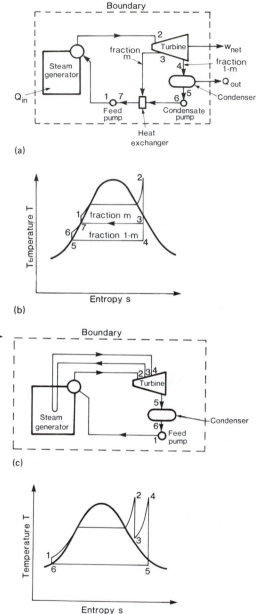

**FIG. A.9.** Rankine cycle diagrams. (a) P-V diagram. Adapted from Faires (1948). (b) T-S diagram. Adapted from Burghardt (1978).

coils in the boiler, the heat input path is extended to a state in the superheated steam region. On the p-V plane of Figure A.9a, of course, the path extension 2–2′ is merely a horizontal line, since the superheated extension is still at constant pressure. In the T-S plane, the path 2–2′ rises to higher temperatures.

The expansion process from the superheated state is nearly isentropic (path 2′–3′). The chief value of the superheated extension (path 2–2′) is an incremental increase of *available energy* for the cycle, which is greater per unit of heat input than the increments leading to the saturated vapor state (point 2). This increase in available energy is represented on the T-S diagram (Figure A.9b) by the area bounded by the points 2, 2′, 3′, and 3. According to the principles set down earlier in this appendix, this will increase the thermal efficiency of the cycle, thereby delivering more useful work per unit of input heat and fuel.

**FIG. A.10.** Rankine cycle efficiency improvements – regeneration and reheating. (a) Schematic diagram for one stage of regenerative heating. (b) T-S diagram for one stage of regenerative heating. (c) Schematic diagram for one stage of reheating. (d) T-S diagram for one stage of reheating. Adapted from Burghardt (1978).

Another significant contribution to thermal efficiency improvement is obtained in the Rankine cycle through the employment of the condenser. The cooling effect of the condenser on the cycle efficiency can be understood through the use of the T-S diagram. On Figure A.9b, it is easy to see that lowering the exhaust temperature $(T_L = T_3 = T_4)$ for a specified input temperature $(T_H = T_2)$ will increase the available energy, because it increases the T-S area bounded by the paths of the cycle. Conversely, power plants operating *without* a condenser would discharge steam at a higher exhaust temperature $(T_L)$ and therefore have less available energy in the cycle. The efficiency benefit of a condenser may be understood by considering pressure, as observed in Chapter 3. There it is noted that an increased pressure differential (reducing *back pressure*) is created across the steam-driven engine by the cooling action of the condenser.

There are several means of increasing the available energy in the Rankine cycle. One important method is regeneration, as shown schematically in Figures A.10a and b, and represents an attempt to approach Carnot cycle efficiency. It does this by recuperating the heat of vaporization of a fraction (labeled $m$ in Figure A.10a) of the working substance to reheat the remainder $(1 - m)$ as feed water back to the boiler. A single stage of regeneration (power plants usually have several) can improve the cycle efficiency by as much as 7 percent.

Reheating is still another means of improving thermal efficiency. Whereas reheating might be understood on a common sense basis, its function in terms of second-law efficiency can be seen explicitly on the T-S diagram Figures A.10c and d. The addition of area in the superheated region (path 3–4–5) results in a gain in available energy (T–S area in Figure A.10d) and a net gain in efficiency.

# Increasing entropy – further ramifications of the second law

The property entropy (S) has another significant interpretation. This interpretation is that entropy is a measure of the state of chaos of a thermodynamic system. Systems of any size are considered in these theoretical speculations, ranging from the microscopic scale to the cosmos. On a cosmic scale, it is said that the state of chaos in the universe is continually growing and hence the entropy of the universe is always increasing.

Lest the scope of these generalizations becomes too great, let us consider a single system as depicted in Figure A.11. For a transfer of heat into the system from its surroundings, as indicated, $dQ > 0$ means heat *into* the system and $T_0 > T$ is required for this heat transfer. Therefore

$$dS_{syst} = \frac{dQ}{T} > 0$$

that is, the change in entropy of the system is positive. The change in entropy of the surroundings for this heat transfer, however, is:

$$dS_{surr} = -\frac{dQ}{T_0}$$

a negative quantity, because the heat transfer $(-dQ)$ is out of the surroundings. The total (system and surroundings) entropy change is therefore:

$$dS_{syst} + dS_{surr} = \frac{dQ}{T} - \frac{dQ}{T_0} > 0$$

a positive quantity, since $l/T > l/T_0$. If the heat transfer were the other way, *out* of the system, then $dQ < 0$ and $T > T_0$ would be required, resulting *again* in a positive total change in entropy. We therefore see that entropy increases no matter which way the heat transfer goes.

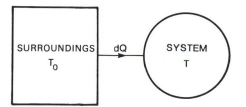

**FIG. A.11.**   Entropy change for a single system.

This result of *increasing entropy* has been generalized widely. Irreversible processes within a system, for example, lead only to increases in entropy. Even an isolated system with no transfers with its surroundings either has a constant entropy or increasing entropy; it will *never* decrease. This is expressed simply in this relation for the entropy of an isolated system $(dS_{isol})$:

$$dS_{isol} \geqslant 0$$

which is called the Clausius relation (after R. J. Clausius, 1822–1888, German physicist). The inequality of this relation is said to apply to the universe as a whole, and embodies the statement that *the entropy of the universe is continually increasing.*

To better appreciate what this all means, let us look at thermal behavior on a detailed microscopic scale. On a scale of individual particles, or small groups of particles, the microscopic view of entropy entails a measure of the randomness of motion and the degree of mixing of particle types, velocities, and motions. Thus, in this microscopic view we would observe increasing randomness and mixing, with a trend toward a statistically uniform sea of chaotic motion. The universe, according to the cosmic interpretation of entropy, is becoming more and more such a sea of chaos, even though its total energy is constant (the first law of conservation of energy). The only way that energy can be made *available*, according to the theory, is for work to be done on a portion of the sea or for there to be a local release of concentrated energy (such as by heat of combustion).

In an isolated system, the microscopic theory of entropy stresses the increase in chaos or disorder that is to be expected if the theory of entropy increase holds true. This view carries over into another area, much more concrete in concept, namely, mineral resources. Economists and others have observed that the world's natural concentrations of minerals[13] are being dispersed as a result of human activities. Thus, concentrations of metal ores (such as copper), precious crystals (such as diamonds), and other useful elements are mined, fed into the manufacturing process, and sold to consumers all over the world. Many of the mineral-based products are discarded and end up in solid waste disposal sites, mixed with many other materials. Even with complete recycling, which requires an input of work, some mass is lost from any metal product due to ordinary processes of wear.

This picture of gradual dispersal of minerals fits the concept of increasing entropy in several ways. First, there is a random dispersal of the material. Next, connected with the dispersal is a mixing of the pure material with many other minerals (and organic materials too). Dispersal and mixing were among the irreversible processes discussed in connection with hot gases in engines in the previous section. In the case of the dispersal of minerals, as in the cases of throttling or mixing of gases, increasing entropy can be taken as a measure of increasing chaos. Therefore, the observation is that mineral use also represents a trend toward increasing entropy, along with all processes of energy conversion, and both appear to be inevitable as a result of human activity.

Indeed, it is the seeming inevitability of the entropy process that leads to another profound concept – *The Arrow of Time* (see Eddington 1958). This notion points out that since actions, reactions, and activity, in general, are continuously taking place in the universe, the ongoing increase of entropy is itself a measure of the passage of time and its irreversibility even carries with it the unidirectional aspect of the passage of time.

A conclusion sometimes drawn from the prediction of inevitable increases of entropy is that ultimately the universe will suffer a *heat death*. According to this extension of the theory, energy and heat will continue to become more diffuse, leading ultimately to a point when it is no longer feasible to concentrate heat, organize matter, or do useful work. By the time of this ultimate catastrophe, of course, life would long have ceased and the universe would be a cold, diffuse, disordered spread of energy and matter approaching the condition of *maximum entropy* (see Georgescu-Roegen 1971).

## Conversion to electricity

In a conventional power plant, the steam turbine drives the electric generator (see Chapter 3). Conventional electric generation takes place according to the principles of electromagnetism, which were formulated in the nineteenth century.[14] The generation of electric voltage, for example, is understood through Faraday's law of induction. The understanding of the creation of magnetic fields by electric currents is attributed mainly to Oersted and Ampère, and the forces on current-carrying conductors in a magnetic field to Ampère and others. These effects are key to a basic understanding of present day electrical generation, but the laws of electromagnetism are also the basis for other major areas of technology, including communications, broadcasting, computers, electromechanical devices, and instrumentation.

Faraday's induction law underlies the creation of electric voltage in the coils of a conventional

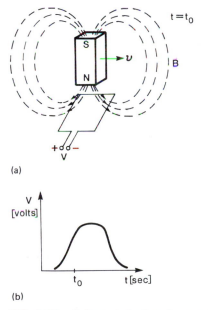

(a)

(b)

**FIG. A.12.** Voltage induction in a coil by a changing magnetic field. (a) Magnet moving by a coil. (b) Induced voltage versus time.

electric generator. It is shown in its simplest form in Figure A.12, where a magnet is depicted being moved past a loop (a one-turn coil) of wire. If a voltmeter or oscilloscope were connected across the two open terminals of the coil, a voltage that varies in time would be measured (see graph in Fig. A.12). This voltage is *induced* by the changing magnetic field.

The discovery of induction is credited mainly to Faraday, who stated that the *electromotive force* (voltage) induced in a coil is proportional to the time-rate of change of the magnetic flux linking the coil. In the configuration of Figure A.12, we can understand Faraday's time-varying flux linkage as follows. As the magnet moves by the coil on a path parallel to the coil, the number of magnetic field lines that pass through the loop varies. The maximum number of field lines passing through the loop occurs, of course, when the magnet is directly opposite the coil. The sum of the field lines passing through the loop is the magnetic flux linking the coil. Therefore, in this configuration the flux linkage varies as the magnet passes by, starting from zero, rising to a maximum and then falling back to zero.

Faraday also stated that the strength of induced voltage in a coil is proportional to the number of turns in the coil. Therefore, if our coil consisted of two turns around the same loop, instead of one, then the same experiment would yield twice the peak voltage in Faraday's law, as expressed in the formula:

$$v = N\frac{d\phi}{dt}$$

where $n$ = number of turns in the coil, $d\phi/dt$ = time rate of change of the magnetic flux linkage in each turn, and $v$ = voltage induced across the open terminals of the coil.

Voltage induction in conventional electric generators takes place much as we have just described, but with several major modifications. One difference is that the motion between the magnetic field and the coil is rotary, not translational (see Figure 3.18). Another important difference is that the magnetic fields are

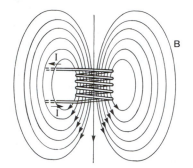

**FIG. A.13.** An electromagnet. Current flows in the coil, thus creating a magnetic field surrounding the coil.

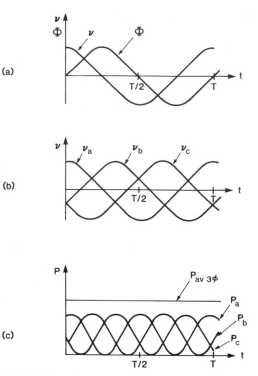

**FIG. A.14.** Alternating-current generator waveforms in time. (a) Magnetic pole flux ($\phi$) versus time. (b) Three-phase voltages versus time. (c) Three-phase powers versus time.

not supplied by permanent magnets but by electromagnets.

Electromagnets are based on Ampère's law, which states that the flow of electric current creates a magnetic field in a plane normal to the direction of the current flow. A simplified picture of an electromagnet is shown in Figure A.13, which depicts a current-carrying coil. The magnetic field lines created by the current flow in the coil suggest that this can substitute for the permanent magnet. Electromagnets in fact are a highly developed technology, and magnetic fields can be created that are far stronger than those that occur naturally with magnetic materials.

The operation of the conventional alternating-current (a.c.) generator takes place as the magnetic poles turn on the rotor past coils embedded in the stator (see the description of the conventional generator in Chapter 3). The voltage induced in each stator coil alternates in polarity (positive and negative) as the passing magnetic poles alternate between north and south.[15] This alternation is consistent with Faraday's law, because the flux linkages reverse themselves as each pole face passes.

If the magnetic field is properly shaped at the pole faces, the flux linkage waveform is very nearly sinusoidal (that is, a sine function in time), as shown in Figure A.14a. Because the induced voltage depends on the time rate of change of the flux linking the coil, the voltage is also sinusoidal, but shifted in time. These sinusoidal variations of flux and voltage repeat themselves in a time period $T$, as indicated on the figure. This time period corresponds to a complete alternation through a pair of poles (for example, N-S-N). In a.c. power systems, this period is commonly $\frac{1}{60}$ second and the frequency of alternation is therefore 60 Hertz (cycles per second).

The basic theory of electricity tells us that the electric power of a generator at any instant[16] is expressed as the product of the voltage times the current at that instant. More precisely, the electric power output of a generator coil is:

$$p = v \cdot i \, [\text{watt}]$$

where $p = p(t) =$ instantaneous power delivered by the generator coil to an external circuit [watts], $v = v(t) =$ the instantaneous voltage across the terminals of the generator coil [volts], $i = i(t) =$ instantaneous current flowing through the terminals of the coil to an external circuit [amperes].

We can understand the basis of this simple formula if we recognize the nature of each of the two terms in the product $v \cdot i$. Voltage represents the potential energy to do work on an electric charge,[17] whereas current represents the rate at which charge flows. Therefore, the product $v \cdot i$ is the rate at which work is supplied or delivered by a generator to an electrical load, such as an electric motor. The rate of work, however, is power, according to our beginning definitions.

In considering the operation of the generator, we conclude that it only supplies energy to an external circuit when a current flows through the terminals of the coil, that is, no power (or energy) is supplied simply by inducing a voltage alone. Since the flow of a current marks the transmission of electric energy out of the generator, we can now ask how that current is related to the conversion of energy in the generator from the prime mover.

The flow of current in a stator coil sets up its own magnetic field, as shown in Figure A.13. This stator magnetic field reacts against the electromagnets of the rotor. These reaction forces on current-carrying conductors are depicted in Figure A.15a. In the case of the a.c. generator, the magnetic field resulting from stator currents causes forces on the current-carrying conductors in the rotor (pole) windings. These forces on the rotor windings always tend to oppose the motion of the rotor pole and thereby set up a countertorque to the mechanical torque supplied by the prime mover (see Figure A.15b). Such a reaction force is observed in all sorts of electromagnetic systems and is generally predicted by Lenz's law.[18] The prime mover (steam turbine) must supply mechanical *work* against such a force (torque) of reaction in electric generation.

## The transformation of electricity

After being generated, electricity must be transmitted and distributed. The conventional means of transmission is either through overhead lines (heavy wires) or underground cables. Either means of transmission results in electrical losses along the conductors.

Early in the evolution of electrical technology, it was recognized that electrical losses are reduced by operating transmission lines at high voltages. The reasons for this can be appreciated from two simple considerations. The first follows from the law of electrical power as just described, namely, that power is equal to voltage multiplied by current. Thus, for any given power to be transmitted, the higher the voltage used, the lower will be the current required in an inverse proportion to the voltage. For example, if we elect to transmit power at twice the voltage ($v$) then we will find that half the current ($i$) is required to transmit any given number of watts ($p$). Second, the losses in conductors vary with the square of the current[19] flowing in the conductor; therefore, losses

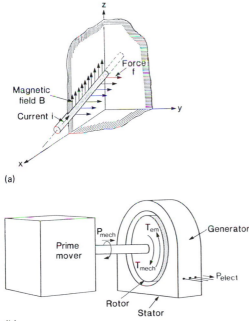

(a)

(b)

**FIG. A.15.** Prime mover/generator reaction. (a) Magnetic force in a current-carrying conductor. (b) Torques and rotation. Adapted from Elgerd (1978).

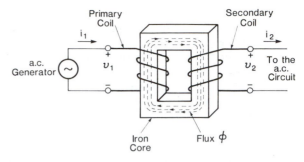

**FIG. A.16.** Idealized electrical transformer. For a.c. power flow: $p_{in} = v_1 i_1$; $p_{out} = v_2 i_2$. For zero losses: $p_{out} = p_{in}$. Adapted from M. El-Hawry, 1983, *Electrical Power Systems – Design and Analysis*. Reston, VA.

decrease *inversely* as the square of the operating voltage of an electrical transmission line. So if the voltage is doubled, the losses are reduced by a factor of four.

The means of achieving high voltages for low-loss transmission was the electrical transformer[20] (Fig. A.16). The transformer converts the voltage and current as generated to a higher voltage and lower current. Aside from small losses in the transformer, the electric power is transmitted *through* the transformer as indicated on Fig. A.16.

Transformers operate on the principle of induction, although applied in a different way than we described for generators. The transformer, for example, has no moving parts. Instead it uses time-varying magnetic fields created by alternating currents to *induce* voltages. Therefore, in order to use transformers, it became necessary to use an a.c. electric system. The voltage magnitude induced across the secondary coil ($v_2$) varies directly with the a.c. primary voltage ($v_1$), and its magnitude depends on the ratio of the number of turns in the secondary coil to the turns in the primary,

as expressed in:

$$v_2 = \frac{N_2}{N_1} v_1.$$

where $N_1$ = number of turns on the primary coil and $N_2$ = number of turns on the secondary coil. This follows, of course, from the law of induction, so that the magnitude of an induced voltage in a coil is proportional to the number of turns in the coil. This formula expresses the *transformation* of voltage, which can be a voltage step-up if:

$$\frac{N_2}{N_1} > 1$$

or a voltage step-down if

$$\frac{N_2}{N_1} < 1$$

In the idealized transformer shown in Figure A.16, we see two coils wound on a closed iron core. The primary coil on the left is connected

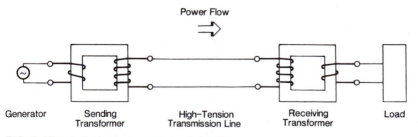

**FIG. A.17.** A single-phase a.c. transmission line.

to the generator that supplies the a.c. voltage ($v_1$) and current ($i_1$). The a.c. primary current ($i_1$) creates an alternating magnetic flux ($\phi$), which is concentrated[21] in the iron core. The alternating magnetic flux links the secondary coil and, according to the induction principle, induces a volatage ($v_2$). If the secondary coil is connected to a circuit, current ($i_2$) then flows out of the secondary coil and a.c. energy flows *through* the transformer. In practice most electric transformers use separate coils that have no direct electrical contact with one another, but rather rely on magnetic induction.

*High-tension* (high-voltage) transmission circuits are illustrated in Figure A.17. The transformer at the sending end (which is fed by the generator) has a larger number of turns on the secondary coil than on the primary coil and therefore provides a step-up in voltage. At the receiving end, on the other hand, there is a step-down in voltage, because the primary side connected to the transmission line has more turns. If the step-up of voltage for this circuit were 10:1, the losses in the transmission lines would be $\frac{1}{100}$ (1 percent) of those with no step-up.

# Notes

1. The MKS unit for charge is the coulomb (the charge of about $2.56 \cdot 10^{18}$ electrons) and for current is the ampere (1 ampere is a charge flow of one coulomb per second).
2. The kilocalorie is based on the heat required to raise the temperature of one kg of water 1°C. Heat energy can also be measured in (metric) joules along with mechanical energy or electrical energy, and so this is a duplication within the MKS system. The equivalence between the two is $1 J = 0.24 \cdot 10^{-3}$ Kcal.
3. Voltage is commonly thought of as a measure of the force pushing the electric current through conductors much as water pressure pushes the flow of water through pipes. Although this is not entirely inaccurate, a better conception of voltage as a measure of electric *potential energy* is given later on in this appendix (see note 17).
4. A mass of 1 slug has a gravitational force of 1 pound at the earth's surface.

5. The BTU is based on the heat required to raise the temperature of 1 pound of water 1°F.
6. One KW-hr is the energy delivered or used at a power of 1 KW ($10^3$ watts) for 1 hour and is the familiar measure used in electric rates. The *heat rate* of a thermally driven electric generator is commonly measured in BTU/KW-hr; 3,412 BTU/KW-hr represents a heat rate at 100 percent conversion efficiency.
7. This approximate conversion gives the average kinetic energy measured in KeV ($10^3$ electron volts see Units of Measure section) of a particle in a hot gas of plasma (see Chapter 12). The exact relation is (kinetic energy) $E = kT$, where $k = 0.862 \times 10^{-7}$ KeV/°K (Boltzman's constant), and $T$ is absolute temperature on the Kelvin scale.
8. The Barrel (bb$\ell$) is the common measure of petroleum.
9. More precisely, heat content is subsumed into the scientific engineering term internal energy, which will be used in the discussion of thermodynamics that follows.
10. In the case of systems capable of storing energy, this equation becomes $E_{in} = E_{out} + E_{stored}$. We review practical systems with storage, including conventional hydroelectric stations, pumped hydroplants, and batteries in Chapter 3 and Appendix C.
11. In the closed path integrals on Fig. A.7b:

$$\int_{S_2}^{S_3} = \int_{S_4}^{S_1} = 0$$

and

$$\int_{S_3}^{S_4} = -\int_{S_4}^{S_3}$$

12. It is possible, physically, to compress the steam–liquid mixture isentropically (vertical T-S path) to a higher temperature/pressure point at or near the saturated liquid line from which the heat input process in the boiler could proceed. Operation with the liquid–vapor mixture causes practical difficulties such as corrosion and pitting of mechanical parts.
13. We are discussing here only durable minerals that do not change in basic composition as a result of human use. Thus we are excluding minerals that are sources of chemicals (for example, calcite and sulfur), which undergo chemical reactions in manufacturing. We also exclude here uranium

and thorium, which undergo the even more fundamental changes of nuclear reactions in their use as fissionable fuels, and fossil fuels, which undergo chemical change in combustion.

14. For a readable account of the development of the theory of electromagnetism see Singer, Holmyard, Hall, and Williams, 1978. Major contributors to these theories were: A. M. Ampère (French physicist, 1775–1836), C. A. deCoulomb (French physicist, 1736–1806), M. Faraday (British physicist and chemist, 1791–1867), K. K. Gauss (German mathematician, 1777–1855), J. Henry (U.S. physicist, 1797–1878), J. C. Maxwell (Scottish physicist, 1831–1879), and C. Oersted (Danish physicist, 1777–1851).

15. Magnetic fields have a direction associated with them. The field direction is customarily chosen to be positive as the lines emanate from a north pole.

16. Electric power is the time rate at which electrical energy is delivered by the generator at a particular instant.

17. The exact definition of a volt is the work done in moving an electric charge against the electric forces of repulsion by other charges. The MKS unit of 1 volt represents the change in electric potential when 1 joule of work is done against electrostatic forces on 1 coulomb of charge.

18. Lenz's law states that when a time-varying magnetic flux induces a voltage in a coil and the voltage is permitted to produce a current in the coil, the current will always flow so as to oppose the magnetic flux changes inducing the voltage (after H. F. E. Lenz, Russian physicist, 1804–1864).

19. Ohm's law tells us that the voltage *drop* (change) along a length of a conductor is

$$v_R = ir$$

where $i$ = the current through the resistance and $r$ = the resistance of a length of conductor. The law of electric power tells us that the power delivered (lost) to the length of conductor is:

$$p_R = v_R \cdot i = i^2 R.$$

20. Faraday applied his theory of induction to both an electric generator and a form of the transformer in 1831. However, early in the electric age, most electric power came from direct-current (d.c.) generators. It was not until later in the century, after the alternating-current generator had been developed by W. Siemens (German engineer, 1816–1892) and others that the idea of using transformers in a.c. circuits was developed. Credit for the introduction of a.c. systems to American goes mostly to George Westinghouse (U.S. engineer and industrialist, 1846–1914).

21. Materials such as iron, which can be magnetized when put in the vicinity of a permanent magnet or a current-carrying coil, are highly *permeable* to magnetic flux. When magnetic field lines flow through a closed path that is composed of such material, magnetic flux becomes concentrated. In transformers, the iron core channels the magnetic lines so that a maximum flux linkage of the secondary coil is achieved. Iron is also used in generators to guide and shape the magnetic flux (see Chapter 3).

## Bibliography

### Thermodynamic principles

These books are ordered in ascending level of sophistication.

Krenz, J. H. 1984. *Energy, Conversion and Utilization*, second edition. Allyn & Bacon, Boston, MA. An elementary-level textbook for undergraduate physical science students.

Severns, W. H., and H. E. Degler. 1948. *Steam, Air and Gas Power*. Wiley, New York.

Faires, V. M. 1948. *Theory and Practice of Heat Engines*. Macmillan, New York. Both textbooks were used by undergraduate engineering students. They were selected because of their readable descriptive material on heat engines and instructive figures. Several of these figures have been adopted here.

Burghardt, M. D. 1978. *Engineering Thermodynamics with Applications*. Harper & Row, New York. A contemporary textbook for undergraduate engineering students.

Van Wylen, G. J., and R. E. Sonntag. 1976. *Fundamentals of Classical Thermodynamics*. Wiley, New York. A modern, graduate-level textbook for engineering students.

### The law of entropy and its philosophical implications

Georgescu-Roegen, N. 1971. *The Entropy Law and the Economic Process*. Harvard University Press, Cambridge, MA.

Eddington, A. 1958. *The Nature of the Physical World.*
    University of Michigan Press, Ann Arbor
    (paperback).
Weinberg, A. M. 1982. "Avoiding the Entropy Trap."
    *The Bulletin of the Atomic Scientists*, October.

## Electromagnetism and electricity

Singer, C. J., E. J. Holmyard, A. R. Hall, and T. I.
    Williams, 1956. *History of Technology*, Vol. 3.
    Oxford University Press. Oxford.
(A summary of the development of the theory of
    electromagnetism in the nineteenth century.)
Elgerd, O. I. 1978. *Basic Electric Power Engineering.*
    Addison-Wesley, Reading, MA.
An introductory textbook about the principles of
    electrical machines for electrical engineering
    students.

# APPENDIX B

# Review of ethical theory

## Introduction

Throughout this book, we discuss the ethical implications of energy issues. In this appendix we will review the concepts and theories of ethics. These few pages cannot, of course, provide a comprehensive analysis of ethical thought, a field that has been developed over literally thousands of years. What follows is intended only to summarize briefly some basic theories in ethics. For more thorough considerations, there are many fine books on theory (for example, Frankena 1973 and Barry 1985), as well as numerous anthologies of some of the great writings on ethics.

Before beginning our survey, we should make a few general points. First, when we use the term *ethics*, we refer to a branch of philosophy concerned with good and bad conduct and the judgment, values, and actions that lead to it. Often, in everyday speech, when we refer to some behavior as ethical, we mean it as a synonym for right or good, and an antonym for bad or wrong. Here, ethics pertains to both right and wrong, to *moral* behavior and thought, and when we say something is an ethical concern that means that it is an issue of morality. Its antonym would be nonethical or nonmoral, with no value judgment attached. So, whether a technical device *can* be built is a nonethical question; whether it *should* be is likely to be an ethical one.

We should also distinguish between ethical principles and statements and ethical theories. There are widely held ethical principles. It is wrong to lie, to cheat, to steal, to kill. It is right to be charitable, to protect those in danger, to tell the truth. But why are these actions right or wrong? Why is it wrong to lie? On what basis can we say it is right to be charitable? Is it always wrong to cheat? The answers to these questions are found in ethical theory, the rationale underlying ethical statements and standards. Not surprisingly, there have been many different approaches to ethics in the history of Western philosophy. No theory is beyond criticism or question, but the ones we will consider have influenced the history of ideas and in one form or another may have proponents today.

Actually, this appendix will be in two parts. The first part will review a few leading theories of the foundations of ethical standards. The second part will summarize theories of justice – a concept that is of particular importance to issues of energy resources and technology.

## Foundations of ethical standards

On what basis do we judge an act or a rule of conduct to be right or wrong? What makes something right? Is one class of actions always right or do circumstance and/or consequence play a part?

There are many answers to these questions in the Western philosophical tradition – many different ways to systematize ethical beliefs. They can, however, be grouped roughly into two separate categories. First, there are systems that assert that ethical acts and rules are inherently right or good regardless of the consequences (sometimes referred to as *deontological* theories, from the Greek – *deon*, for duty and *logos* for science). Second, some theories assert that right or wrong is determined not by the act or rule itself but by its nonmoral consequences (*teleological* theories, from the Greek – *teleo*, for end or purpose). We will consider several examples.

## Deontological theories

### Divine command

According to this view, an act is right because God has told us to do it, and the primary justification is obedience to the will of God. Indeed, right and wrong mean God commands one to do the former and not do the latter. These rules are eternal and binding for the believing. Typically, the commands appear as revealed word in sacred texts such as the Bible or the Koran.

Philosophers have questioned such theories by pointing out inconsistencies within sacred texts as well as disagreements among texts and religions. From the standpoint of reason, no one set of divine commands can be proven to take precedence over any other; thus the disagreements cannot be resolved. Of course, to believers only one text and religion is true, the others false. But this is an opinion from faith, not from any logical deduction.

For the philosopher, the most compelling question about divine command theory was posed by Plato (427–347 B.C.) in his dialogue *Euthyphro*. In the dialogue, Socrates asks, "Is something right because God commands it, or does God command it because it is right?" If the answer is the latter, then divine command

is no longer the basis for the ethical argument; something besides God's command makes an act right or wrong. If the answer is that something is right solely because God commands it, then God's will is arbitrary and so are right and wrong.

### Natural law

A way around Socrates' question might be to maintain, as St. Thomas Aquinas (1225–1274) did, that God imbues people with a sense of moral obligation in the form of natural inclinations that people can ascertain through the exercise of their powers of reason. Ethical standards, then, emerge from a kind of natural imperative or natural law.

A natural law approach need not have a theological basis. It can be based simply on a notion that people possess some inherent inclination (or inherent need) to act in accord with a universal principle, such as nature or the harmony of the universe. Even without God, the standards derived from such a theory must be immutable or they will not be inherent law. The major question then becomes: Why, if the laws are in our nature, are they so hard to discern? In fact, why do some people discern them differently?

### Kant's categorical imperative

One of the most influential deontological theories was prosposed by the German philosopher Immanuel Kant (1724–1804). According to Kant, moral standards have to pass two tests, they have to be consistent and they have to be universal. He concluded, therefore, that we must act only on a rule of ethical conduct – called a *maxim* – that we can freely and deliberately choose to apply universally or, as Kant put it, "will to be universal law" (often called the *categorical imperative*). Every part of this formulation is important. Our rules have to be freely chosen and choosable. That means that they have to be objectively possible and subjectively acceptable, as well as possible for everyone to follow. In fact, Kant believed that a truly universal ethical maxim would be followed by any rational person looking at the same situation.

Obviously any maxim that uses the phrasing "I feel" is not universal, but Kant argued that even some maxims that were couched in more general terms still did not pass his test. Take, for instance, the maxim that everybody should break promises to suit his or her interests. Kant argued that one could not choose such a maxim to be universal law. Promise breaking implies an institution of promise keeping so if you will everyone to break promises, the concept has no meaning. The maxim is thus not simply wrong, it is logically impossible. Other maxims, such as declaring that no one should be charitable, are logically possible, but Kant argued that it was hardly believable that people could truly condemn charity without being guilty of self-contradiction (because nearly everyone is sometimes in need).

Kant's arguments raise numerous complications and difficulties. One problem is that there may be conflicts between rules. For instance, we can say we must protect the environment and help the poor, but what if doing the latter hurts the former? Kant's theory does not give us an answer to such hierarchical questions. Also, Kant's test does not clearly show what maxims are specifically ethical ones. As one philosopher has noted, "Tie your left shoestring first," could pass Kant's test, but it is not an ethical rule (see Frankena 1973).

Kant's theory is worth considering if only because of its rigorous philosophical approach. The theory reminds us that ethical statements need to be logically consistent and capable of being generalized. The latter is especially important in discussions of energy issues. In some instances regarding energy, we may feel that our society applies one standard of conduct for ourselves and a different one for others.

# Teleological theories

## Ethical egoism

Ethical egoism is the belief that the individual's sole moral requirement is to promote his or her own long-term welfare. Or put another way, the individual should do those things that lead to the greatest amount of good in a nonmoral sense (that is happiness, pleasure, and so on) for the individual.

At first glance, this hardly seems the basis for an acceptable ethical system. It would appear to condone conduct that would be widely construed as immoral, or at least selfish or hedonistic. Although some advocates of ethical egoism have argued for hedonism (for example, the Greek philosopher Epicurus), such a theory need not be construed that way. An individual's long-term self-interest may mean living in harmony with others and might argue for generosity rather than selfishness. There might even be occasions *requiring* generosity, especially if a person decides that reciprocal goodwill is in his or her interest. At the same time, an ethical egoist could not assert that generosity was universally good or right or a moral obligation.

Ethical egoism has been proposed as a serious theory of ethics and has even gained support from some modern psychological theories that claim that moral judgments are often made on this basis. Not surprisingly, however, it raises many questions as a philosophical theory. One important problem involves moral judgment and advice by people not directly involved in a moral question. It can be demonstrated that such third-party advice is untrustworthy if we assume that everyone operates from a framework of ethical egoism (see Frankena 1973). Thus, ethical egoism cannot resolve moral conflicts. Also, it is hard to see how ethical egoism can lead to a harmonious social order, which is presumably in everyone's long-term interest. If everyone acts egoistically, then unless everyone has identical interests, conflicts are not only inevitable, they are likely to lead to social disintegration.

## Utilitarianism

Like ethical egoists, utilitarians look at the nonmoral benefit derived from moral acts and rules. But to utilitarians the benefits must be general, not individual. In fact, the test of the moral goodness of an act or proposition is whether it produces a greater balance of good over evil or the greatest possible happiness (utility) distributed as widely in society as possible or the greatest good for the greatest number. According to early proponents of the theory, good could in some way be quantified, and whatever ethical act or rule that produced the most good was the best.

But of course, this is problematic to say the least. Not only is there a problem of how to quantify good, but there is the more fundamental question: What is good? One early utilitarian, Jeremy Bentham (1748–1832), believed that the measure of utility was the intensity of the pleasure it produced, and he tried to work out a specific rating system of pleasures and pains. His approach gave a particularly hedonistic slant to the utilitarian theory. Others have dropped the requirement of quantification and have maintained that the quality of pleasure counted as much as the quantity or intensity. Although this seems an improvement, it does make it difficult, if not impossible, to weigh good and bad because a statement of quality is subjective. At the same time, philosophers have not seen that problem as grounds for outright rejection of utilitarian ethics because the basic requirement of promoting good in the world and fighting evil is at the heart of ethical behavior.

## Two utilitarian approaches

Just as there is disagreement about how one measures good among utilitarians, there are also fundamental differences in formulation of the theory itself. Two approaches are especially noteworthy.

The first, *act utilitarianism*, asserts that every act must promote the greatest balance of good over evil. That is, in every situation, the individual must examine the consequences of a course of action. If it is likely to lead to a greater ratio of good over evil (by whatever measure) than any other course, it is the appropriate course of action. It is important to recognize that according to this view actions and consequences are seen in isolation. Although an act utilitarian might agree that keeping promises is

useful (although only because past experience has shown its utility), it would not mean that he or she would assert that everyone should keep promises. Or indeed that the individual should keep his or her next promise. The decision would depend solely on the situation. The problem with such an approach, as philosophers have demonstrated, is that it is easy to construct scenarios in which immoral behavior is condoned because narrowly construed consequences are preferred. (For example, a person promises to pay a worker for a job but does not because he or she sees better uses for the money.)

An alternative is *rule utilitarianism*. This approach argues that it is not the individual act but the rules governing kinds of acts that matter. The rule that on balance leads to the greatest ratio of good is morally preferable. We can readily see how this improves on act utilitarianism. A rule such as "Always keep promises" can be asserted even if individual cases produce more evil than good (for instance, a promise of vengeance). But on balance, the rule produces more good over evil. This way, we can have a code of ethical conduct that is based on the nonmoral consequences of every rule.

At the same time, it is hardly perfect. First, it is difficult to measure all of the consequences of any rule or act. This is especially true where consequences are intergenerational, as they so often are with energy technology and policy. Also, even with rule utilitarianism it is hard to escape instances when acts are condoned that are morally unsupportable. For example, building a nuclear plant on an earthquake fault may put a very few people at a small statistical risk and lead to benefit – jobs, assurance of electric supply, and so on – for hundreds of thousands of people. There may be no way to calculate benefits that would lead to any other conclusion, yet the imposition of risk on some people by others seems unfair and unjust. John Stuart Mill (1806–1873), a leading utilitarian thinker, argued that you could not impose the risk because an underlying sense of justice is implicit "in the very meaning of utility." As Frankena (1973) points out, there is no basis for this assertion in utilitarian theory. At the same time, as he also indicates, a separate principle of justice may be required to complement an ethical theory. Justice certainly is a key concern in many of the issues we raise in this book, and there are many theories that attempt to derive principles of justice.

## Justice

There are two types of justice, which is usually defined as fairness or correct treatment. The first, *retributive justice*, is concerned with the punishment of crimes. In this book, however, we are more concerned with the second form of justice, *distributive justice*. Distributive justice is concerned with the relative treatment of people, and whether and how people can obtain wealth and benefits including material wealth, power, prestige, and education. How should the world's wealth and resources be distributed? What is the fairest, most morally acceptable means of distribution of those things people value? How should we deal with the distributions of those things people find threatening? The relevance of these questions to energy resources, technology, and policy is self-evident, but the criteria used to answer these questions are not. We will consider four approaches to distributive justice.

### Merit

The view of distributive justice based on merit says that the greatest benefits should go to those who are most deserving. Such a formulation, however, must come to grips with a crucial and difficult question: What makes one person more deserving than another? Over the course of human history, some people have offered reasons why one person should get superior benefit over another that are clearly unjust – for example, race and sex. But other merit-oriented ideas bear greater scrutiny. For instance, some philosophers have argued that the most deserving individual is the one who displays the greatest virtue. But what is virtue and how is it recognized? As has been pointed out (see

Wenz 1988), virtue has often been identified with material success – as a sign of hard work – although given that people can obtain wealth unscrupulously, this hardly seems a tenable basis to deem an individual virtuous.

Even assuming that we could agree on a criterion based on merit, we would need to answer another question: Are people equally able to attain virtue? In other words, do people by birth or wealth or education have an unfair advantage in gaining recognition for their virtue? If the answer is yes, then their gain of public benefit as a result would be unjust.

### Social utility

Most other theories of justice begin with the basic premise that all people are inherently equal with equal rights to pursue social and personal benefit. However, the reality is that people do not now enjoy, and never have enjoyed, benefits equally. That does not mean that a society is necessarily unjust. The question becomes, however: On what basis can it be just for there to be an unequal distribution of wealth and benefit (as well as risk) among equal individuals?

John Stuart Mill argued that if social benefit as a whole increased then it was just to distribute benefits unequally. In other words, if there were a gasoline shortage, it could be just to give the gasoline to a business to create new jobs and increase economic growth and at the same time to deny gasoline to individuals who want to take automobile trips for pleasure. It would be argued that there is less social benefit in the latter case. Distributive justice then can be thought of as social expedience; whatever is more socially expedient carries greater weight.

Just what determines the more socially expedient course, Mill recognized, is a matter of opinion. Thus, Mill seems to contradict himself. As we note in Chapter 10, his view of liberty requires that no individual be placed at risk against his or her will. Yet Mill's view of social utility would seem to permit society to place a few people at risk if that risk was deemed socially expedient. Indeed, if opinion determines expediency then what is expedient to one person may well be unjust to another.

## Equality

A modern theory by philosopher John Rawls (1971) begins with the same basic assumptions of equality as the social utility theory. People are inherently equal, but do not have equal benefits. Rawls, however, gives more specific rules for distributive justice. The cornerstone of his theory is what is called the *maximin* principle. An unequal distribution is acceptable provided it maximizes the position of those who are worst off. It maximizes the minimum, hence maximin.

Such a theory might be used to justify a reverse income tax for those below the poverty level or a special tax on luxury gas-guzzling cars, with the money to be used to subsidize poor people who buy economy cars. Possibly, it would also argue against levying a tax that would be so prohibitive it might prevent, say, a coal company from exploiting a mine, thus limiting jobs and growth. Yet at the same time, a tax or an insurance fund to protect or compensate those who might suffer from land subsidence or pollution (maximizing the position of those worst off) might be mandatory.

It should be noted that Rawls applies his theory not only to inequality of wealth but to social and political inequality as well. His *difference principle* states that social as well as economic "inequalities are to be arranged so that they are both a) to the greatest benefit of the least advantaged and b) attached to offices and positions open to all under conditions of fair equality and opportunity" (see Rawls 1971).

Other philosophers, of course, have raised questions and objections (some too complex to be covered here) to this theory. For example, the perspective of an individual becomes important in determining what is just, since the theory sees everything from the position of the least advantaged. Another question arises over whether Rawls's concept requires a forced and patterned distribution of wealth, which may itself be unjust. Indeed, proponents of the next

concept of distributive justice would regard forced redistribution as insupportable.

## Liberty

In the libertarian view, individual liberty takes precedence over all else. People should have the equal right to acquire goods, use property freely, and enjoy other benefits to the extent that they are able, provided that in doing so they do not infringe the liberty of others. Obviously, acquisition by fraud or force is disallowed because it would necessarily mean the infringement of another's liberty. The role of the state – its laws and coercive power – is to ensure that liberty is maintained for all. Patterned redistribution of wealth, however, would be coercive, an invasion of basic freedom. It would, in other words, be unjust to make a wealthy person less well off by force. Or more generally, the libertarian theory of justice would regard any act as unjust that makes anyone less well off so long as equality of rights is maintained. Distribution of wealth could only be effected freely through unfettered markets.

Libertarian theories appear problematic, especially with regard to environmental issues. It would, for example, be unjust to interfere with someone's use of justly gained property, say a coal mine. But if it pollutes the air and water, then it might impinge on other people's freedom to breathe clean air. It can even be argued that all pollution has to be proscribed, although this would seem a huge infringement on the right of people to acquire and use their property. If outlawing pollution is too extreme, a libertarian might permit it on the basis of compensation of those affected. But who are the victims of any given quantity of air or water pollution? Those living now with lung diseases? Everyone living now and in the future? Any decision on compensation would inevitably be arbitrary. The libertarian approach leaves the question unsettled, yet the right to use property is a fundamental issue in our discussion of energy. Social utility or equality theories may be more functional for such questions; whether their answers would be entirely just, however, is a matter for interpretation.

## Conclusion

Despite the many fundamental differences between these and other theories of justice, it should not be forgotten that they all have at least one thing in common: They all attempt to find a path toward optimal fairness and equity among people throughout the world. This has been at the heart of attempts to systematize concepts such as justice through history. But we should note, too, that in recent years a few thinkers have criticized all traditional concepts of justice because of their anthropocentricity. In place of traditional concepts, they have proposed ideas of justice that include fairness and equity to all sentient life and even the entire ecosystem on the earth – arguing that anything less will not suffice in a technological world.

## Bibliography

Barry, V. 1985. *Applying Ethics*. Wadsworth, Belmont, CA.

Brandt, R. B. 1959. *Ethical Theory*. Prentice-Hall, Englewood Cliffs, NJ.

Frankena, W. K. 1973. *Ethics*. Prentice-Hall, Englewood Cliffs, NJ.

Nozick, R. 1974. *Anarchy, State, and Utopia*. Basic Books, New York.

Rawls, J. 1971. *A Theory of Justice*. Harvard University Press, Cambridge, MA.

Wenz, P. S. 1988. *Environmental Justice*. State University of New York Press, Albany, NY.

# APPENDIX C

# Synopsis of new technologies

## Technical efficiency

Energy conservation, as we noted in Chapter 4, can be understood to mean various resource substitutions, doing without, or *technical efficiency*, which we will focus on here.

Technical efficiency is best discussed with reference to the laws of thermodynamics (Appendix A). The First Law of Thermodynamics states that energy can neither be created nor destroyed; it can only change form or be stored temporarily. According to the Second Law, energy tends to be transformed from forms available for useful work and transfer of heat to forms that are less available or unavailable.

The efficiency of a thermal plant is usually defined according to the First Law. A thermal plant supplies useful heat either for a stage of industrial operation or for a heat engine, such as the modern turbine. Plant efficiency merely compares the *useful output* (useful heat or electricity) to the *fuel input* as a practical measure of the best use of the fuel.

Energy inputs and outputs for an industrial process heat (IPH) plant are shown in Figure C.1. This figure may be compared with Figure A.1, which is a similar diagram for an electric power plant. For both plants, we can identify a fuel input, useful energy output, and losses. Both thermal plants can reduce the nonuseful heat output and thereby make better use of the fuel input. This is done with heat-recovery

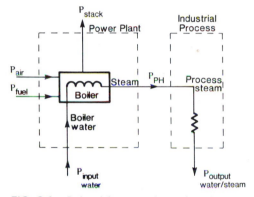

**FIG. C.1.** Industrial process heat plant (no heat recovery). $P_{ph}$ = process heat, rate of heat transfer; $P_{input} = P_{fuel} + P_{in} + P_{input\,water} \simeq P_{fuel}$; $P_{output} = P_{ph} + P_{stack} + P_{output\,water/steam}$; plant efficiency: $\eta = P_{ph}/P_{fuel}$.

technology, as is indicated in Figure C.2. A *recuperator* is used on the process water/steam output, and an *economizer* is used on the stack (hot flue gas) output. Each of these devices is a *heat exchanger*, which transfers part of the outgoing flow of heat back to preheat the air or water flowing in. (Such devices are not new, but were neglected until the advent of high fuel prices.) Using heat recovery, the useful output energy can be increased for the same fuel input or, alternatively, the same useful output can be obtained with less fuel input. Either type of recovery leads to improved plant efficiency.

There are limits to efficiency, however, as a

**297**

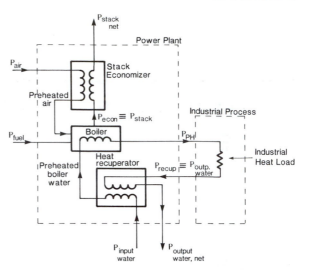

**FIG. C.2.** Technical efficiency with industrial process heat. $p_{econ}$ = stack economizer input, rate of heat transfer; $p_{recup}$ = heat recuperation input, rate of heat transfer.

consequence of the Second Law of Thermodynamics. As nonoutput steam has more and more heat exchanged out of it, we would expect the temperature of the effluent to decrease. But the output temperature could not go *below* the temperature of the surrounding air or water of the plant, because we could not then transfer any more heat to it. We therefore realize that the outside temperature places a theoretical limit on the amount of low-temperature energy that can be recovered. Nonetheless, typical heat-recovery systems can reclaim at least 8 percent more useful heat, depending on the

efficiency of the boiler and the system. The efficiency of a basic IPH system can usually be raised from about 87 percent to close to 95 percent.

These thermodynamic considerations also give us insight into the conversion of thermal energy into mechanical work and electricity (Fig. A.1). The mechanical work is, of course, the action of high-pressure steam on the turbine blades, which in turn spin the shaft of the electric generator. The efficiency with which the thermal plant delivers such useful work is likewise limited by the upper and lower temperatures of

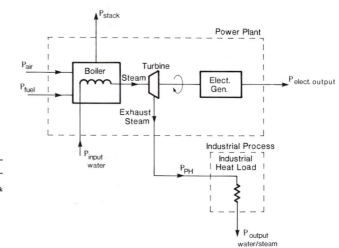

**FIG. C.3.** Cogeneration – technical efficiency in electrical generation. $P_{output} = P_{elect\,out} + P_{ph} + P_{stack} + P_{output\,water};$

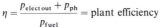

$$\eta = \frac{P_{elect\,out} + P_{ph}}{P_{fuel}} = \text{plant efficiency}$$

the plant, as expressed in the Carnot limit. Correspondingly, the purpose of the cooling-water loop is to reduce the lower temperature ($T_L$) of the power-plant cycle. Remember from Appendix A that the Carnot efficiency is given by $(T_H - T_L)/T_H$; thus if $T_L$ falls, efficiency rises. Similarly, the higher the starting temperature, the higher the ideal (Carnot) limiting efficiency of the heat engine. The practical upper limits on temperature are determined by the quality of the materials and their ability to withstand the high-temperature/high-pressure conditions and higher heat losses.

A major gain in useful energy output is achieved by *cogeneration* (Fig. C.3). Both electricity and IPH are outputs. In the particular version in Figure C.3, the lower-temperature exhaust steam is used for IPH instead of being discharged as waste heat.

A typical steam cogeneration system can result in a fuel saving of about 16 percent over the separate production of the same amount of IPH and electricity. This is depicted in the bar graph in Figure C.4, where the steam-turbine plant is compared with gas turbine and diesel cogeneration systems that can save 27 percent

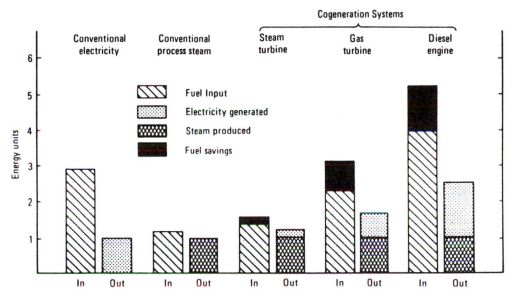

**FIG. C.4.** Cogeneration systems – performance comparisons. Energy inputs and outputs for sample congeneration technologies. The fuel savings (solid-black areas) represent the difference between fuel consumption by each cogeneration system and the fuel that would be required if separate equipment were used to produce the same amount of steam and electricity. Shown from left to right are the inputs and outputs of each of two conventional energy systems and three cogeneration systems. The input/output pair on the left is the conventional generation of electricity by means of steam, using a boiler and a steam turbine, with an output of 1 energy unit and an input of about 2.9 units ($\eta = 0.34$). Next is shown the input/output for conventional process steam using a boiler only with an output of 1 energy unit and an input of 1.15 units ($\eta = 0.87$). The next three pairs are for cogeneration systems, steam turbine, gas turbine, and diesel engine, respectively. In each cogeneration case, the output process steam is the same – 1 energy unit. Added to each cogeneration steam output is the electricity output, which varies according to the technology – steam turbines providing the least and diesel engines the most. The corresponding fuel input energy for each combined output is shown for each of the three cases in cross hatching and the fuel savings are shown in black for producing the same outputs separately and conventionally. The fuel savings shown are, from left to right, 16 percent, 27 percent, and 24 percent, for the three cogeneration technologies. Reproduced, with permission, from the ANNUAL REVIEW OF ENERGY, vol. 3, © 1978 by Annual Reviews Inc., article by R. H. Williams, "Industrial Cogeneration."

and 24 percent respectively. Whether or not a cogeneration system will be a worthwhile investment will depend on the cost (including financing) of the equipment and fuel. It should be noted that in many instances – for example with large industrial energy users – cogeneration has already proven to be a cost-effective technology.

## Small hydroelectric generation

Small-scale hydroelectric plants have an output capacity of 30 $MW_e$ or under, typically a range of 5 $KW_e$ to 5 $MW_e$. As we noted in Chapter 3, most small hydroplants utilize low heads, operating to a significant degree on the flow of the river as well as the pressure of the height of the head behind the dam.

A significant expansion in hydrocapacity is possible in the United States if the development of small hydrosites is considered. In the Northeast states alone, the potential capacity of small hydroplants equals about twelve 1,000 $MW_e$ nuclear plants. Throughout the 48 contiguous states, hydropower could provide an estimated 95,000 $MW_e$ of additional electric capacity (see Fig. 3.12), mostly using small hydroplants. In regions where increased electric system capacity is needed on a relatively short-term basis, small hydroplants may be particularly useful because they only take two or three years to construct.

A major factor in the adoption (or readoption) of small hydropower generation is, of course, its cost. Small hydroplants have capital costs that are higher than thermal plants, but because of low operating/maintenance cost, long plant lives, and, of course, zero fuel costs, small hydroplants often enjoy a cost advantage over thermal plants. If we consider the total generating cost of small hydropower – reduced to a per kilowatt-hour basis – we find the costs range from 1.5 to 6.4¢/KW-hr. As we saw in Chapter 9 (Table 9.2), that compares with about 4¢ for coal and 5¢ (or often more) for nuclear plants for total generating costs (in constant $1979).

Environmental degradation may be a barrier to hydroelectric development where construction means disruption of natural habitats or recreational areas. However, small hydroplants present few environmental concerns because the impacts are on a small scale at each site and because many sites are already dammed. In some cases small hydroplant operations improve the environment by sprucing up facilities that were abandoned and left in disrepair.

## Enhanced oil recovery

Enhanced oil recovery (EOR) is a means of extracting additional resources from already discovered and exploited oil fields. As discussed in Chapter 2, the amount of the resource that is *recoverable* depends on the technology of extraction as well as the amount in place.

Recovering oil is not as simple as drawing water from a well. Petroleum usually is embedded in porous rock formations. These formations are under pressure from overlying and surrounding geologic formations and, often, from accompanying natural gas formations. When the oil-bearing formation is first drilled, a pressure differential is created at the point of drilling, where the pressure is atmospheric (whereas the surrounding formation has higher geologic pressures). This pressure differential causes a migration of the oil through the rock formations toward the drilled well shaft. Once at the well, the oil comes to the surface under natural pressure. When the natural pressure is sufficiently high, the well erupts as a legendary *gusher*.

If the natural pressure is not sufficient to bring the oil to the surface through the shaft, as usually happens with older wells, then other methods must be employed to force the oil out. Petroleum extraction techniques are classified into primary, secondary, and tertiary practices. Primary extraction is where natural geologic pressures are sufficient to cause the flow of oil to the well. It also applies to simple pumping, which becomes necessary when the pressure becomes too low.

As pressures fall further, secondary practices

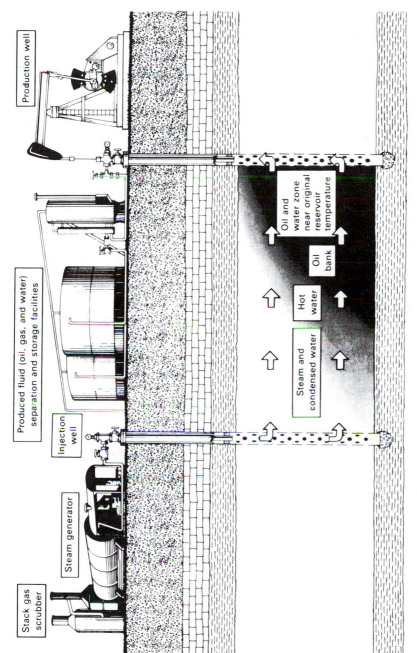

**FIG. C.5.** Steam-injection enhanced oil recovery. *Source:* Energy Technologies and the Environment, June 1981, U.S. Department of Energy, Report No. DOE/EP0026. Available from National Technical Information Service, Springfield, VA 22161.

need to be initiated. The reservoir may be flooded with water to maintain pressures and to displace oil toward the well. Typically, about one-third of the oil in a given field can be recovered by primary and secondary techniques. In the earlier era of petroleum exploitation, extraction was only rarely taken beyond the secondary stage. But with declining oil reserves in the United States and rising prices of imported oil, tertiary methods – EOR – have become more attractive.

Application of EOR technologies increases the amount of the resource that is recoverable. The U.S. Department of Energy has set a goal to extend ultimate recovery of U.S. oil resources by an additional 18 to 52 billion barrels.

EOR technologies variously apply heat, chemicals, and external pressure to extract additional oil. These applications are carried out by a variety of techniques, including steam injection, in situ combustion, and chemical injection. We will review only the first technique.

Figure C.5 shows a diagram of steam-injection EOR. On the left is a steam generator, an ordinary boiler. The steam is injected into the petroleum formation through a perforated pipe. The steam, under pressure, causes hot water to press against the oil, moving it toward the well bore. The movement underground is aided by the heat, which thins the petroleum and makes it less viscous. After the oil reaches

the well, it is pumped to the surface, processed to separate the water, and then shipped or piped to a refinery.

Because tertiary methods require burning fuel – usually some of the petroleum that is produced – the cost of every barrel burned must be charged against the costs of production, thus raising the net cost of the product. The added production costs of EOR made petroleum produced by these techniques uncompetitive in the world oil market in the mid- and late 1980s. Oil produced by tertiary techniques, however, can be expected to become competitive in the world market as world resources move toward exhaustion over the decades to come. This trend will be manifested by declining finding rates or escalating costs of discovery (see Chapter 2). Then tertiary exploitation of the older oil fields will be economically viable.

## Fluid-bed combustion

In Chapter 6 we noted current methods for reduction of air pollution from coal, including coal washing, fabric filters, and flue-gas desulfurization (scrubbers). Some new concepts focus not on cleaning the coal or coal smoke, but on burning it more efficiently, particularly through what is called fluid-bed combustion (FBC). This technology is already in the process of passing

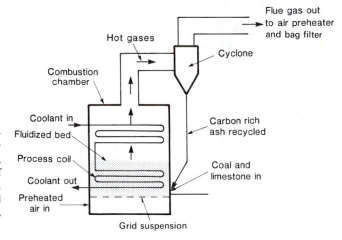

**FIG. C.6.** Fluid-bed combustion – a schematic diagram. *Source*: Energy Technologies and the Environment, June 1981, U.S. Department of Energy, Report No. DOE/EP0026. Available from National Technical Information Service, Springfield, VA 22161.

from the demonstration to the commercialization stage (Chapter 11).

The operation of an FBC boiler is depicted in Figure C.6. Combustion takes place in a bed of moving coal particles, which are suspended by both the natural convection of hot gases from combustion and from the forced movement of air through the bed region. The creation of a region of suspended particles has several advantages:

1. Combustion is more uniform and complete.
2. The temperature of combustion is lower than in a static bed, resulting in less chemical reaction of atmospheric nitrogen into nitrogen oxides.
3. Limestone is added to produce a more complete reaction of sulphur with (alkali) coal ash, thus greatly reducing sulphur dioxide emissions.
4. Unburned coal particles (fines) can be separated from the flue gas and reinjected for combustion.

The overall FBC system (Fig. C.7) operates in the following manner:

1. Coal is conveyed from storage, crushed to particle sizes less than one-quarter of an inch in diameter, and forced into the combustion chamber.
2. Preheated air is forced upward into the combustion chamber through the gridlike structure from below.
3. Crushed limestone is forced into the chamber along with the crushed coal to react with the sulphur in the coal.
4. Combustion of coal particles takes place over the vertical range of the bed region, above and below the feed point.
5. Some heavier particles of ash fall to the bottom through the grid (Fig. C.6) and are removed, but many small particles are carried out with the flue gases.
6. Unburned, fine coal particles can be separated from the flue gas in a mechanical separator (often a cyclone device like the

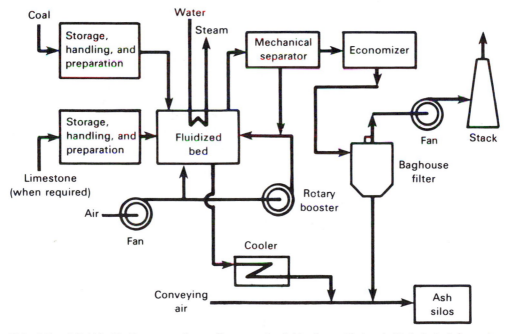

**FIG. C.7.** A fluid-bed boiler system. *Source*: Energy Technologies and the Environment, June 1981, U.S. Department of Energy, Report No. DOE/EP0026. Available from National Technical Information Service, Springfield, VA 22161.

one in Fig. C.6, which works on centrifugal forces) and fed back for more complete combustion.

7. Additional fine particles, including ash, are collected from the flue gas by a fabric filter or by electrostatic precipitators before the gas flows (or is forced) up the smokestack.

Demonstration FBC operations (without scrubbers) have met EPA New Source Performance standards for emission of pollutants such as sulphur dioxide and nitrogen oxide. Removal of sulphur emission, for example, is close to 90 percent and limits the emission to 1.2 pounds or less of sulphur dioxide for each MBtu of heat generated. Nitrogen oxide emissions have been one-half of the maximum federal standard. Only the emission of fine particles remains a problem for FBC, as it does largely for other combustion technologies.

Adoption of FBC boilers is expected to increase in the 1990s for both industrial and electric utility applications. As of the late 1980s, several utilities were running large demonstration projects. It is a potentially cost-effective alternative to the present scrubber technology because it pays for itself, at least in part, through higher thermal efficiency. FBC may also make an important technological contribution to resolution of the acid rain controversy, if further reductions are deemed necessary in sulphur and nitrogen.

## Resource recovery

In the 1970s, municipal solid waste (MSW) resource recovery was thought to be approaching commercialization. Pilot plants had been operating in more than twenty cities, providing energy and materials recovery from municipal wastes. Municipal waste is customarily defined as refuse from households and commercial buildings. The quantities of MSW generated in the United States are immense (over $150 \times 10^6$ tons/year) and increasing. The increase has been due not only to population growth, but also to increasing waste generation per capita.

**Table C.1**. Municipal solid waste composition (by weight)

| | Percent composition as received (dry weight basis) | |
| --- | --- | --- |
| | anticipated range | nominal |
| Paper | 37–60 | 55 |
| Newsprint | 7–15 | 12 |
| Cardboard | 4–18 | 11 |
| Other | 26–37 | 32 |
| Metallics | 7–10 | 9 |
| Ferrous | 6–8 | 7.5 |
| Nonferrous | 1–2 | 1.5 |
| Food | 12–18 | 14 |
| Yard | 4–10 | 5 |
| Wood | 1–4 | 4 |
| Glass | 6–12 | 9 |
| Plastic | 1–3 | 1 |
| Miscellaneous | <3 | 3 |
| Moisture content range: 20–40% nominal: 30% | | |

*Source*: Derived from *Recovery and Utilization of Municipal Solid Waste*, U.S. Environmental Protection Agency, Washington, DC, 1971.

The percentage breakdown of typical municipal refuse is shown in Table C.1. Approximately one-third of MSW is combustible, with a heating value of about 10MBTU/ton. Thus, if fully exploited as a fuel, MSW represents a potential source of thermal energy for the country approaching 1 quad/year. If we include the potential energy savings from materials in waste that can be recycled (metals, glass, and so on), then the MSW recovery can have an even greater impact on national conservation.

The process of MSW collection, processing, and either disposal or recovery, as it has been envisaged, is shown in Figure C.8. In resource recovery plants, wastes are divided into either recovered energy or recovered materials. The products of the materials separation process are:

1. Ferrous metals – mostly from tin cans.
2. Nonferrous metals – mostly aluminum goods.

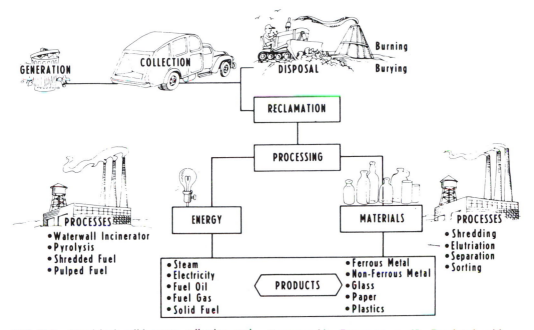

**FIG. C.8.** Municipal solid waste collection and processing. *Source*: S. L. Blum, in P. H. Abelson and A. L. Hammord, eds., *Materials: Renewable and Nonrenewable Resources.* p. 48. Reprinted with permission. Copyright © 1976 American Association for the Advancement of Science.

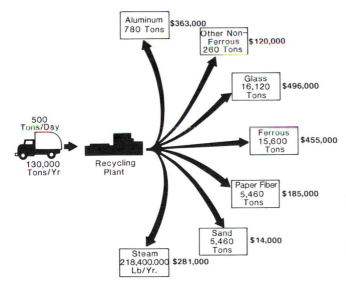

**FIG. C.9.** Resource recovery products and expected revenues (in $1986). Adapted from O. J. Hagerty, 1973, in J. Pavoni, ed., *Handbook of Solid Waste Disposal – Materials and Energy Recovery.* R. E. Krieger, Co., Melbourne, FL.

3. Glass – from bottles.
4. Paper fiber – from newsprint, if separated from combustible components.

In Figure C.9, we see the product breakdown for a hypothetical MSW resource recovery plant with a processing capacity of 500 tons/day. The materials can be sold, where markets exist.

Following materials separation, energy recovery would take place. This energy has usually been derived by burning the reduced MSW directly. The refuse is fed as a fuel supplement for coal or another conventional fuel to generate steam for industrial process heat, district heating, or electricity generation. Such use of waste fuel would cut the cost of steam generation through reduction of conventional fuel purchases.

Emission controls are especially important for resource recovery plants (or ordinary incinerator plants), because they are often in close proximity to residential areas. A major concern in emissions is with particulates and the products of burning plastics (mainly hydrogen chloride from incineration of polyvinyl chlorides, or PVCs). In fact, this concern has caused sufficient public opposition to stop construction of resource recovery plants and to shut down plants already operating. MSW, on the other hand, is lower in sulphur content than many conventional fuels, such as oil and coal, and the use of an MSW fuel supplement reduces sulphur-related emissions.

Resource recovery, like several of the other near-term alternative technologies discussed in Chapter 11, has faced barriers to adoption. These barriers occur within the institutional setting of municipalities. A municipality faces the responsibility to collect and dispose of MSW, come what may. This overriding requirement on municipalities, and hence on the practicality of any MSW technology, molds the way in which its feasibility and economics are evaluated. The most obvious factor in feasibility is the reliability of the technology. Will it function consistently? Will it operate in an environmentally accepted way? Frequent or lengthy operational outages, without alter-

native means of disposal, can create intolerable buildups of waste, with attendant health and esthetics problems. Thus, the *trialability* factor in the process of innovation (see Chapter 11) becomes critical, and many municipalities have been reluctant to take the risk of an unproven technology.

In fact, resource recovery plants, as defined here, ceased being built by the mid-1980s. In their place came *mass burn* incinerators that, at most, produced usable heat or electricity, but performed little or no processing to separate glass, metals, and newsprint. The trend seemed to be toward *source separation*, in other words municipalities shifted the job to the home owner, often under (mandatory) local ordinance.

Actually, resource recovery foundered for both technological and institutional reasons. On the technical side, many pilot plants experienced problems with processing or environmental controls. Even though solutions to these problems were matters of straightforward engineering, the municipalities and their contractors seemed incapable of overcoming them.

## Heat pumps

The conventional use of electricity for space heating is *resistance heating*, which is just the ohmic dissipation of electric energy in a resistance (see Appendix A). Heat pumps represent an instance of conservation through *technical efficiency*, as discussed in the first section of this appendix. A heat pump operates like an air conditioner in reverse – it takes heat from the outside and transfers it inside. The process provides electric space heating much more efficiently than conventional electric heating.

A heat pump, like a refrigerator or an air conditioner, is composed basically of a *compressor*, an *evaporator*, a *condenser*, and an *expansion valve* (see Fig. C.10). The compressor is a necessity – thermodynamically – because if we are to transfer heat from a low temperature body to a higher temperature body, mechanical

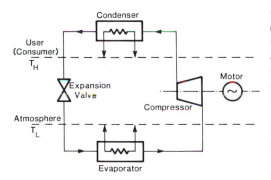

**FIG. C.10.** Heat pump – schematic diagram. Adapted from E. Camatini, ed., (1976), *Heat Pumps: Their Contribution to Energy Conservation.* Kluver Academic Publishers Norwell, MA.

work is required (see Appendix A). The compressor does the work by pumping the refrigerant (working fluid) from a low-pressure, vaporized state to a higher-pressure, liquid state.

The refrigerant is hot as a result of the compression, and heat exchange to the room or building takes place from the condenser. After passing through the condenser, the refrigerant goes through an expansion valve into a low-pressure evaporator. In the process of expansion, the refrigerant drops in temperature and a cooling action takes place, drawing heat from the air volume surrounding the evaporator (even though the outside air is cold, it still contains latent heat that can be transferred). Work is again done on the refrigerant by the compressor and the cycle continues.

The effectiveness of a heat pump (or a refrigerator) is measured by its coefficient of performance (COP), defined as the ratio of the quantity of heat to be transferred to the quantity of work required for the transfer from lower ($T_L$) to higher temperature ($T_H$) shown in Figure C.10. The COPs of working heat pumps today approach a factor of 2, meaning that it takes about one unit of energy expended as work to deliver two units of heat energy. Conventional electric heating has a COP of one, meaning that *twice* as much electric energy is used for the same heating than through a heat pump.

Heat pumps have been available commercially in the United States and Europe since the 1950s. But the process of commercialization (Chapter 11) had not progressed very far prior to the 1970s. Part of the reason was that the capital investment requirement for a heat pump installation was significantly higher than for a natural gas or oil furnace of comparable size. Even during the energy crisis period of the 1970s, the cost savings were not always obvious. However, this technology has gained increasing market penetration in recent years.

## Wind generation

### Wind power basics

The modern aerodynamic theory, upon which a theoretical understanding of power conversion from the wind is based, dates back to the classic work of Daniel Bernoulli (Swiss physicist, 1700–1782) and others. It is based on principles of conservation of energy and momentum of each volume of air moving over a blade, wing, or sail – the so-called *thrust* surface. The output power of a wind generator varies in direct proportion to the area of interception of the windmill, which is approximately the projected area swept out by the rotor blades. The maximum power from a modern wind turbine ranges up to about 10 MW$_e$ for a 300-foot-diameter rotor. This corresponds to converting the energy in a column of air moving at 40 miles per hour by an interception cross section of about 70,000 square feet.

As we might expect, it is not possible to convert all of the available power that exists in a given moving column of air. The aerodynamic theory, for example, shows that the maximum power that can be extracted from such a wind stream is 59 percent of the power in that cross section. The maximum power output of a windmill is furthermore limited by its size, as determined by its rotor diameter.

Output power varies as the square of the rotor diameter and, thus, rapid increases are made for larger and larger diameters. As the diameter of the rotor is increased, so also must the height of the supporting pedestal (Fig. C.11) be increased. Ultimately the size of the entire windmill depends on structural considerations

**FIG. C.11.** A wind turbine generator. Courtesy of the American Wind Energy Association.

and other factors such as TV interference and aesthetics. Thus far, the size limit appears to be about a 300-foot diameter for a propeller-type wind machine (there are other novel types, also having limitations of size). This size limitation implies an upper limit on output power of about 5 MW$_e$ for a wind speed of 17 mph.

So far in our discussion we have either related output power to specific wind speeds or to a rated power capacity. However, wind speeds are variable, indeed in many locations highly variable. The measures of wind machine energy production, like measures of solar energy systems, must therefore be in terms of an expected average rate of energy production (expressed in KW). This information is often presented as the average power per unit of *cross-sectional area* – measured in average watts per square meter (W/m$^2$) for various locations.

Another way to measure the expected output of windmills, however, is to indicate the number of hours per year they are available to deliver a certain power output. This measure is essential because of the variability of the wind – not only are there periods when the wind does not blow, there are times when it blows too hard.

All such machines must be shut down during hurricane or gale force winds. When either the wind is calm or is blowing too hard, the machine is not available to deliver power. Conversely, when the wind is blowing in the usable range of velocities, the machine is said to be available and the number of hours per year that the wind machine is available is known as its *availability*.

Figure C.12 shows the worldwide regions where a wind machine will deliver its upper-limit-rated power output. The numbers shown in the key, starting at values under 750 KW-hr/KW and going over 5,000 KW-hr/KW, are equivalent to *hours per year* in which the machine can produce its rated output power. Thus, the best performance shown is over 5,000 hours (57 percent of the 8,760 hours per year) and the worst shown is under 750 hours (8.6 percent). No matter what measure is used, however, it is clear that only certain regions in the world are promising for wind generation.

There is another consideration. The wind data given above are on a broad *regional* basis. Actually, wind speeds depend on the local conditions, such as terrain, buildings, and vegetation, and thus the performance of a wind machine is said to be very *site specific*.

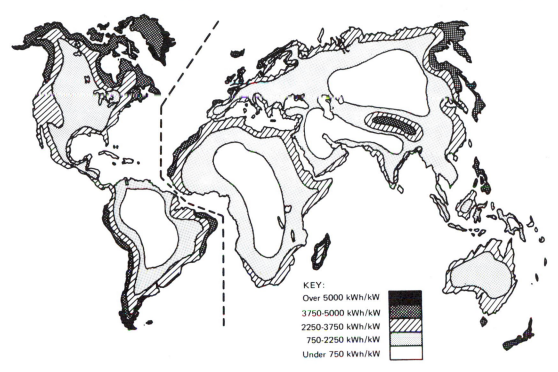

**FIG. C.12.** Worldwide wind availability. *Source:* Mitre Corporation (1975), for the National Science Foundation, *Wind Machines*, NSF-RA-N-75-051. U.S. Government Printing Office, Washington, DC.

KEY:
Over 5000 kWh/kW
3750-5000 kWh/kW
2250-3750 kWh/kW
750-2250 kWh/kW
Under 750 kWh/kW

## Current applications

In their present state of development, wind turbines appear to be appropriate for two modes of operation: (1) remote-site, stand-alone operation, and (2) dispersed-site operation on the electric grid. Remote-site operation refers to the generation of electricity in rural and wilderness locations where there are no transmission lines from electric utility grids. Dispersed-site operation entails feeding electric power from wind machines into the utility grid. When remote-site operation is used, there must be battery storage or some backup source of electricity, such as a gasoline-driven generator.

At the present time the capital costs of wind generators are high compared with conventional generation. Using cost per unit of output capacity, wind generators run about $4,000/KW or about 75 percent above the median present day cost of nuclear plants and 100 percent above the average for small hydroplants. Of course, in comparing the costs of these different plant types, we should account for the fact that wind generation, like solar generation, has zero fuel costs and may therefore have a competitive advantage despite a high capital cost. However, like solar, wind generation must have its economic viability assessed with respect to its availability. When the availability is low, the charge per KW-hr must be higher to pay off the investment.

The net effect of these cost factors is that wind generation is not competitive with most conventional generation. Advocates of wind power, however, contend that design improvements, economies of size, and large-scale manufacture will lower the capital costs in the near- to medium-term future. In addition, developments in energy storage technologies, such as fuel cells and batteries, could improve the availability of wind-generated electricity without adding prohibitively to the capital cost.

## Biomass synfuels

Synthetic fuels from biomass sources are mainly liquid alcohols and gaseous methane. Present day sources of organic materials for commercial production of biomass fuels are wood, agricultural residues, animal wastes, energy crops (grown for conversion), municipal wastes, and sewage.

### Alcohol fuels

Alcohols, although typically associated with other uses, can also be used as fuels in the forms of *ethanol* and *methanol* – respectively derived from grain and wood. Alcohols are derivatives of water, wherein a hydrocarbon radical replaces a hydrogen atom in water to combine with the hydroxyl radical (OH). The simplest example is methanol, having the chemical formula $CH_3OH$, in which the ethyl radical $(CH_3)$ has replaced the hydroxyl radical. Alcohol vapors explode when mixed with air, thus making them useful as fuels for properly designed internal combustion engines.

Alcohols occur naturally in organic materials, but not in sufficient concentrations to make extraction at the high volumes needed for the commercial production of fuels. However, synthesis processes are well known and with further development can be turned to fuel production. These processes include: (1) *fermentation* – for ethanol, (2) *anaerobic digestion* – for methanol, and (3) *biomass gasification* – for methane.

Fermentation to ethanol (ethyl alcohol) results from the decomposition of glucose (sugar) by the action of enzymes. Molasses, for example, contains over 50 percent glucose, and with malt added to supply enzymes, molasses will convert directly to ethyl alcohol. When a carbohydrate (such as corn) is the feedstock, however, intermediate enzymic reactions must take place, first converting the starches to

sucrose and then converting the sucrose to glucose. Sucrose $(C_{12}H_{22}O_{11})$ is common table sugar, usually derived from plants such as sugar cane or sugar beets. Glucose $(C_6H_{12}O_6)$ is the sugar usually produced by fermentation.

The fermentation process itself takes several days to complete and the fermented mash is then distilled. The products of distillation are 96 percent ethanol, a 4 percent water faction and a solid residue (*stillage*). The ethanol-water mixture must be dehydrated before it can be used as a fuel, in a separate step indicated in the flow diagram in Figure C.13.

The overall ethanol-fermentation process requires support processes that add to the costs of production. A major support requirement is boiler heat for distillation, alcohol dehydration, and stillage drying. But portions of the feedstock, such as combustible factions of agricultural products, can be used as a captive fuel, avoiding the cost of purchasing coal or oil. Also ethanol fermentation leaves by-products that can be sold to offset partially the costs of production.

Biomass synthesis of *methanol* requires processes similar to the indirect liquefaction of coal-derived fuels (see Chapter 12). However, in the initial gasification stage for methanol two possible methods can be used. The first, anaerobic digestion, is a biological process that produces methane, and the second, gasification, is simply a chemical reaction producing hydrogen and carbon monoxide. Anaerobic digestion is shown in Figure C.14. Bacteria in a heated, closed, air-deficient atmosphere break down the various organic feedstock components into other organic compounds. Anaerobic digestion must take place in an airtight *digester*, where the temperature and the conditions of acidity are constant, and continuous stirring must take place to optimize the reaction rates.

The alternative means of generating the methane synthesis gas for methanol is the *gasifier* shown in Figure C.15. The process, using wood feedstocks, is quite similar to those described in Chapter 12 for coal. The initial gasifier, for example, operates on wood chips in an oxygen-deficient atmosphere at high temperatures and pressures to produce a

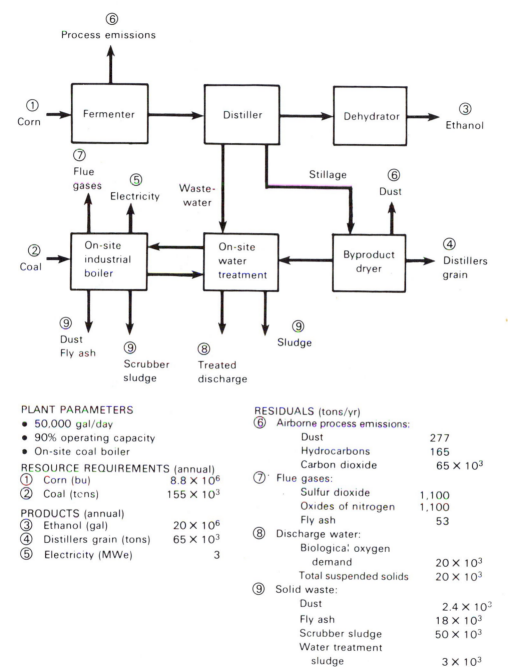

**PLANT PARAMETERS**
- 50,000 gal/day
- 90% operating capacity
- On-site coal boiler

**RESOURCE REQUIREMENTS (annual)**

| | | |
|---|---|---|
| ① | Corn (bu) | $8.8 \times 10^6$ |
| ② | Coal (tons) | $155 \times 10^3$ |

**PRODUCTS (annual)**

| | | |
|---|---|---|
| ③ | Ethanol (gal) | $20 \times 10^6$ |
| ④ | Distillers grain (tons) | $65 \times 10^3$ |
| ⑤ | Electricity (MWe) | 3 |

**RESIDUALS (tons/yr)**

| | | |
|---|---|---|
| ⑥ | Airborne process emissions: | |
| | Dust | 277 |
| | Hydrocarbons | 165 |
| | Carbon dioxide | $65 \times 10^3$ |
| ⑦ | Flue gases: | |
| | Sulfur dioxide | 1,100 |
| | Oxides of nitrogen | 1,100 |
| | Fly ash | 53 |
| ⑧ | Discharge water: | |
| | Biological oxygen demand | $20 \times 10^3$ |
| | Total suspended solids | $20 \times 10^3$ |
| ⑨ | Solid waste: | |
| | Dust | $2.4 \times 10^3$ |
| | Fly ash | $18 \times 10^3$ |
| | Scrubber sludge | $50 \times 10^3$ |
| | Water treatment sludge | $3 \times 10^3$ |

**FIG. C.13.** Process flow diagram for a corn-to-ethanol fermentation plant. *Source:* Energy Technologies and the Environment, June 1981, U.S. Department of Energy, Report No. DOE/EP0026. Available from National Technical Information Service, Springfield, VA 22161.

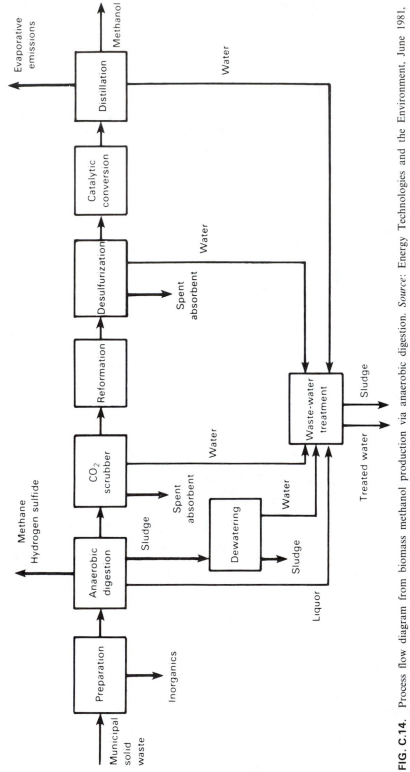

**FIG. C.14.** Process flow diagram from biomass methanol production via anaerobic digestion. *Source:* Energy Technologies and the Environment, June 1981, U.S. Department of Energy, Report No. DOE/EP0026. Available from National Technical Information Service, Springfield, VA 22161.

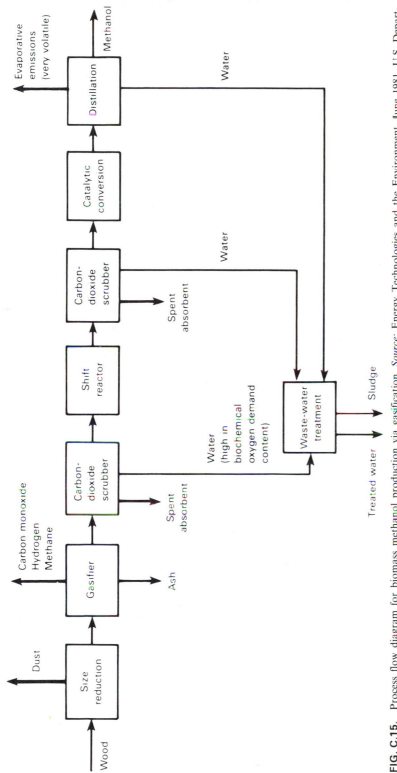

**FIG. C.15.** Process flow diagram for biomass methanol production via gasification. *Source:* Energy Technologies and the Environment, June 1981, U.S. Department of Energy, Report No. DOE/EP0026. Available from National Technical Information Service, Springfield, VA 22161.

synthesis gas consisting mainly of hydrogen and carbon dioxide. This gas, after the scrubbing of carbon dioxide, is subjected to a water-gas shift reaction (see Chapter 12) that produces a hydrogen-carbon monoxide mixture to feed the next stage: catalytic conversion (Fig. C.15). Catalytic conversion in this case produces liquid methanol, water, and impurities, which are distilled to give a final output of methanol.

## Market potential

Methanol is sometimes used as a fuel in racing cars. Ethanol has been sold in a mixture with gasoline for automotive use, especially in Brazil where a major development effort was undertaken. But for the most part, biomass synfuels suffer from the same disadvantage as most alternative technologies; they currently cost too much. Their market potential depends on several factors, a few of which are technical. An important consideration is the performance of alcohols as fuels for internal combustion engines. In general, alcohols have a lower heat value (per unit of volume) than petroleum-derived gasoline. For example, ethanol has only about 60 percent of the heating value of gasoline, and methanol only about 50 percent. However, the alcohols have a higher octane rating than gasoline, somewhat compensating for the loss of fuel efficiency (see Chapter 3).

Also, alcohol fuels, either pure or in high percent blends, are corrosive to the type of piping and fixtures used for petroleum products. The fuels themselves are prone to degradation by hydration (they pick up water in solution) and they separate when in long-time storage. These problems of shipping and storage could be avoided in the case of methanol by converting it to a synthetic gasoline with one additional step of chemical processing. This product would have shipping, storage, and ignition properties very similar to petroleum-derived gasoline.

## Shale oil

Shale oil represents a natural energy resource comparable to, or larger than, petroleum. It is derived from a sedimentary rock known as marlstone or oil shale found concentrated in particular geologic formations on all continents of the world. In the United States the largest formations are in Utah, Wyoming, and Colorado, but there are smaller resources in the Eastern states.

The oil is locked into the rock in a form called *kerogen*, a solid organic substance that is mostly insoluble. Kerogen is a large, complex organic chain that, like other organic compounds, can be broken down by *cracking*. Cracking, in this case, means subjecting pulverized shale to temperatures in the 800° to 1,000°F range. This produces an oily vapor that condenses into shale oil and is then purified by removing sulphur and nitrogen. Raw shale oil is more viscous than crude petroleum, thus presenting some shipping problems until refining techniques can be found that create an oil product more physically similar to high-grade petroleum.

Like coal, shale deposits are mined either underground or at the surface, depending on the depth of the deposits. Both the mining techniques and the problems of coal mining (see Chapter 6) apply to shale as well. Large-scale operations can lead to subsidence, land degradation, and other problems. Shale also creates large volumes of waste material for disposal. In a high-grade formation, about 25 gallons of shale oil can be produced per ton of marlstone. About 85 percent of the marlstone is waste. The spent shale is less dense after crushing and thus the volumes of material for disposal are larger than the original deposits by as much as 30 percent. Consequently, spent shale represents a greater disposal problem than coal.

After the shale is mined, it must be pulverized and then fed into a retort or kiln for cracking. In the retort, the crushed shale is heated by direct or indirect means. About 75 percent removal of the kerogen has been obtained at the 800° to 1,000°F retort temperatures of pilot plants.

Alternatively, cracking can take place in situ. The shale is first fractured by explosive charges or high-pressure hydraulic pumping. The fractured zone in the shale seam underground is then ignited using a combustible gas and partial

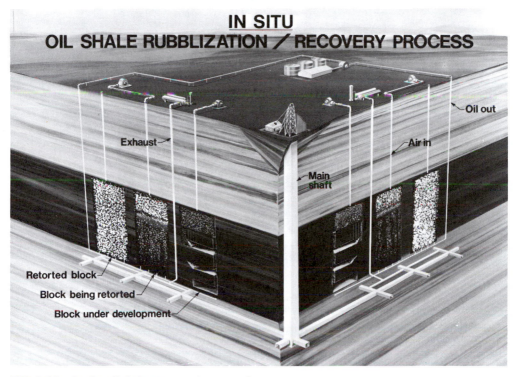

**IN SITU
OIL SHALE RUBBLIZATION / RECOVERY PROCESS**

Oil out

Exhaust

Air in

Main shaft

Retorted block

Block being retorted

Block under development

**FIG. C.16.** In situ oil shale recovery. Courtesy of Lawrence Livermore Laboratory.

combustion of the shale sustained by compressed air. Then, instead of extracting rock, raw shale oil alone is removed. This technique works when about 30 percent air spaces have been created by fracturing. A cross-sectional conception of in situ recovery is shown in Figure C.16. Only about 30 percent kerogen recovery has been obtained with the in situ technique.

The in situ method has the obvious advantages of eliminating spent shale disposal requirements, and unlike the retorting process, it does not generate air pollution. But there are environmental impacts. One of the major concerns is the pollution of underground aquifers, an especially serious question in regions of scarce water resources in the Western states. The shale processing itself requires large amounts of water, running close to 0.2 million acre feet annually to produce 1 million barrels per day of shale oil. (An acre foot of water is

the volume required to fill a basin of 1 acre area to a depth of 1 foot.) The water resources of the arid Western states could be taxed beyond their limits in a multimillion barrel per day shale industry.

## Unconventional fossil resources

### Tar sands

Tar sands are natural beds of sand, water, and bitumen. The bitumen is a tarlike hydrocarbon that can be separated from the sand by non-chemical means to give the equivalent of crude oil.

The largest ongoing tar sands operation, in the Athabasca sands in Alberta, Canada, requires the mining of hundreds of thousands of tons of sands each day using giant bucket-

**FIG. C.17.** Tar Sand extraction in an Athabasca deposit in Canada. Courtesy of Alberta Public Affairs Bureau.

shell excavators and conveyor belts (see Figure C.17). The highly abrasive sand is mostly composed of quartz particles, leading to wear of shovel bits at more than twice the rate of ordinary use. A rich tar sand contains only about 10 percent bitumen by weight. As many as five tons of tar sands are needed to extract one barrel of crude, tar-sand oil.

Tar sand processing consists basically of heating, sifting, and floating the bitumen tar to separate it from the sand particles. Following separation, the bitumen is upgraded in a process that corresponds roughly to the refining of crude petroleum. Upgrading also includes purification of the product and the creation of useful hydrocarbons such as butane, naphtha, and kerosene. The prime concern in purification is the removal of sulphur, which typically composes about 4 percent by weight (a level considered unacceptably high in petroleum) of the tar sand.

As we have noted elsewhere, tar sand deposits are huge worldwide. However, the cost of

exploiting them is very high. The capital investment at the Athabasca facility started at about $2.5 billion and grew to $3.5 billion by 1984. Production costs, meanwhile, were expected to run about $9.50 per barrel of crude, which is competitive with the world oil prices of the 1980s. However, the costs have proven to be higher, making production costs at best marginally competitive with petroleum. As we saw for synfuels and other alternative fossil fuels, oil market fluctuations have caused perceptions of the market competitiveness of tar sands to change.

## Unconventional natural gas

Extraction of unconventional gaslike tar sands does not present obstacles due to lack of basic conceptual knowledge. Rather, the cost of extraction must first be proven competitive with conventional fuels before unconventional gas can be a viable alternative.

Unconventional gas has the same geological

origins as other fossil fuels, including conventional natural gas; it is simply more difficult to reach and extract. Much of it is simply natural gas that resides in highly impermeable geological strata, so that the gas does not flow at sufficient rates using ordinary natural gas drilling and pumping techniques. Examples of such impermeable deposits are Devonian shale formations in the Eastern states, deep tight sand deposits in the Western states (see Fig. C.18), and methane embedded in coal deposits.

A more speculative source is methane believed to be trapped under deep aquifers in regions near the Gulf of Mexico (Fig. C.18) called *geopressurized domes*. These deposits, if they exist, would be found at tremendous depths (10,000 to 30,000 feet), under huge pressures (2,000 to 5,000 psi) due to the domelike geological formations above them, and at high temperatures (300° to 500°F). These conditions illustrate an extreme case of what was stated above; there is no conceptual obstacle to

recovering such a resource, but it would be an extremely difficult engineering task and probably would be prohibitively expensive.

Several techniques have been investigated, however, to recover tight sand and Devonian shale deposits. These include hydraulic fracturing and explosive fracturing, both of which are designed to open fissures in the sand or shale formations to allow the flow of gas to a bored well. Flow may also be enhanced by the use of emulsifiers and compressed gas following the fracture.

In hydraulic fracturing, a fluid is forced under high pressure into underground strata from a drilled well operated at the surface. An advanced technique called massive hydraulic fracturing (MHF) uses hundreds of thousands of gallons of the fracture fluid at pressures in the range of 15,000 psi. After the high-pressure fluid has caused fissures throughout the stratum, the fractured layers are propped open by injecting gelling agents so that the gas can continue to

**FIG. C.18.** Unconventional gas resources in the United States. *Source*: U.S. Department of Energy (1981).

flow through the system of fissures to an extraction well.

Explosive fracturing involves pumping the components of an explosive mix slurry into the stratum of sand or shale. This mixture is forced under pressure into existing fissures to penetrate deep into the gas-bearing seam. The entire explosive volume can then be detonated, extending the fractured zone more than 10,000 feet from the bore hole.

Unconventional gas recovery techniques on the whole require further development before they can be technically feasible and economic. Progress on technical feasibility has been made on tight sands as well as on coal bed degasification. But the engineering feasibility of deep-well gas extraction from geopressurized domes has not been demonstrated.

Before unconventional gas approaches economic viability, total extraction costs (including exploration, capital, and operating costs) must be less than $4/scf to be competitive with conventional gas and petroleum.

## Geothermal energy

The core of the earth is a vast thermal reservoir that is effectively inexhaustible. Yet only a tiny fraction of geothermal energy has been tapped. The principal use of geothermal heat has been with *hydrothermal* techniques, which take advantage of water that has been heated in rock formations deep in the earth's crust. These rock formations lie just above the molten magma, thousands of feet below the surface.

A typical geothermal reservoir is depicted in Figure C.19. In the figure, we are able to follow the path of water flow originating from rain that works its way down through fractures in rock until it reaches the depth where it is heated by the magma. Once hot, the water rises from natural convection. But often its path upward will be blocked by low-permeability strata of rock, trapping it in a reservoir thousands of feet below the earth's surface. As a result of heating and confinement, both temperatures and pressure in the hydrothermal reservoir build up.

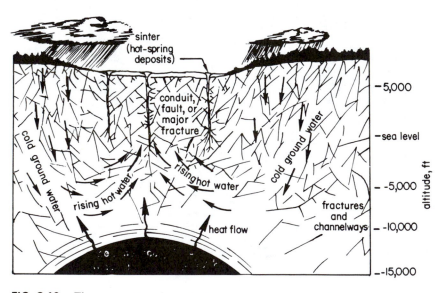

**FIG. C.19.** The structure of a hydrothermal reservoir. *Source*: U.S. Geodetic Survey, 1973, *Geothermal Resources*. U.S. Government Printing Office, Washington, DC.

The temperatures typically exceed 200°C and pressures run into the range of 100 psi (about seven times atmospheric pressure). Depending on the temperature and pressure, the water can be turned to pure or dry steam or become a wet mixture of steam and water. The steam or hot water of a hydrothermal reservoir can be brought to the surface only through a natural fracture in the overlying rock or through a drilled conduit (Fig. C.19). (Natural fractures are the sites of natural geysers and hot springs.)

The location of sites for drilled wells is a prospecting process similar to that for oil and natural gas. Only particular geological formations are known to be likely sites. When a successful well is found, it can supply steam for a thermal electrical plant, replacing the furnace boiler in a fossil-fired plant. In the case of a well with dry steam, the thermal system can have a direct input of the steam (Fig. C.20), which is required only to pass through a separator and filter before entering the steam turbine. A typical geothermal plant has a plant efficiency around 20 percent, compared with a conventional plant's 35 percent. This alone, however, would not be the critical factor in the viability of this technology, because these plants operate on cost-free fuel once the well is found and successfully drilled.

However, hydrothermal energy is not without potential problems. For example, the steam and water often contain salts and minerals that corrode and pit turbine blades. Further development in corrosion-resistant metals is needed to make the technology less susceptible to breakdown and replacements.

The main barrier to geothermal energy, however, is the limited scope of the resource. The sites with the largest potential generating capacities are located in Central America, Italy, New Zealand, the Philippines, and the Western United States. But the total world capacity is likely to remain small, and there is at present only about 2 $GW_e$ installed capacity worldwide.

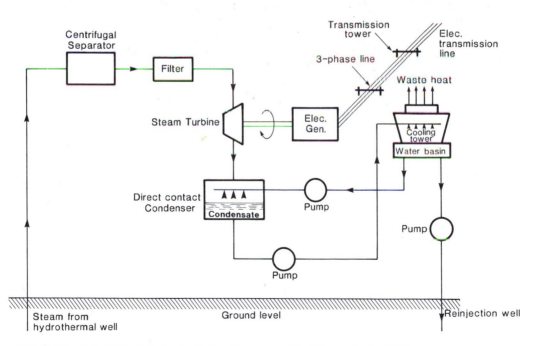

**FIG. C.20.** A hydrothermal electric plant – direct, dry-steam type. Adapted from DiPippo, 1980, *Geothermal Energy as a Source of Electricity, A Worldwide Survey,* for the U.S. Department of Energy. U.S. Government Printing Office. Washington, DC.

## Direct solar conversion

Direct conversion of sunlight to electricity is a very attractive technological prospect. Like solar thermal energy, it is pollution free and uses the renewable energy of the sun. It would also eliminate all of the intermediate processes of thermal plant generation, which are necessary if the solar energy is first collected in the form of heat.

Direct conversion can occur through several different physical mechanisms, including photochemical processes, the thermoelectric effect, and the photovoltaic efffect. Of these alternatives, *photovoltaics* (PV) have received the most attention and development. The photovoltaic effect was first observed over a hundred

years ago by A.C. Becquerel in 1839, but was not fabricated into a useful device until the 1950s, when it was first used in the space program. Today, photovoltaic conversion is a working technology, but its major barrier to widespread adoption is its cost, as we noted in Chapter 11.

Photovoltaic conversion cells are semiconductor, junction-type devices. They are able to supply d.c. electric power when illuminated by sunlight, converting the incident solar energy directly with efficiencies ranging up to a theoretical limit of about 30 percent (which corresponds to the Carnot limiting efficiency for thermal conversion). Working arrays of PV cells are shown in Figure C.21. Each cell in an array develops a small voltage and supplies

**FIG. C.21**. A solar photovoltaic installation. Courtesy of the Electric Power Research Institute.

current. The cells are interconnected to supply an overall voltage and current for direct use or for storage in batteries.

Photovoltaic electricity is created in semiconductor materials when a photon of sunlight is absorbed into the material and energizes a charge carrier such as an electron. This energy enables the carrier to overcome an electric potential barrier at the semiconductor junction in the cell. Junctions are a feature of many semiconductor devices, including transistors. They are created by *doping* two adjacent regions of the material with different impurities. On one side, the impurities create an excess of negative carriers (mobile electrons), while on the other side an excess of positively charged carriers (so-called holes) is created. The transition region between the two sides is called the *junction* region and it has a potential (voltage) barrier caused by fixed ions of opposite-charged polarities in the two adjacent materials.

If a mobile charge carrier gains sufficient energy from a solar photon to overcome the potential barrier at the junction, then the carrier is in a position to do useful electrical work. This follows from the physical principles outlined in Appendix A and has its mechanical analog in hydropower, where a mass of water is recognized to have gravitational potential energy if it is raised to a height.

The efficiency of PV conversion depends on two major factors: optical losses and the height of the junction barrier (that is, the electrical potential that rises across the junction region, usually measured in electron volts; see Appendix A). Optical losses, such as reflections from the cell's surface, can be reduced by proper optical design, whereas the conversion efficiency of carriers across the junction barriers presents a fundamental limitation.

These limitations on PV conversion efficiency are due to an inability of the semiconductor junction to provide for the conversion of the entire spectrum of incoming sunlight. An optimum photovoltaic conversion takes place when the energy of an incident photon is just equal to the energy difference of the barrier junction, thus absorbing all of the photon's energy in taking the charge carrier over the barrier potential. If the photon energy is less than the barrier energy, then no conversion takes place. If it is greater, then part of the photon energy is wasted.

A new form of solar cell called the *tandem junction* has several junctions fabricated together with varying barrier potentials. These tandem junctions are able to cover more of the solar spectrum, but the basic limitation is still present. In addition, tandem cells are more expensive to fabricate.

The entire spectrum of sunlight photon energies spans a 10-to-1 range. It is apparent that PV conversion only will be optimal for one narrow range within that spectral spread. Mass-produced, single-layer silicon PV cells – the cheapest and easiest to manufacture – currently convert less than 10 percent of incident solar energy into electricity, about half their limit. R&D efforts sponsored both by government and industry have been directed toward improving the efficiency of silicon devices and searching for other semiconductors with better efficiencies. Some semiconductors, such as gallium arsenide, have somewhat higher theoretical efficiencies.

Present-day applications of the PV technology, like those of solar thermal, are limited by capital costs and solar availability (see Chapter 12). The costs of PV cells are determined by the current state of semiconductor technology and manufacture. New developments are needed – and may be possible – to permit the manufacture of large quantities of PV cells cheaply at efficiencies close to the limit. It is generally thought that these objectives can ultimately be reached by the development of new semiconductor materials and techniques of manufacture. The prospects, however, are far from certain for the near term (see Chapter 11).

It should be noted that there is a technique that raises efficiencies significantly: focusing. The idea is to focus sunlight to anywhere from 100 to 10,000 times its incident intensity. By such focusing, it is argued, the cost of the cell per unit of area can be much higher, because the amount of energy converted will be greater. But the fabrication of such cells is far too expensive at present for commercial viability. The per unit cost of mass production would have to fall significantly.

# Advanced storage

## Conventional storage

For many energy technologies, storage is crucial. If we consider the storage of fuels as the storage of the energy embedded in them, then oil is an excellent example. The massive amounts of petroleum stored worldwide are necessary for the reliable, economical availability of gasoline, fuel oil, and petrochemicals.

Electric utilities also store energy using a scheme called *pumped storage*. Electricity generated by thermal power plants drives large electric motors to pump water uphill to elevated reservoirs during periods of low electric demand – the off-peak hours. During the periods of peak demand, the water is allowed to flow back downhill to redeliver the energy from hydroelectric generation.

Also, electricity can be stored in batteries. However, the present-day automobile battery, to use an important and common example, is used for starting the internal combustion engine and not for locomotion.

Finally, energy storage includes heat storage. In thermodynamic terms (Appendix A), we are considering the storage of *transferred heat* before it is put to useful purposes. A long-standing conventional practice is hot water storage in homes and industry. Such heat storage smoothes out the delivery of hot water or steam, but it is not usually considered for a term longer than one day.

*Advanced* storage – *new* technology not in conventional use – is an integral part of other new technologies that will be made more feasible by innovations in storage. Wind and solar technologies especially will benefit from advances in storage. Also, new storage technologies may lead to an electric-powered automobile.

## Advanced batteries

New developments in batteries are important for the stand-alone operation of solar PV and wind generators, as we have already seen. They are also needed for utilities in place of pumped hydropower and for electric vehicles. But all of these applications require major improvements in technical characteristics and cost reduction of the storage technique.

Advanced battery concepts are typically a variation on the conventional lead-acid storage battery. This operates on the principle of the galvanic cell, whose discovery goes back to the eighteenth century. A single-cell battery consists of two electrodes immersed in an electrolyte (Fig. C.22). Chemical reactions of the electrolyte with the substances at each of the two electrodes release electrons with the potential to do electrical work. In the lead-acid battery, one electrode is lead (Pb), the other electrode is made of lead oxide ($PbO_2$), and the electrolyte is sulphuric acid ($H_2SO_4$). For the battery to deliver electricity, electrons are supplied to the lead oxide electrode from the lead electrode. But as the energy is discharged, the chemical composition changes. We can see this by writing the electrochemical equation:

$$PbO_2 + Pb + 2H_2SO_4 \rightarrow 2PbSO_2 + 2H_2O$$

which indicates the products of the discharge on the right side. The lead sulphate ($PbSO_2$) product is deposited on the electrodes, while water goes into the solution of the electrolyte.

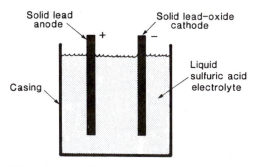

FIG. C.22.   Lead-acid battery.

These products build up, of course, as the discharge progresses and eventually the battery no longer produces electricity.

However, the reaction is reversible. A storage battery can be recharged by a d.c. source (generator). The charge flow in the cell reverses and so do the reactions – from $PbSO_2$ back to $PbO_2$ on the positive electrode, to lead on the negative electrode, to sulphuric acid in the electrolyte. The cycle of discharge and recharge can only be repeated a limited number of times due to electrode and electrolyte deterioration. This determines the life cycles of the battery. For lead-acid batteries in automobiles, this is around 300 cycles; if these batteries were used in a solar application with daily cycling, they would last less than a year. Industrial grade lead-acid batteries have over five times the number of life cycles, but cost more than twice as much and would not be economically viable for such service. The lead-acid storage battery is too short-lived for the new applications envisaged. It is also too heavy and bulky, especially for electric vehicle application.

Although proposed advanced batteries operate on the same galvanic principle as conventional batteries, major innovations have been tried in their construction and operation. In an effort to get away from deposits and corrosion of solid electrodes, liquid electrodes have been designed. An example of this (Fig. C.23) is a sodium-sulphur battery invented at the Ford Motor Company in the 1960s. In this experimental battery, the two electrodes are comprised of molten sodium and a molten sulphur-carbon mixture. The use of the liquid electrodes in the sodium-sulphur battery not only reduced corrosion, it extended the life cycle of the electrodes.

But this battery has many drawbacks. It requires operation at temperatures in the range 300° to 350°C, which can present a safety risk in electric vehicles. And it is still too heavy and too costly. Other experimental batteries, such as zinc-chlorine and lithium-iron sulphide have thus far exhibited the same shortcomings.

Thus, the cost of electric storage systems is uncompetitive at the present. For example, battery storage may cost up to 50 percent more than pumped hydropower. The cost goal of the developers of advanced batteries for utilities is about half that of present heavy duty lead-acid batteries.

One way to make storage batteries more cost competitive would be to extend their lifetimes. Developers of advanced batteries hope to attain life cycles in excess of 2,000 and lifetimes of over 10 years. (It should be noted that daily cycling would require 3,650 cycles for a 10-year lifetime.)

For comparison, the lifetime of a conventional pumped hydroplant can be as high as 50 years and the number of charge-discharge cycles presents no additional limitation to that technology. This has important economic ramifications because the capital cost of equipment is usually amortized over periods no longer than the working life of the equipment. The short lifetimes now expected of batteries means that the projected costs of *delivered* energy will be more than 50 percent higher than those of pumped hydropower. Thus, the cost to the utility ratepayer would be correspondingly higher for that fraction of the electric energy delivered by batteries. In stand-alone operation for batteries, such as remote wind generation, cost considerations can be made in a similar manner, but with different operational parameters if high availability is the objective.

It is clear, however, that achieving economic

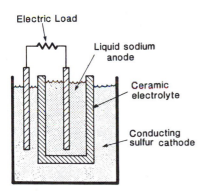

**FIG. C.23.** Schematic diagram of a sodium-sulfur battery.

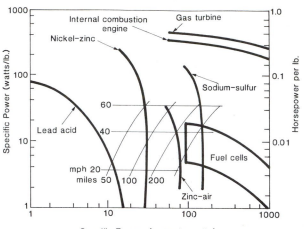

**FIG. C.24.** Weight-specific properties of energy storage systems. Adapted from Dorf, 1978 (see Chapter 2, Bibliography).

viability for large-scale wind or solar generation will require significantly greater energy storage capacities. That is, the system must be capable of storing several days' equivalent usage of energy so that the system can deliver the energy at the normal rate of daily consumption during periods of limited sunlight or wind.

Finally, the weight and volume requirements for batteries must be able to compete technically with certain conventional technologies. Weight is by far the most important technical requirement, if batteries are to be the means of storing energy in an electric vehicle and thus are to be the equivalent to a tank of gasoline in an automobile or truck. Actually, there are two parameters relative to weight: the *specific energy* and the *specific power*. The specific energy gives the amount of energy stored per unit of weight of a vehicle using a particular storage technology and is measured in watt-hours per pound. The specific power gives the power rate of delivery of that energy per pound of vehicle as measured in watts per pound. The driving range of a vehicle is roughly proportional to its specific energy, whereas the speed capability is nearly proportional to its specific power.

Figure C.24 shows weight-specific characteristics of several batteries, each assumed to be installed in an electric vehicle. The lead-acid (battery) curve illustrates the meaning of the curves. As the power capacity is increased, the storage capacity drops and vice versa. The lead-acid curve shows, for example, that an electric vehicle could have about a 50-mile range at a speed of 25 mph as one possible operational combination.

The curves for some proposed advanced batteries, such as nickel-zinc, zinc-air, and sodium-sulphur, show significant improvements over lead-acid. Each displacement to the right means increased driving range, so the sodium-sulphur could have a range over 250 miles at 40 mph.

When we compare batteries against the other vehicular technologies, we see that only the advanced batteries begin to be technically competitive. In general, the best performance corresponds to the upper-right-hand corner of the weight-specific plot. In that region, of course, is the internal combustion engine. This means that the portability of electric vehicles will have to be developed further if they are going to compete with the existing technology of petroleum-based fuels and heat engines (see Chapter 12).

## Advanced heat storage

The conventional technique of storing heat in hot water tanks is an example of the use of the physical property of *sensible heat* storage. In this form of storage, thermal energy is transferred (or "heat is added"; see Appendix A) to a medium and the increase in temperature of the material is proportional to the heat added, as long as the material does not change state. Accordingly, water is a material for sensible heat storage as long as it remains a liquid.

Heat can also be stored by hot air systems, for example, by forcing the air through a porous bed of pebbles or crushed stone. Once the stones are warmed to the higher temperature, the heat has been stored – to be extracted later by passing cooler air through the bed.

However, water is the best sensible heat medium per unit of weight of material used (Table C.2). Specific heat is the constant of proportionality for sensible heat, expressed here in units of BTUs per pound of material per °F. With water, it takes 1 BTU per pound to raise the temperature 1°F.

Solar-thermal systems probably represent the best examples of advanced storage applications. When water is used as the storage medium, storage is accomplished by heat exchange coils in the tank (see Fig. 12.3a) through which the hot coolant liquid is circulated from the solar collector. When the rate of heat exchange out of the tank is less than the rate of input from the collector, then the water temperature in the tank rises and thermal energy is stored. Conversely, when the heat output rate is greater than the input rate, energy is supplied from storage and the temperature of the tank water drops. These heat exchanges only work if the coolant from the collector is hotter than the tank water. The exchanges are aided if the input coil is placed in the lower – hence cooler – part of the tank and the output is placed in the hotter top part.

Typical daily residential hot water use is cyclical, rising to a daytime or early evening peak and dropping late at night. Industrial hot water and steam would also have daily cycles, but most often would be confined to daytime work hours and drop to zero on weekends and holidays. The use of short-time storage can extend the heat supply into the evening or possibly to the next morning.

But any attempt to increase short-time storage capacity requires an ever-increasing amount of collector area, which rapidly boosts the capital cost. Because solar energy costs are at best marginally competitive where the collector area is sufficient only for one day's supply, an attempt to collect and store heat for several days is at present prohibitively expensive.

Storage tanks for solar-heated hot water are already quite large. Take the case of residential hot water. A family of four typically might use 75,000 BTU per day for hot water. A typical solar collector would have an area of about 150 square feet and a storage tank (at 2.5 gallons storage per square foot of collector) of 375

**Table C.2.** Sensible heat materials

| Sensible heat material | Specific heat (BTU/lb°F) | True density (lb/ft³) | Heat capacity (BTU/ft³°F) | |
|---|---|---|---|---|
| | | | No voids | 30% voids |
| Water | 1.00 | 62 | 62 | — |
| Scrap iron | 0.12* | 490 | 59 | 41 |
| Magnetite (Fe₃O₄) | 0.18* | 320 | 57 | 40 |
| Scrap aluminum | 0.23* | 170 | 39 | 27 |
| Stone | 0.21 | ~170 | 36 | 25 |
| Sodium (to 208°F) | 0.23 | 59 | 14 | — |

*From 77° to 600°F.
*Source*: Adapted from Hottel & Howard (1971). See Fig. C.3 caption.

**Table C.3.** Latent heat materials

| Latent heat material | Melting point (°F) | Density (lb/cu ft) | Heat of fusion | |
|---|---|---|---|---|
| | | | BTU/lb | BTU/cu ft |
| Calcium chloride hexahydrate | 84–102 | 102 | 75 | 7,900 |
| Glauber's salt | 90 | 92 | 105 | 9,700 |
| Sodium metal | 208 | 59 | 42 | 2,500 |
| Lithium nitrate | 482 | 149 | 158 | 23,500 |
| Lithium hydride | 1,260 | 51 | 1800 | 92,000 |

*Source*: Adapted from H. C. Hottel and J. B. Howard 1971, *New Energy Technology – Some Facts and Assessments*, MIT Press, Cambridge, MA.

gallons. This is a volume of about 2,800 cubic feet, which would require a 12-foot-diameter, 25-foot-long cylindrical tank – quite large for a modest sized home. A solar system that also supplies space heat would need a tank over ten times larger. (An even larger space requirement for storage would hold for hot air solar home space heating.)

The size of storage volumes could be reduced considerably by the use of latent heat materials. High-pressure steam storage, for example, has been considered for large high-temperature industrial or utility operations. For lower-temperature applications, such as residential hot water or space heating, solid materials that change phase have been considered. Table C.3 shows some common solid latent heat materials that melt at elevated temperatures and thus absorb latent heat in changing phase from solid to liquid. The heat capacities per unit volume of these materials may be compared with the sensible heat materials.

However, latent heat materials have not been adopted for solar and other thermal applications for storage. A technical reason is the long time required for changes of phase and hence long charge-discharge times. Also, materials like Glauber's salt were found to decompose into a simple sodium sulphate after repeated cycling and then settle out of the liquid solution. Further research on these compounds is being conducted in government laboratories.

Finally, there is a form of long-term heat storage that has found limited adoption. This is called *seasonal storage* and involves the storage of heat, or the *storage of cold*, underground – usually in aquifers. Both heat pumps and air conditioning are *heat exchangers* and as such rely on temperature differentials to function efficiently. Heat pumps, for example, can function by exchanging thermal energy from the outside ambient air, but their efficiency drops rapidly as the outside temperature falls below freezing. For subfreezing weather, heat pumps can be made to function by exchanging energy from coils placed underground, where the temperatures remain well above freezing. Summer air conditioning, of course, uses a reservoir of lower-temperature material underground. Seasonal storage has not found wide use, however, mostly because of added investment and operating costs.

## Magnetohydrodynamics

Magnetohydrodynamics (MHD) is a means of *direct conversion* of the thermal energy of fossil fuel to electricity that omits the intermediate step of conversion into mechanical energy. MHD offers the prospect of higher thermal efficiencies for power plants, and it is one of several possible advanced fossil fuel technologies that can lower air pollution.

Although MHD is a novel concept, it is based on classical principles of thermodynamics and electricity generation. The MHD thermodynamic cycle is quite similar to that of the combustion turbine (see Chapter 3). Figure C.25 shows an open cycle, in which a compressor raises the pressure and temperature of incoming air to be mixed with fuel for combustion, as in the combustion turbine. However, the MHD nozzle and duct replace the turbine for the expansion part of the cycle. In the MHD duct, electricity is generated directly through the motion in the expanding combustion gases of the electrons and ions as they pass through an applied magnetic field.

High-velocity motion of the gas in the MHD duct is created by passing it through a nozzle at high pressure and temperature (2,500° to 3,000°C). The gases gain speed as they pass through the nozzle, which widens toward its output. A large electromagnet is applied to both sides of the MHD duct, so that a magnetic field crosses the duct at right angles to the direction of gas flow (see Figure C.26).

An electric potential is created across the MHD channel by the flow of the charged particles through the magnetic field. The principle by which this occurs is basically the

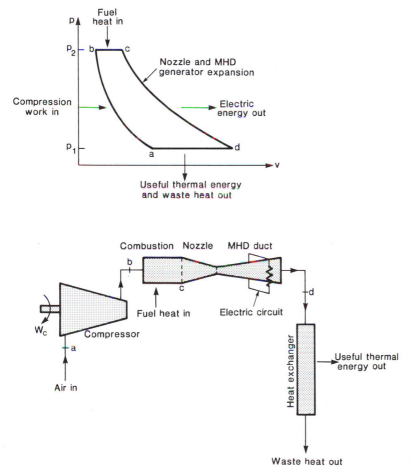

**FIG. C.25.** An open-cycle magnetohydrodynamic (MHD) generator. Adapted from J. H. Krenz, 1984, *Energy Conversion and Utilization.* Allyn and Bacon, Boston, MA.

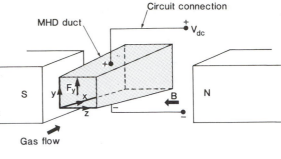

**FIG. C.26.** Magnetohydrodynamic duct section with magnet pole pieces and circuit connection. Adapted from J. H. Krenz, 1984, *Energy Conversion and Utilization.* Allyn and Bacon, Boston, MA.

same as that for a current-carrying conductor when it passes through a magnetic field in a conventional generator. Here, the magnetic deflection of the charges is equivalent to an electric field force in a direction at right angles to both the gas flow and the magnetic field (Fig. C.26). This force ($F_Y$ in the figure) is called the Lorentz force (named after H. A. Lorentz 1853–1928, Dutch physicist). It causes an accumulation of positive charge at one wall and negative charge at the opposite wall of the MHD duct. The charges build up at the walls and cause an electric field opposing the Lorentz field, and thus an equilibrium is established between them. Of course, the two opposing walls are insulated electrically from one another, thus sustaining a voltage between them. When an electric circuit is connected across the two electrode walls, current flows.

But our description thus far has overlooked some practical considerations. High conversion efficiencies require high temperatures at the start of the cycle and this, in turn, requires the lining of the nozzle and duct with refractory materials capable of withstanding several-thousand-degree temperatures. These materials have not yet been fully developed for long-lived service, and continuous operation of more than a few thousand hours has not been achieved.

MHD generators have thus far been conceived as operating in conjunction with conventional fossil-fueled operations. In this form of *combined-cycle operation*, the MHD generator would utilize the combustion gases

at their top temperatures, while the MHD waste heat would be supplied to a conventional steam system. In such an operation, the combined thermal efficiency has already been shown to reach 50 percent and 60 percent is thought possible (remember, conventional steam plants have thermal efficiencies generally less than 40 percent).

Prototype MHD generators have been built in the United States since the 1960s. There was a heightened effort at MHD development during the energy crisis and a slackening of effort thereafter. Still, development efforts in MHD have been sustained in the USSR and are also being carried out in Japan, China, Israel, and some Western European countries.

## Inertial confinement fusion

An artist's sketch of laser-driven inertial confinement fusion is shown on Figure C.27. The figure shows a small pellet of fusion fuel (such as DT, see Chapter 12) that is centered at the point of intersection of several laser beams. The pellet has been timed to drop into this position such that its arrival coincides with a high-power pulse of laser light simultaneously emitted through all laser beam transport tubes (four are shown). The incident laser light heats and compresses the fuel pellet, causing fusion ignition within an extremely short time interval. When the fuel pellet is ignited, the liquid lithium just outside the pellet chamber absorbs the

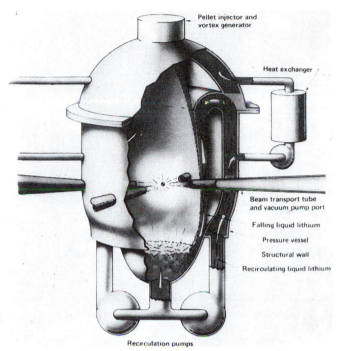

Pellet injector and vortex generator

Heat exchanger

Beam transport tube and vacuum pump port

Falling liquid lithium

Pressure vessel

Structural wall

Recirculating liquid lithium

Recirculation pumps

**FIG. C.27.** Inertial confinement fusion reactor concept. Courtesy of the Lawrence Livermore Laboratory.

neutrons emitted by the fusion reactions. The lithium acts as a coolant and potentially a means of breeding tritium.

When fusion schemes such as this were first proposed in the late 1960s, they were received with great enthusiasm in the scientific community. The concept was elegantly simple compared to the magnetic confinement schemes. Furthermore, with the rapid advances being made with lasers during that period, it was thought that quick advances could be made this way to a fusion break-even experiment. The new laser-fusion projects appealed to the public as futuristic marvels because of their exotic features.

Multibeam laser systems capable of huge power outputs were built in the 1970s. For example, the HELIOS laser built at the Los Alamos Scientific Laboratory emits infrared pulses in eight beams directing 10,000 joules of light energy onto the target in 1 nanosecond ($10^{-9}$ seconds). This rate of energy delivery equals 1 terawatt ($10^{12}$ watts). The density of the light flux flowing onto the target during the 1-nanosecond interval can also be huge. For example, if the HELIOS laser beams were all focused into a 1-mm-diameter spot, the light flux intensity would be over $10^{14}$ W/cm$^2$ (compare this with the average solar flux density of 245 W/m$^2$ = 0.0245 W/cm$^2$, mentioned in Chapter 12). With the construction of the new ANTARES CO$_2$ laser at Los Alamos, it was hoped that target densities could be increased over ten times higher. Another laser, the SHIVA laser at the Livermore Laboratory, can deliver the same energy in one-tenth the time, thus increasing the power rate by another factor of ten.

With these extremely high concentrations of power in laser light, violent dynamics take place within the pellet. An artist's conception of these dynamics is shown on Figure C.28. They actually evolve in stages. The first stage is the formation of a plasma (see Chapter 12), which is *ablated* directly from the solid pellet material by the rapid laser heating. This plasma surrounding the solid pellet itself gets heated even more and expands violently away from the

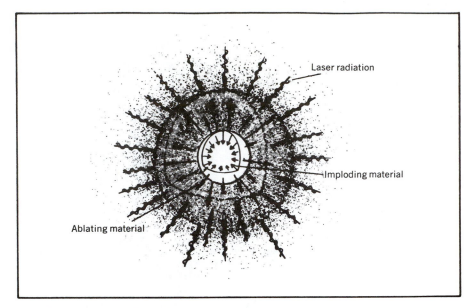

**Laser implosion of a pellet.** The atmosphere extends to several pellet radii and is formed before the main laser pulse by a prepulse that ablates some of the pellet surface. Absorption of the laser light in the outer atmosphere generates hot electrons. As the electrons move inward heating the atmosphere and pellet surface, scattering and solid-angle affects greatly increase the spherical symmetry. Violent ablation and blowoff of the pellet surface generates the pressures that implode the pellet; the effect is similar to a spherical rocket. The pellet core then undergoes thermonuclear burn.

**FIG. C.28.** Laser fusion – compression stage, implosion of the pellet. *Source*: *Physics Today* (August 1973). Courtesy of the University of California, Lawrence Livermore Laboratory under the auspices of the U.S. Department of Energy.

pellet in a blowoff – the second stage. The force of this blowoff is so great that it acts like a rocket thrust, causing a compression wave to propagate into the pellet. This compression is calculated to lead to densities in the isotope nuclei of at least 1,000 times the density of liquid hydrogen, pressures at $10^{10}$ times atmospheric pressure, and a temperature within the compressed pellet of millions of degrees centigrade.

Ideally, a further deposition of heat from incoming laser light accompanies this implosion, serving to raise the temperatures even further within the pellet (to hundreds of millions of degrees centigrade). This should lead to the ignition condition – the third stage of the dynamic sequence. The entire sequence of three stages has been theorized to take place within

less than 1 nanosecond. Whereas much of this dynamic sequence has been observed experimentally, the ignition condition has not yet been attained. If it is attained ultimately, then and only then will the final stage of thermonuclear burn be observed, which is required to demonstrate the *scientific feasibility* of this version of fusion technology.

The progress toward the achievement of ignition is indicated on the graph of compression ratio versus peak fuel temperature in Figure C.29. Each cross on the chart represents an experimental result already attained by 1982 at a major research center. As with magnetic confinement research, the inertial program has both failed to reach break-even and has encountered unexpected difficulties – instabilities similar to those mentioned for magnetic con-

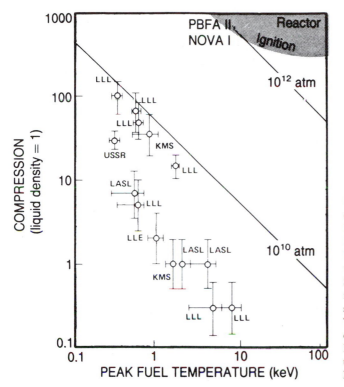

FIG. C.29. Inertial confinement experiment results. LLL = Lawrence Livermore Laboratory; KMS = KMS Fusion Inc., Ann Arbor, MI; USSR = Lebedev Physical Institute, USSR; LASL = Los Alamos Scientific Laboratory; LLE = Laboratory for Laser Energetics, University of Rochester. *Source*: Physics Today (September 1982). Courtesy of the University of California, Lawrence Livermore Laboratory under the auspices of the U.S. Department of Energy.

finement and unwanted heating of electrons – all of which still require solutions. The inertial confinement fusion effort appears to have lost much of its glamour. The emphasis in the U.S. Congress since 1980 has been on funding magnetic confinement research.

## Fuel cells

The construction and operation of a fuel cell appears, at first glance, to be similar to a battery. Both convert chemical energy directly into electricity using a pair of elecrodes immersed in an electrolyte. However, the fuel cell is supplied continually by chemical energy, and in the fuel cell, unlike the battery, the electrode material is not depleted as it supplies electricity.

The fuel cell is in some ways an ideal electric generator. Operation is quiet because the cell has no moving parts, and it is clean. These qualities make it especially promising for electric generation in urban areas.

A diagram of the structure of a hydrogen-oxygen fuel cell is shown in Figure C.30. Hydrogen gas ($H_2$) flows in on the upper left and oxygen ($O_2$) on the upper right. These two gases penetrate the porous electrodes, each of which is in contact with the liquid electrolyte. In this example, the electrolyte is alkaline (potassium hydroxide, $KOH \cdot H_2O$), although an acid, such as phosphoric acid, can also be used.

When using an alkaline electrolyte, hydroxyl ions are the charge carriers in the solution. The hydroxyl ions ($OH^-$) are created in the porous cathode on the right in Figure C.30a by the reaction of water and oxygen in the presence of excess electrons in the cathode solution. These ions then diffuse toward the anode. The reaction at the anode liberates electrons (e), which flow through the electrical load and, in turn, feed more reactions at the cathode. Also hydroxyl

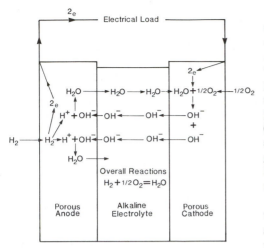

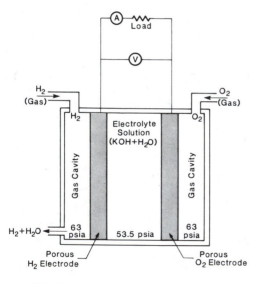

a) Basic operation of the fuel cell.

b) Construction.

**FIG. C.30.** The fuel cell. (a) Basic operation. (b) Construction. *Source*: B. J. Crowe, 1973, *Fuel Cells – A Survey*. National Aeronautics and Space Administration (NASA), U.S. Government Printing Office, Washington, DC.

ions pass from the electrolyte and react with any excess hydrogen there to produce water. An electrical potential exists across a fuel cell, just as it does across a battery, and the electrons created can thus do *electrical work*.

Practical operation of a fuel cell requires either a catalyst or elevated temperatures to enhance the electrode reactions. These temperatures typically exceed the boiling point of water and the cells are therefore operated under pressure to prevent the water from turning to steam. The example in Figure C.30 has a pressure in the electrolytic chamber of 53.5 psi (3.6 times atmospheric pressure). Note also a somewhat higher pressure in each of the two gas cavities; this is designed to force the hydrogen and oxygen into the porous electrodes.

Although oxygen can easily be supplied to the cell by air, the supply of hydrogen requires an additional process. The most successful method to date has been steam reforming of either naphtha or natural gas. This reaction is similar to some of those shown in Chapter 12 for coal gasification. If methane (from natural gas) is the input, for example, the reaction is:

$$CH_4 + H_2O \rightarrow CO + 3H_2$$

where the water is actually steam at high temperatures (1,500°F). In addition to the high temperature, the reaction also requires catalysts, such as nickel oxide and silicon.

A working prototype fuel cell has been operated by the Consolidated Edison Company in New York City, in a demonstration cosponsored by the U.S. Department of Energy and the Electric Power Research Institute. The plant has used the steam reforming method on naphtha to supply the hydrogen and phosphoric acid as the electrolyte. The entire unit contains over 9,000 individual fuel cells, which with series and parallel connections give a d.c. output of 2,800 volts and a total power output of 4.8 $MW_e$.

Presently, the unit capital costs of fuel cells are estimated to be less than $1,000/KW_e$, which if accurate, would make this technology competitive with combustion turbines, the present conventional technology for small, peak-load generators. But a fuel cell, like a turbine, uses hydrocarbon fuels and thus the overall economics of the two – including capital and operating costs – will be similar. Even assuming that fuel cell stations are not quite cost competitive with combustion turbines, they might still be the choice in urban areas because they are low in noise and air pollution.

Fuel cells have also been considered as the means to power an electric vehicle. Here, as for a battery-powered vehicle, storage of fuel and weight become major concerns. A hydrogen-oxygen cell, for instance, might require the storage of liquid hydrogen; the hydrogen would be prepared separately, just as is gasoline. Another technical possibility, however, is storage in metal hydrides, where the hydrogen enters into a chemical compound for storage and then is released for use later.

The weight-specific properties of fuel-cell-powered vehicles have been estimated and are shown in Figure C.24 along with those of battery systems. It can be seen that fuel cell vehicles characteristically have high specific energies but low specific powers, meaning they would have long driving ranges, but have low acceleration and speeds.

The other barrier to adoption of fuel cells for automotive use is their costs, as compared to the conventional internal combustion automobile. Further development and engineering of mass production techniques would still be needed for this technology, even if its technical characteristics compared favorably with conventional automobiles and trucks.

## The hydrogen economy

In the previous section, we have seen the technical possibility of the use of hydrogen in a fuel cell for producing electricity. Hydrogen is considered the ideal fuel, because the product of combustion or the fuel cell reactions with oxygen is simply water. If useful thermal energy and electricity could be created in this ideal process, then we would solve several disturbing dilemmas, notably: air pollution, acid rain, and the greenhouse effect.

For this reason, some futuristic thinkers have proposed the creation of a *hydrogen economy*, in which hydrogen would be synthesized on a mass scale for a variety of uses. Not only could it replace natural gas for all thermal applications in homes and industry, but it could conceivably be used for automotive applications, replacing gasoline and diesel fuel. Although this could mean that electric vehicles would be powered by a fuel cell, an internal combustion engine burning hydrogen is also a possibility.

The major technical difficulties with such schemes for mass conversion to hydrogen center around the fact that hydrogen does not occur in nature as a basic energy source. Instead, a significant energy input is needed to create hydrogen from its natural state in water. Conversion from water, rather than from a hydrocarbon such as naphtha, is essential if the hydrogen economy is to free us from dependence on fossil fuels. The converted hydrogen, therefore, is more in the nature of a fuel *medium* in which energy can be stored and transmitted.

The best known means of hydrogen conversion is the hydrolyzer, which is a fuel cell in reverse. That is, the separation of hydrogen from water requires a source of electricity, such as a conventional thermal electric power plant. The process, furthermore, is only 60 to 70 percent efficient and thus requires more energy from the electrical source than is regained when the hydrogen is burned or used in a fuel cell.

Hydrogen production in this futuristic scheme probably would require large nuclear fusion power plants. Thus, the conception is a set of massive electric power plants whose basic energy is derived from the superabundant supply of hydrogen isotopes (see Chapter 12) and that use relatively safe nuclear fusion reactions. The quantities of electricity would be so large and the resources effectively unlimited that the inefficiencies of the hydrolyzers would be of no concern.

There are several reasons why the hydrogen economy is unrealistic at this time. The principle one, of course, is that nuclear fusion has not been proven technically feasible, nor are there alternative mass means for hydrogen conversion under development. In addition, it would not be entirely pollution free. It has been pointed out that the combustion of hydrogen in *air* (not pure oxygen) generates some of the same pollutants that concern us today, such as the nitrogen oxides.

In all, the hydrogen economy appears to be a technical dream that may be realized but only in the far future.

# Index